Automating Manufacturing Systems with PLCs

Version 7.0, April, 2010

by Hugh Jack

ISBN 978-0-557-34425-3

Additional materials and updates for this work will be available at http://engineeronadisk.com

TABLE OF CONTENTS

PREFACE

Designing software for control systems is difficult. Experienced controls engineers have learned many techniques that allow them to solve problems. This book was written to present methods for designing controls software using Programmable Logic Controllers (PLCs). It is my personal hope that by employing the knowledge in the book that you will be able to quickly write controls programs that work as expected (and avoid having to learn by costly mistakes.)

This book has been designed for students with some knowledge of technology, including limited electricity, who wish to learn the discipline of practical control system design on commonly used hardware. To this end the book will use the Allen Bradley ControlLogix processors to allow depth. Although the chapters will focus on specific hardware, the techniques are portable to other PLCs. Whenever possible the IEC 61131 programming standards will be used to help in the use of other PLCs.

In some cases the material will build upon the content found in a linear controls course. But, a heavy emphasis is placed on discrete control systems. Figure .1 crudely shows some of the basic categories of control system problems.

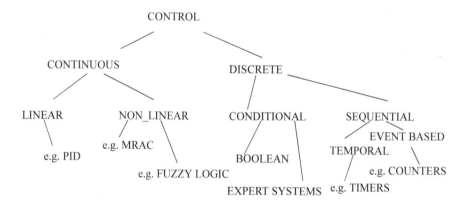

Figure .1   *Control Dichotomy*

- Continuous - The values to be controlled change smoothly. e.g. the speed of a car.
- Logical/Discrete - The value to be controlled are easily described as on-off. e.g. the car motor is on-off. NOTE: all systems are continuous but they can be treated as logical for simplicity.
     - e.g. "When I do this, that always happens!" For example, when the power is turned on, the press closes!
- Linear - Can be described with a simple differential equation. This is the preferred starting point for simplicity, and a common approximation for real world problems.
     - e.g. A car can be driving around a track and can pass same the same spot at a constant velocity. But, the longer the car runs, the mass decreases, and it travels faster, but requires less gas, etc. Basically, the math gets tougher, and the problem becomes non-linear.
     - e.g. We are driving the perfect car with no friction, with no drag, and can predict how it will work perfectly.
- Non-Linear - Not Linear. This is how the world works and the mathematics become much more complex.
     - e.g. As rocket approaches sun, gravity increases, so control must change.
- Sequential - A logical controller that will keep track of time and previous events.

The difference between these control systems can be emphasized by considering a simple elevator. An elevator is a car that travels between floors, stopping at precise heights. There are certain logical constraints used for safety and convenience. The points below emphasize different types of control problems in the elevator.

Logical:

1. The elevator must move towards a floor when a button is pushed.
2. The elevator must open a door when it is at a floor.
3. It must have the door closed before it moves.
   etc.

Linear:

1. If the desired position changes to a new value, accelerate quickly towards the new position.
2. As the elevator approaches the correct position, slow down.

Non-linear:

1 Accelerate slowly to start.
2. Decelerate as you approach the final position.
3. Allow faster motion while moving.
4. Compensate for cable stretch, and changing spring constant, etc.

Logical and sequential control is preferred for system design. These systems are more stable, and often lower cost. Most continuous systems can be controlled logically. But, some times we will encounter a system that must be controlled continuously. When this occurs the control system design becomes more demanding. When improperly controlled, continuous systems may be unstable and become dangerous.

When a system is well behaved we say it is self regulating. These systems don't need to be closely monitored, and we use open loop control. An open loop controller will set a desired position for a system, but no sensors are used to verify the position. When a system must be constantly monitored and the control output adjusted we say it is closed loop. A cruise control in a car is an excellent example. This will monitor the actual speed of a car, and adjust the speed to meet a set target speed.

Many control technologies are available for control. Early control systems relied upon mechanisms and electronics to build controlled. Most modern controllers use a computer to achieve control. The most flexible of these controllers is the PLC (Programmable Logic Controller).

The book has been set up to aid the reader, as outlined below.

> Sections labeled *Aside:* are for topics that would be of interest to one discipline, such as electrical or mechanical.
> Sections labeled *Note:* are for clarification, to provide hints, or to add explanation.
> Each chapter supports about 1-4 lecture hours depending upon students background and level in the curriculum.
> Topics are organized to allow students to start laboratory work earlier in the semester.
> Sections begin with a topic list to help set thoughts.
> Objective given at the beginning of each chapter.
> Summary at the end of each chapter to give big picture.
> Significant use of figures to emphasize physical implementations.
> Worked examples and case studies.
> Problems at ends of chapters with solutions.

1PROGRAMMABLE LOGIC CONTROLLERS

Topics:

- PLC History
- Ladder Logic and Relays
- PLC Programming
- PLC Operation
- An Example

Objectives:

- Know general PLC issues
- To be able to write simple ladder logic programs
- Understand the operation of a PLC

Control engineering has evolved over time. In the past humans were the main method for controlling a system. More recently electricity has been used for control and early electrical control was based on relays. These relays allow power to be switched on and off without a mechanical switch. It is common to use relays to make simple logical control decisions. The development of low cost computer has brought the most recent revolution, the Programmable Logic Controller (PLC). The advent of the PLC began in the 1970s, and has become the most common choice for manufacturing controls.

PLCs have been gaining popularity on the factory floor and will probably remain predominant for some time to come. Most of this is because of the advantages they offer.

- Cost effective for controlling complex systems.
- Flexible and can be reapplied to control other systems quickly and easily.
- Computational abilities allow more sophisticated control.
- Trouble shooting aids make programming easier and reduce downtime.
- Reliable components make these likely to operate for years before failure.

1.1 LADDER LOGIC

Ladder logic is the main programming method used for PLCs. As mentioned before, ladder logic has been developed to mimic relay logic. The decision to use the relay logic diagrams was a strategic one. By selecting ladder logic as the main programming method, the amount of retraining needed for engineers and tradespeople was greatly reduced.

Modern control systems still include relays, but these are rarely used for logic. A relay is a simple device that uses a magnetic field to control a switch, as pictured in Figure 1.1. When a voltage is applied to the input coil, the resulting current creates a magnetic field. The magnetic field pulls a metal switch (or reed) towards it and the contacts touch, closing the switch. The contact that closes when the coil is energized is called normally open. The normally closed contacts touch when the input coil is not energized. Relays are normally drawn in schematic form using a circle to represent the input coil. The output contacts are shown with two parallel lines. Normally open contacts are shown as two lines, and will be open (non-conducting) when the input is not energized. Normally closed contacts are shown with two lines with a diagonal line through them. When the input coil is not energized the normally closed contacts will be closed (conducting).

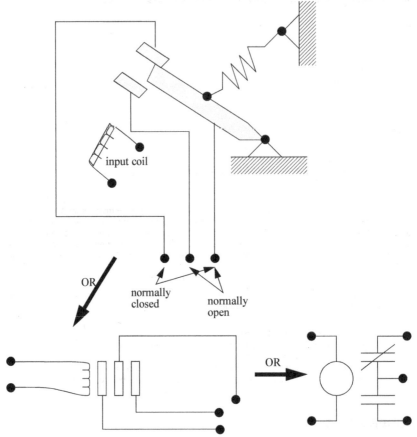

input coil

OR

normally
closed

normally
open

OR

Figure 1.1   *Simple Relay Layouts and Schematics*

Relays are used to let one power source close a switch for another (often high current) power source, while keeping them isolated. An example of a relay in a simple control application is shown in Figure 1.1. In this system the first relay on the left is used as normally closed, and will allow current to flow until a voltage is applied to the input A. The second relay is normally open and will not allow current to flow until a voltage is applied to the input B. If current is flowing through the first two relays then current will flow through the coil in the third relay, and close the switch for output C. This circuit would normally be drawn in the ladder logic form. This can be read logically as C will be on if A is off and B is on.

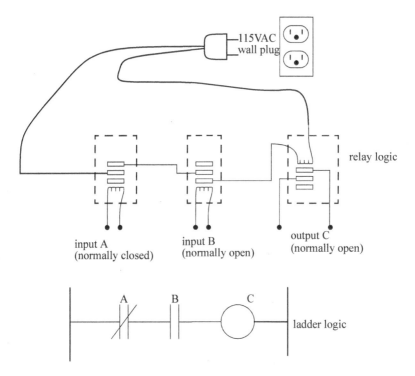

Figure 1.2 **A Simple Relay Controller**

The example in Figure 1.1 does not show the entire control system, but only the logic. When we consider a PLC there are inputs, outputs, and the logic. Figure 1.1 shows a more complete representation of the PLC. Here there are two inputs from push buttons. We can imagine the inputs as activating 24V DC relay coils in the PLC. This in turn drives an output relay that switches 115V AC, that will turn on a light. Note, in actual PLCs inputs are never relays, but outputs are often relays. The ladder logic in the PLC is actually a computer program that the user can enter and change. Notice that both of the input push buttons are normally open, but the ladder logic inside the PLC has one normally open contact, and one normally closed contact. Do not think that the ladder logic in the PLC needs to match the inputs or outputs. Many beginners will get caught trying to make the ladder logic match the input types.

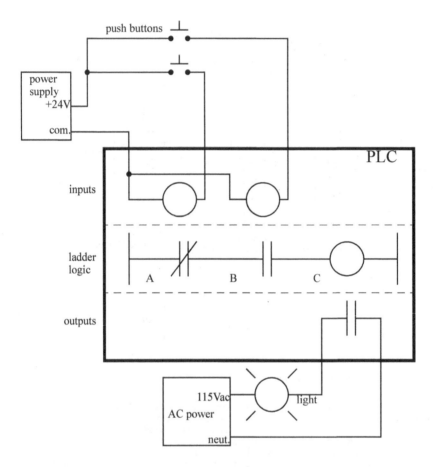

Figure 1.3   **A PLC Illustrated With Relays**

Many relays also have multiple outputs (throws) and this allows an output relay to also be an input simultaneously. The circuit shown in Figure 1.1 is an example of this, it is called a seal in circuit. In this circuit the current can flow through either branch of the circuit, through the contacts labelled A or B. The input B will only be on when the output B is on. If B is off, and A is energized, then B will turn on. If B turns on then the input B will turn on, and keep output B on even if input A goes off. After B is turned on the output B will not turn off.

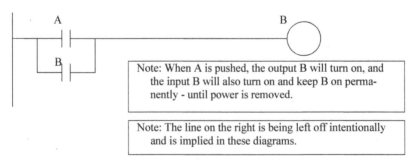

Note: When A is pushed, the output B will turn on, and the input B will also turn on and keep B on permanently - until power is removed.

Note: The line on the right is being left off intentionally and is implied in these diagrams.

4

Figure 1.4   **A Seal-in Circuit**

1.1.1 Programming

The first PLCs were programmed with a technique that was based on relay logic wiring schematics. This eliminated the need to teach the electricians, technicians and engineers how to *program* a computer - but, this method has stuck and it is the most common technique for programming PLCs today. An example of ladder logic can be seen in Figure 1.1. To interpret this diagram imagine that the power is on the vertical line on the left hand side, we call this the hot rail. On the right hand side is the neutral rail. In the figure there are two rungs, and on each rung there are combinations of inputs (two vertical lines) and outputs (circles). If the inputs are opened or closed in the right combination the power can flow from the hot rail, through the inputs, to power the outputs, and finally to the neutral rail. An input can come from a sensor, switch, or any other type of sensor. An output will be some device outside the PLC that is switched on or off, such as lights or motors. In the top rung the contacts are normally open and normally closed. Which means if input *A* is on and input *B* is off, then power will flow through the output and activate it. Any other combination of input values will result in the output *X* being off.

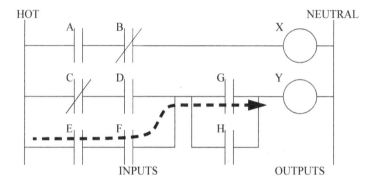

Note: Power needs to flow through some combination of the inputs (A,B,C,D,E,F,G,H) to turn on outputs (X,Y).

Figure 1.5   **A Simple Ladder Logic Diagram**

The second rung of Figure 1.1 is more complex, there are actually multiple combinations of inputs that will result in the output *Y* turning on. On the left most part of the rung, power could flow through the top if *C* is off and *D* is on. Power could also (and simultaneously) flow through the bottom if both *E* and *F* are true. This would get power half way across the rung, and then if *G* or *H* is true the power will be delivered to output *Y*. In later chapters we will examine how to interpret and construct these diagrams.

There are other methods for programming PLCs. One of the earliest techniques involved mnemonic instructions. These instructions can be derived directly from the ladder logic diagrams and entered into the PLC through a simple programming terminal. An example of mnemonics is shown in Figure 1.1. In this example the instructions are read one line at a time from top to bottom. The first line *00000* has the instruction *LDN* (input load and not) for input *A*. This will examine the input to the PLC and if it is off it will remember a *1* (or true), if it is on it will remember a *0* (or false). The next line uses an *LD* (input load) statement to look at the input. If the input is off it remembers a *0*, if the input is on it remembers a *1* (note: this is the reverse of the *LDN*). The *AND* statement recalls the last two numbers remembered and if the are both true the result is a *1*, otherwise the result is a *0*. This result now replaces the two numbers that were recalled, and there is only one number remembered. The process is repeated for lines *00003* and *00004*, but when these are done there are now three numbers remembered. The oldest number is from the *AND*, the newer numbers are from the two *LD* instructions. The *AND* in line *00005* combines the results from the last *LD* instructions and now there are two numbers remembered. The *OR* instruction takes the two numbers now remaining and if either one is a *1* the result is a *1*, otherwise the result is a *0*. This result replaces the two numbers, and there is now a single number there. The last instruction is the *ST* (store output) that will look at the last value stored and if it is *1*, the output will be turned on, if it is *0* the output will be turned off.

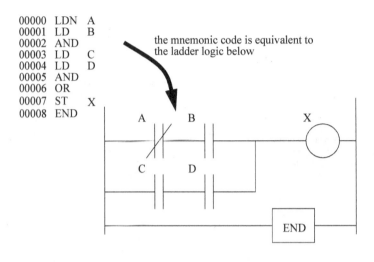

```
00000  LDN  A
00001  LD   B
00002  AND
00003  LD   C
00004  LD   D
00005  AND
00006  OR
00007  ST   X
00008  END
```

the mnemonic code is equivalent to
the ladder logic below

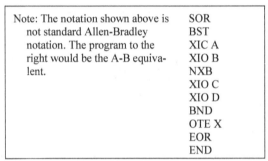

Note: The notation shown above is not standard Allen-Bradley notation. The program to the right would be the A-B equivalent.	SOR BST XIC A XIO B NXB XIO C XIO D BND OTE X EOR END

Figure 1.6   ***An Example of a Mnemonic Program and Equivalent Ladder Logic***

The ladder logic program in Figure 1.1, is equivalent to the mnemonic program. Even if you have programmed a PLC with ladder logic, it will be converted to mnemonic form before being used by the PLC. In the past mnemonic programming was the most common, but now it is uncommon for users to even see mnemonic programs.

Sequential Function Charts (SFCs) have been developed to accommodate the programming of more advanced systems. These are similar to flowcharts, but much more powerful. The example seen in Figure 1.1 is doing two different things. To read the chart, start at the top where is says *start*. Below this there is the double horizontal line that says follow both paths. As a result the PLC will start to follow the branch on the left and right hand sides separately and simultaneously. On the left there are two functions the first one is the *power up* function. This function will run until it decides it is done, and the *power down* function will come after. On the right hand side is the *flash* function, this will run until it is done. These functions look unexplained, but each function, such as *power up* will be a small ladder logic program. This method is much different from flowcharts because it does not have to follow a single path through the flowchart.

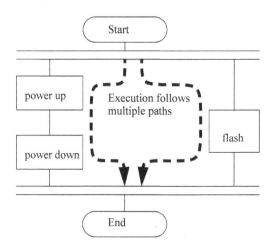

Figure 1.7   *An Example of a Sequential Function Chart*

Structured Text programming has been developed as a more modern programming language. It is quite similar to languages such as BASIC. A simple example is shown in Figure 1.1. This example uses a PLC memory location *i*. This memory location is for an integer, as will be explained later in the book. The first line of the program sets the value to 0. The next line begins a loop, and will be where the loop returns to. The next line recalls the value in location *i*, adds 1 to it and returns it to the same location. The next line checks to see if the loop should quit. If *i* is greater than or equal to 10, then the loop will quit, otherwise the computer will go back up to the *REPEAT* statement continue from there. Each time the program goes through this loop *i* will increase by 1 until the value reaches 10.

```
i := 0;
REPEAT
  i := i + 1;
UNTIL i >= 10
END_REPEAT;
```

Figure 1.8   *An Example of a Structured Text Program*

1.1.2 PLC Connections

When a process is controlled by a PLC it uses inputs from sensors to make decisions and update outputs to drive actuators, as shown in Figure 1.1. The process is a real process that will change over time. Actuators will drive the system to new states (or modes of operation). This means that the controller is limited by the sensors available, if an input is not available, the controller will have no way to detect a condition.

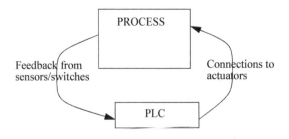

Figure 1.9   *The Separation of Controller and Process*

The control loop is a continuous cycle of the PLC reading inputs, solving the ladder logic, and then chang-ing the outputs. Like any computer this does not happen instantly. Figure 1.1 shows the basic operation cycle of a PLC. When power is turned on initially the PLC does a quick *sanity check* to ensure that the hardware is working properly. If there is a problem the PLC will halt and indicate there is an error. For example, if the PLC power is drop-ping and about to go off this will result in one type of fault. If the PLC passes the sanity check it will then scan (read) all the inputs. After the inputs values are stored in memory the ladder logic will be scanned (solved) using the stored values - not the current values. This is done to prevent logic problems when inputs change during the ladder logic scan. When the ladder logic scan is complete the outputs will be scanned (the output values will be changed). After this the system goes back to do a sanity check, and the loop continues indefinitely. Unlike normal computers, the entire program will be *run* every scan. Typical times for each of the stages is in the order of milliseconds.

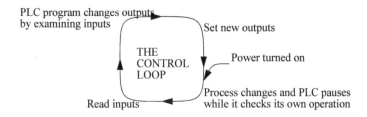

Figure 1.10   *The Scan Cycle of a PLC*

1.1.3 Ladder Logic Inputs

PLC inputs are easily represented in ladder logic. In Figure 1.1 there are three types of inputs shown. The first two are normally open and normally closed inputs, discussed previously. The *IIT* (Immediate InpuT) function allows inputs to be read after the input scan, while the ladder logic is being scanned. This allows ladder logic to examine input values more often than once every cycle. (Note: This instruction is not available on the ControlLogix processors, but is still available on older models.)

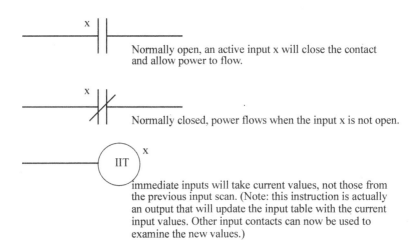

X

Normally open, an active input x will close the contact and allow power to flow.

X

Normally closed, power flows when the input x is not open.

IIT X

immediate inputs will take current values, not those from the previous input scan. (Note: this instruction is actually an output that will update the input table with the current input values. Other input contacts can now be used to examine the new values.)

Figure 1.11   **Ladder Logic Inputs**

1.1.4 Ladder Logic Outputs

In ladder logic there are multiple types of outputs, but these are not consistently available on all PLCs. Some of the outputs will be externally connected to devices outside the PLC, but it is also possible to use internal memory locations in the PLC. Six types of outputs are shown in Figure 1.1. The first is a normal output, when energized the output will turn on, and energize an output. The circle with a diagonal line through is a normally on output. When energized the output will turn off. This type of output is not available on all PLC types. When initially energized the *OSR* (One Shot Relay) instruction will turn on for one scan, but then be off for all scans after, until it is turned off. The *L* (latch) and *U* (unlatch) instructions can be used to lock outputs on. When an *L* output is energized the output will turn on indefinitely, even when the output coil is deenergized. The output can only be turned off using a *U* output. The last instruction is the *IOT* (Immediate OutpuT) that will allow outputs to be updated without having to wait for the ladder logic scan to be completed.

When power is applied (on) the output x is activated for the left output, but turned off for the output on the right.

An input transition on will cause the output x to go on for one scan (this is also known as a one shot relay)

When the L coil is energized, x will be toggled on, it will stay on until the U coil is energized. This is like a flip-flop and stays set even when the PLC is turned off.

Some PLCs will allow immediate outputs that do not wait for the program scan to end before setting an output. (Note: This instruction will only update the outputs using the output table, other instruction must change the individual outputs.)

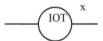

Note: Outputs are also commonly shown using parentheses -( )- instead of the circle. This is because many of the programming systems are text based and circles cannot be drawn.

Figure 1.12   **Ladder Logic Outputs**

1.2 A CASE STUDY

Problem: Try to develop (without looking at the solution) a relay based controller that will allow three switches in a room to control a single light.

Solution: There are two possible approaches to this problem. The first assumes that any one of the switches on will turn on the light, but all three switches must be off for the light to be off.

The second solution assumes that each switch can turn the light on or off, regardless of the states of the other switches. This method is more complex and involves thinking through all of the possible combinations of switch positions. You might recognize this problem as an exclusive or problem.

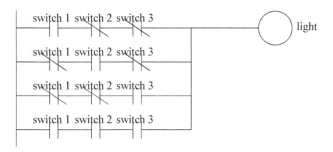

Note: It is important to get a clear understanding of how the controls are expected to work. In this example two radically different solutions were obtained based upon a simple difference in the operation.

1.3 SUMMARY

- Normally open and closed contacts.
- Relays and their relationship to ladder logic.
- PLC outputs can be inputs, as shown by the seal in circuit.
- Programming can be done with ladder logic, mnemonics, SFCs, and structured text.
- There are multiple ways to write a PLC program.

1.4 PRACTICE PROBLEMS

(Note: Problem solutions are available at http://sites.google.com/site/automatedmanufacturingsystems/)

1. Give an example of where a PLC could be used.

2. Why would relays be used in place of PLCs?

3. Give a concise description of a PLC.

4. List the advantages of a PLC over relays.

5. A PLC can effectively replace a number of components. Give examples and discuss some good and bad applications of PLCs.

6. Explain why ladder logic outputs are coils?

7. In the figure below, will the power for the output on the first rung normally be on or off? Would the output on the second rung normally be on or off?

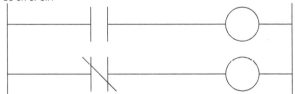

8. Write the mnemonic program for the Ladder Logic below.

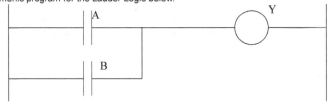

1.5 ASSIGNMENT PROBLEMS

1. Explain the trade-offs between relays and PLCs for contro applications.

2. Develop a simple ladder logic program that will turn on an output X if inputs A and B, or input C is on.

1.1 PRACTICE PROBLEM SOLUTIONS

1. To control a conveyor system

2. For simple designs

3. A PLC is a computer based controller that uses inputs to monitor a process, and uses outputs to control a process using a program.

4. Less expensive for complex processes, debugging tools, reliable, flexible, easy to expand, etc.

5. A PLC could replace a few relays. In this case the relays might be easier to install and less expensive. To control a more complex system the controller might need timing, counting and other mathematical calculations. In this case a PLC would be a better choice.

6. The ladder logic outputs were modelled on relay logic diagrams. The output in a relay ladder diagram is a relay coil that switches a set of output contacts.

7. off, on

8. Generic: LD A, LD B, OR, ST Y, END; Allen Bradley: SOR, BST, XIO A, NXB, XIO B, BND, OTE Y, EOR, END

2PLC HARDWARE

Topics:

- PLC hardware configurations
- Input and outputs types
- Electrical wiring for inputs and outputs
- Relays
- Electrical Ladder Diagrams and JIC wiring symbols

Objectives:

- Be able to understand and design basic input and output wiring.
- Be able to produce industrial wiring diagrams.

Many PLC configurations are available, even from a single vendor. But, in each of these there are common components and concepts. The most essential components are:

    Power Supply - This can be built into the PLC or be an external unit. Common voltage levels required by the PLC (with and without the power supply) are 24Vdc, 120Vac, 220Vac.

    CPU (Central Processing Unit) - This is a computer where ladder logic is stored and processed.

    I/O (Input/Output) - A number of input/output terminals must be provided so that the PLC can monitor the process and initiate actions.

    Indicator lights - These indicate the status of the PLC including power on, program running, and a fault. These are essential when diagnosing problems.

The configuration of the PLC refers to the packaging of the components. Typical configurations are listed below from largest to smallest as shown in Figure 2.1.

    Rack - A rack is often large (up to 18" by 30" by 10") and can hold multiple cards. When necessary, multiple racks can be connected together. These tend to be the highest cost, but also the most flexible and easy to maintain.

    Mini - These are smaller than full sized PLC racks, but can have the same IO capacity.

    Micro - These units can be as small as a deck of cards. They tend to have fixed quantities of I/O and limited abilities, but costs will be the lowest.

    Software - A software based PLC requires a computer with an interface card, but allows the PLC to be connected to sensors and other PLCs across a network.

Figure 2.1   *Typical Configurations for PLC*

2.1 INPUTS AND OUTPUTS

Inputs to, and outputs from, a PLC are necessary to monitor and control a process. Both inputs and outputs can be categorized into two basic types: logical or continuous. Consider the example of a light bulb. If it can only be turned on or off, it is logical control. If the light can be dimmed to different levels, it is continuous. Continuous values seem more intuitive, but logical values are preferred because they allow more certainty, and simplify control. As a result most controls applications (and PLCs) use logical inputs and outputs for most applications. Hence, we will discuss logical I/O and leave continuous I/O for later.

Outputs to actuators allow a PLC to cause something to happen in a process. A short list of popular actuators is given below in order of relative popularity.

> Solenoid Valves - logical outputs that can switch a hydraulic or pneumatic flow.
> Lights - logical outputs that can often be powered directly from PLC output boards.
> Motor Starters - motors often draw a large amount of current when started, so they require motor starters, which are basically large relays.
> Servo Motors - a continuous output from the PLC can command a variable speed or position.

Outputs from PLCs are often relays, but they can also be solid state electronics such as transistors for DC outputs or Triacs for AC outputs. Continuous outputs require special output cards with digital to analog converters.

Inputs come from sensors that translate physical phenomena into electrical signals. Typical examples of sensors are listed below in relative order of popularity.

> Proximity Switches - use inductance, capacitance or light to detect an object logically.
> Switches - mechanical mechanisms will open or close electrical contacts for a logical signal.
> Potentiometer - measures angular positions continuously, using resistance.
> LVDT (linear variable differential transformer) - measures linear displacement continuously using magnetic coupling.

Inputs for a PLC come in a few basic varieties, the simplest are AC and DC inputs. Sourcing and sinking inputs are also popular. This output method dictates that a device does not supply any power. Instead, the device only switches current on or off, like a simple switch.

> Sinking - When active the output allows current to flow to a common ground. This is best selected when different voltages are supplied.
> Sourcing - When active, current flows from a supply, through the output device and to ground. This method is best used when all devices use a single supply voltage.

This is also referred to as NPN (sinking) and PNP (sourcing). PNP is more popular. This will be covered in detail in the chapter on sensors.

2.1.1 Inputs

In smaller PLCs the inputs are normally built in and are specified when purchasing the PLC. For larger PLCs the inputs are purchased as modules, or cards, with 8 or 16 inputs of the same type on each card. For discussion purposes we will discuss all inputs as if they have been purchased as cards. The list below shows typical ranges for input voltages, and is roughly in order of popularity.

> 12-24 Vdc
> 100-120 Vac
> 10-60 Vdc
> 12-24 Vac/dc
> 5 Vdc (TTL)
> 200-240 Vac
> 48 Vdc
> 24 Vac

PLC input cards rarely supply power, this means that an external power supply is needed to supply power for the inputs and sensors. The example in Figure 2.1 shows how to connect an AC input card.

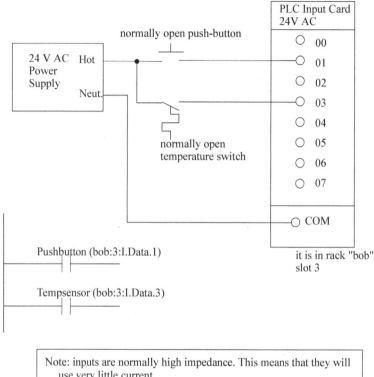

Note: inputs are normally high impedance. This means that they will use very little current.

Figure 2.2   *An AC Input Card and Ladder Logic*

In the example there are two inputs, one is a normally open push button, and the second is a temperature switch, or thermal relay. (NOTE: These symbols are standard and will be discussed later in this chapter.) Both of the switches are powered by the positive/hot output of the 24Vac power supply - this is like the positive terminal on a DC supply. Power is supplied to the left side of both of the switches. When the switches are open there is no voltage passed to the input card. If either of the switches are closed power will be supplied to the input card. In this case inputs 1 and 3 are used - notice that the inputs start at 0. The input card compares these voltages to the common. If the input voltage is within a given tolerance range the inputs will switch on. Ladder logic is shown in the figure for the inputs. Here it uses Allen Bradley notation for ControlLogix. At the top is the tag (variable name) for the rack. The input card ('I') is in slot 3, so the address for the card is bob:3.I.Data.x, where 'x' is the input bit number. These addresses can also be given alias tags to make the ladder logic less confusing.

NOTE: The design process will be much easier if the inputs and outputs are planned first, and the tags are entered before the ladder logic. Then the program is entered using the much simpler tag names.

Many beginners become confused about where connections are needed in the circuit above. The key word to remember is *circuit*, which means that there is a full loop that the voltage must be able to follow. In Figure 2.1 we can start following the circuit (loop) at the power supply. The path goes *through* the switches, *through* the input card, and back to the power supply where it flows back *through* to the start. In a full PLC implementation there will be many circuits that must each be complete.

A second important concept is the common. Here the neutral on the power supply is the common, or reference voltage. In effect we have chosen this to be our 0V reference, and all other voltages are measured relative to it. If we had a second power supply, we would also need to connect the neutral so that both neutrals would be connected to the same common. Often common and ground will be confused. The common is a reference, or datum voltage that is used for 0V, but the ground is used to prevent shocks and damage to equipment. The ground is connected under a building to a metal pipe or grid in the ground. This is connected to the electrical system of a building, to the power outlets, where the metal cases of electrical equipment are connected. When power flows through the ground it is bad. Unfortunately many engineers, and manufacturers mix up ground and common. It is very common to find a power supply with the ground and common mislabeled.

Remember - Don't mix up the ground and common. Don't connect them together if the common of your device is connected to a common on another device.

One final concept that tends to trap beginners is that each input card is isolated. This means that if you have connected a common to only one card, then the other cards are not connected. When this happens the other cards will not work properly. You must connect a common for each of the output cards.

There are many trade-offs when deciding which type of input cards to use.

- DC voltages are usually lower, and therefore safer (i.e., 12-24V).
- DC inputs are very fast, AC inputs require a longer on-time. For example, a 60Hz wave may require up to 1/60sec for reasonable recognition.
- DC voltages can be connected to larger variety of electrical systems.
- AC signals are more immune to noise than DC, so they are suited to long distances, and noisy (magnetic) environments.
- AC power is easier and less expensive to supply to equipment.
- AC signals are very common in many existing automation devices.

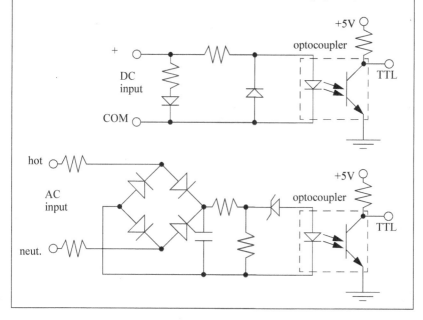

ASIDE: PLC inputs must convert a variety of logic levels to the 5Vdc logic levels used on the data bus. This can be done with circuits similar to those shown below. Basically the circuits condition the input to drive an optocoupler. This electrically isolates the external electrical circuitry from the internal circuitry. Other circuit components are used to guard against excess or reversed voltage polarity.

Figure 2.3   *Aside: PLC Input Circuits*

2.1.2 Output Modules

WARNING - ALWAYS CHECK RATED VOLTAGES AND CURRENTS FOR PLC's AND NEVER EXCEED!

As with input modules, output modules rarely supply any power, but instead act as switches. External power supplies are connected to the output card and the card will switch the power on or off for each output. Typical output voltages are listed below, and roughly ordered by popularity.

120 Vac
24 Vdc
12-48 Vac
12-48 Vdc
5Vdc (TTL)
230 Vac

These cards typically have 8 to 16 outputs of the same type and can be purchased with different current ratings. A common choice when purchasing output cards is relays, transistors or triacs. Relays are the most flexible output devices. They are capable of switching both AC and DC outputs. But, they are slower (about 10ms switching is typical), they are bulkier, they cost more, and they will wear out after millions of cycles. Relay outputs are often called dry contacts. Transistors are limited to DC outputs, and Triacs are limited to AC outputs. Transistor and triac outputs are called switched outputs.

> Dry contacts - a separate relay is dedicated to each output. This allows mixed voltages (AC or DC and voltage levels up to the maximum), as well as isolated outputs to protect other outputs and the PLC. Response times are often greater than 10ms. This method is the least sensitive to voltage variations and spikes.
> Switched outputs - a voltage is supplied to the PLC card, and the card switches it to different outputs using solid state circuitry (transistors, triacs, etc.) Triacs are well suited to AC devices requiring less than 1A. Transistor outputs use NPN or PNP transistors up to 1A typically. Their response time is well under 1ms.

ASIDE: PLC outputs must convert the 5Vdc logic levels on the PLC data bus to external voltage levels. This can be done with circuits similar to those shown below. Basically the circuits use an optocoupler to switch external circuitry. This electrically isolates the external electrical circuitry from the internal circuitry. Other circuit components are used to guard against excess or reversed voltage polarity.

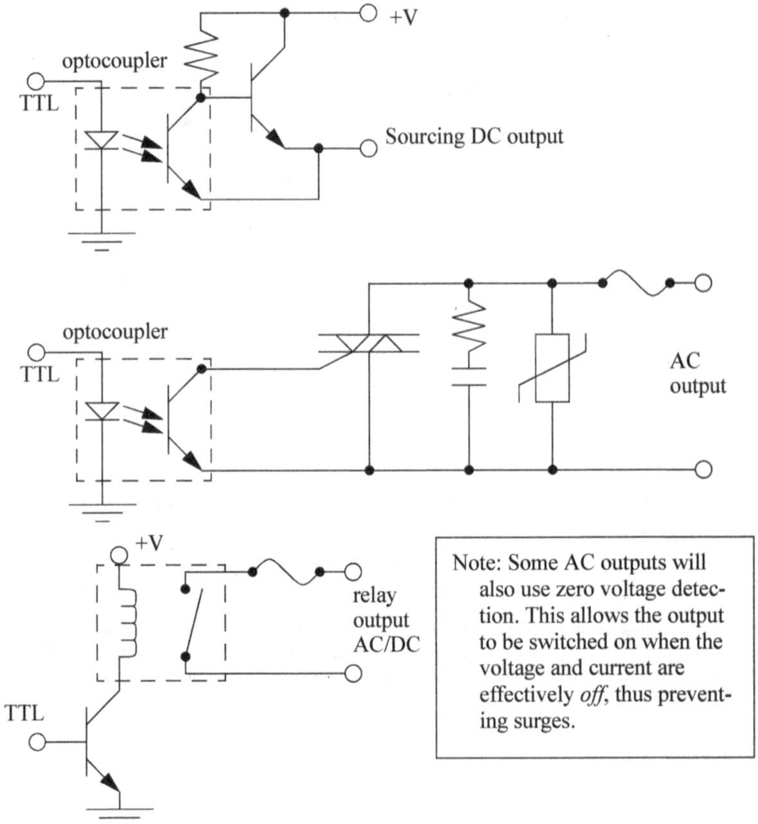

Note: Some AC outputs will also use zero voltage detection. This allows the output to be switched on when the voltage and current are effectively *off*, thus preventing surges.

Figure 2.4   **Aside: PLC Output Circuits**

20

Caution is required when building a system with both AC and DC outputs. If AC is accidentally connected to a DC transistor output it will only be on for the positive half of the cycle, and appear to be working with a diminished voltage. If DC is connected to an AC triac output it will turn on and appear to work, but you will not be able to turn it off without turning off the entire PLC.

ASIDE: A transistor is a semiconductor based device that can act as an adjustable valve. When switched off it will block current flow in both directions. While switched on it will allow current flow in one direction only. There is normally a loss of a couple of volts across the transistor. A triac is like two SCRs (or imagine transistors) connected together so that current can flow in both directions, which is good for AC current. One major difference for a triac is that if it has been switched on so that current flows, and then switched off, it will not turn off until the current stops flowing. This is fine with AC current because the current stops and reverses every 1/2 cycle, but this does not happen with DC current, and so the triac will remain on.

A major issue with outputs is mixed power sources. It is good practice to isolate all power supplies and keep their commons separate, but this is not always feasible. Some output modules, such as relays, allow each output to have its own common. Other output cards require that multiple, or all, outputs on each card share the same common. Each output card will be isolated from the rest, so each common will have to be connected. It is common for beginners to only connect the common to one card, and forget the other cards - then only one card seems to work!

The output card shown in Figure 2.1 is an example of a 24Vdc output card that has a shared common. This type of output card would typically use transistors for the outputs.

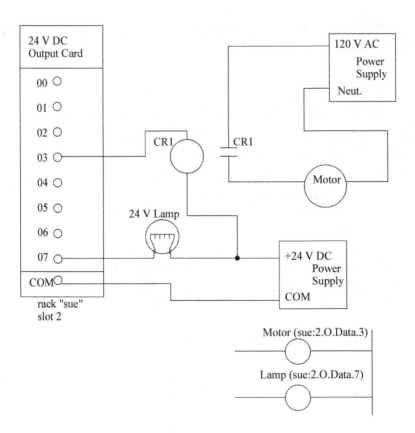

Figure 2.5   *An Example of a 24Vdc Output Card (Sinking)*

In this example the outputs are connected to a low current light bulb (lamp) and a relay coil. Consider the circuit through the lamp, starting at the 24Vdc supply. When the output *07* is on, current can flow in *07* to the *COM*, thus completing the circuit, and allowing the light to turn on. If the output is off the current cannot flow, and the light will not turn on. The output *03* for the relay is connected in a similar way. When the output *03* is on, current will flow through the relay coil to close the contacts and supply 120Vac to the motor. Ladder logic for the outputs is shown in the bottom right of the figure. The notation is for an Allen Bradley ControlLogix. The output card ('O') is in a rack labelled 'sue' in slot 2. As indicated for the input card, it is good practice to define and use an alias tag for an output (e.g. Motor) instead of using the full description (e.g. sue:2.O.Data.3). This card could have many different voltages applied from different sources, but all the power supplies would need a single shared common.

The circuits in Figure 2.1 had the sequence of power supply, then device, then PLC card, then power supply. This requires that the output card have a common. Some output schemes reverse the device and PLC card, thereby replacing the common with a voltage input. The example in Figure 2.1 is repeated in Figure 2.1 for a voltage supply card.

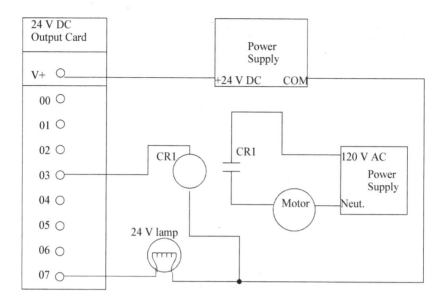

Figure 2.6   *An Example of a 24Vdc Output Card With a Voltage Input (Sourcing)*

In this example the positive terminal of the 24Vdc supply is connected to the output card directly. When an output is on power will be supplied to that output. For example, if output *07* is on then the supply voltage will be output to the lamp. Current will flow through the lamp and back to the common on the power supply. The operation is very similar for the relay switching the motor. Notice that the ladder logic (shown in the bottom right of the figure) is identical to that in Figure 2.1. With this type of output card only one power supply can be used.

We can also use relay outputs to switch the outputs. The example shown in Figure 2.1 and Figure 2.1 is repeated yet again in Figure 2.1 for relay output.

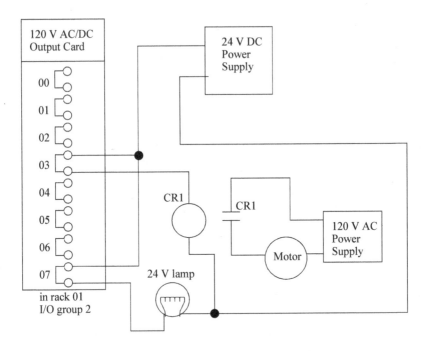

Figure 2.7   *An Example of a Relay Output Card*

In this example the 24Vdc supply is connected directly to both relays (note that this requires 2 connections now, whereas the previous example only required one.) When an output is activated the output switches on and power is delivered to the output devices. This layout is more similar to Figure 2.1 with the outputs supplying voltage, but the relays could also be used to connect outputs to grounds, as in Figure 2.1. When using relay outputs it is possible to have each output isolated from the next. A relay output card could have AC and DC outputs beside each other.

2.2 RELAYS

Although relays are rarely used for control logic, they are still essential for switching large power loads. Some important terminology for relays is given below.

Contactor - Special relays for switching large current loads.
Motor Starter - Basically a contactor in series with an overload relay to cut off when too much current is drawn.
Arc Suppression - when any relay is opened or closed an arc will jump. This becomes a major problem with large relays. On relays switching AC this problem can be overcome by opening the relay when the voltage goes to zero (while crossing between negative and positive). When switching DC loads this problem can be minimized by blowing pressurized gas across during opening to suppress the arc formation.
AC coils - If a normal coil is driven by AC power the contacts will vibrate open and closed at the frequency of the AC power. This problem is overcome by relay manufacturers by adding a shading pole to the internal construction of the relay.

The most important consideration when selecting relays, or relay outputs on a PLC, is the rated current and voltage. If the rated voltage is exceeded, the contacts will wear out prematurely, or if the voltage is too high fire is possible. The rated current is the maximum current that should be used. When this is exceeded the device will become too hot, and it will fail sooner. The rated values are typically given for both AC and DC, although DC ratings are lower than AC. If the actual loads used are below the rated values the relays should work well indefinitely. If the values are exceeded a small amount the life of the relay will be shortened accordingly. Exceeding the values significantly may lead to immediate failure and permanent damage. Please note that relays may also include minimum ratings that should also be observed to ensure proper operation and long life.

- Rated Voltage - The suggested operation voltage for the coil. Lower levels can result in failure to operate, voltages above shorten life.
- Rated Current - The maximum current before contact damage occurs (welding or melting).

2.3 A CASE STUDY

(Try the following case without looking at the solution in Figure 2.1.) An electrical layout is needed for a hydraulic press. The press uses a 24Vdc double actuated solenoid valve to advance and retract the press. This device has a single common and two input wires. Putting 24Vdc on one wire will cause the press to advance, putting 24Vdc on the second wire will cause it to retract. The press is driven by a large hydraulic pump that requires 220Vac rated at 20A, this should be running as long as the press is on. The press is outfitted with three push buttons, one is a NC stop button, the other is a NO manual retract button, and the third is a NO start automatic cycle button. There are limit switches at the top and bottom of the press travels that must also be connected.

SOLUTION

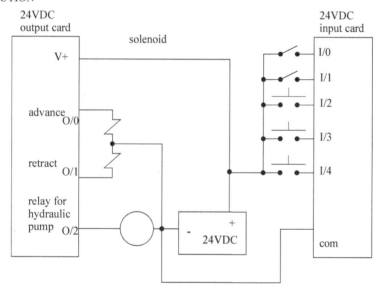

Figure 2.8   *Case Study for Press Wiring*

The input and output cards were both selected to be 24Vdc so that they may share a single 24Vdc power supply. In this case the solenoid valve was wired directly to the output card, while the hydraulic pump was connected indirectly using a relay (only the coil is shown for simplicity). This decision was primarily made because the hydraulic pump requires more current than any PLC can handle, but a relay would be relatively easy to purchase and install for that load. All of the input switches are connected to the same supply and to the inputs.

2.4 ELECTRICAL WIRING DIAGRAMS

When a controls cabinet is designed and constructed ladder diagrams are used to document the wiring. A basic wiring diagram is shown in Figure 2.1. In this example the system would be supplied with AC power (L1 is 120Vac or 220Vac) between the left and right (neutral or 0V) rails. The lines of these diagrams are numbered, and these numbers are typically used to number wires when building the electrical system. The switch before line 010 is a master disconnect for the power to the entire system. A fuse is used after the disconnect to limit the maximum current drawn by the system. Line 020 of the diagram is used to control power to the outputs of the system. The stop button is normally closed, while the start button is normally open. The branch, and output of the rung are CR1, which is a master control relay. The PLC receives power on line 30 of the diagram.

The inputs to the PLC are all AC, and are shown on lines 050 to 090. Notice that Input I/0 is a set of contacts on the MCR *CR1*. The three other inputs are a normally open push button (line 060), a limit switch (070) and a normally closed push button (080). A DC power supply is shown on line 100, to supply 24Vdc to the outputs. This powers the relay outputs of the PLC to control a green indicator light (200), a red indicator light (210), a solenoid (220), and another relay (230). The relay on 230 switches a set of contacts (040) that turn on the *drill station*.

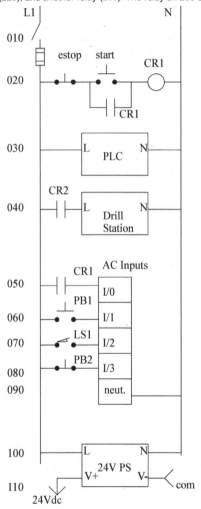

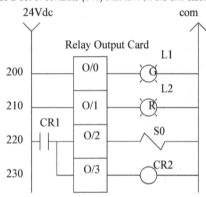

Figure 2.9   *A Ladder Wiring Diagram*

In the wiring diagram the choice of a normally close stop button and a normally open start button are intentional. Consider line 020 in the wiring diagram. If the stop button is pushed it will open the switch, and power will not be able to flow to the control relay and output power will shut off. If the stop button is damaged, say by a wire falling off, the power will also be lost and the system will shut down - safely. If the stop button used was normally open and this happened the system would continue to operate while the stop button was unable to shut down the power. Now consider the start button. If the button was damaged, say a wire was disconnected, it would be unable to start the system, thus leaving the system unstarted and safe. In summary, all buttons that stop a system should be normally closed, while all buttons that start a system should be normally open.

2.4.1 JIC Wiring Symbols

To standardize electrical schematics, the Joint International Committee (JIC) symbols were developed, these are shown in Figure 2.1, Figure 2.1 and Figure 2.1.

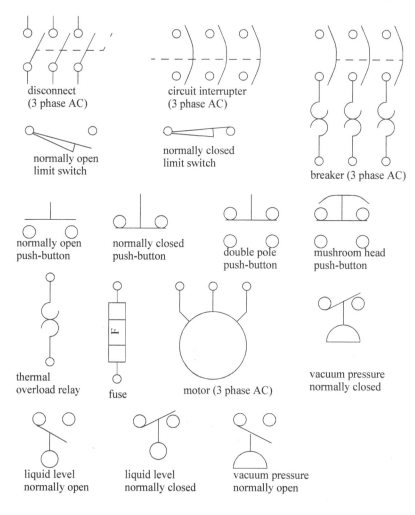

Figure 2.10   **JIC Schematic Symbols**

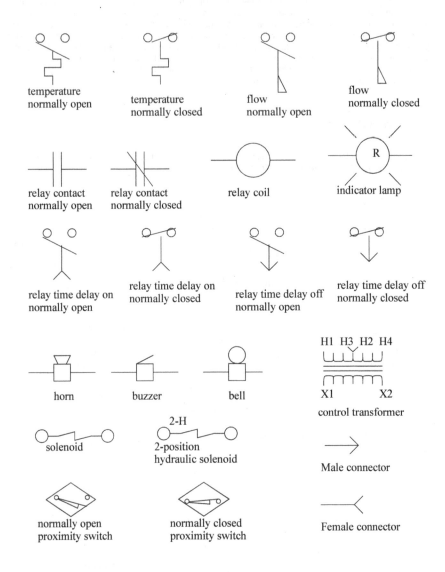

temperature
normally open

temperature
normally closed

flow
normally open

flow
normally closed

relay contact
normally open

relay contact
normally closed

relay coil

indicator lamp

relay time delay on
normally open

relay time delay on
normally closed

relay time delay off
normally open

relay time delay off
normally closed

horn

buzzer

bell

H1 H3 H2 H4

X1   X2

control transformer

solenoid

2-H

2-position
hydraulic solenoid

Male connector

normally open
proximity switch

normally closed
proximity switch

Female connector

Figure 2.11   **JIC Schematic Symbols**

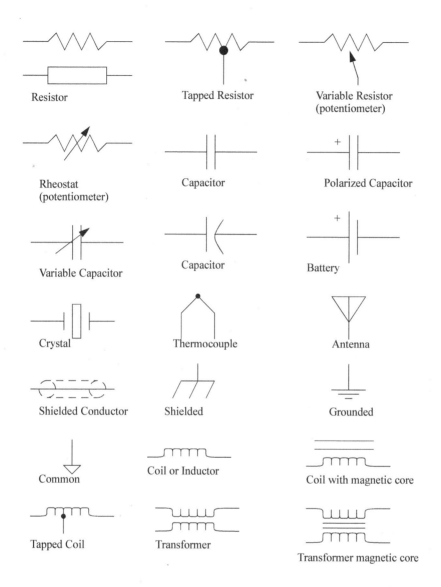

Resistor

Tapped Resistor

Variable Resistor
(potentiometer)

Rheostat
(potentiometer)

Capacitor

Polarized Capacitor

Variable Capacitor

Capacitor

Battery

Crystal

Thermocouple

Antenna

Shielded Conductor

Shielded

Grounded

Common

Coil or Inductor

Coil with magnetic core

Tapped Coil

Transformer

Transformer magnetic core

Figure 2.12   *JIC Schematic Symbols*

2.5 SUMMARY

- PLC inputs condition AC or DC inputs to be detected by the logic of the PLC.
- Outputs are transistors (DC), triacs (AC) or relays (AC and DC).
- Input and output addresses are a function of the card location/tag name and input bit number.
- Electrical system schematics are documented with diagrams that look like ladder logic.

2.6 PRACTICE PROBLEMS

(Note: Problem solutions are available at http://sites.google.com/site/automatedmanufacturingsystems/)

1. Can a PLC input switch a relay coil to control a motor?

2. How do input and output cards act as an interface between the PLC and external devices?

3. What is the difference between wiring a sourcing and sinking output?

4. What is the difference between a motor starter and a contactor?

5. Is AC or DC easier to interrupt?

6. What can happen if the rated voltage on a device is exceeded?

7. What are the benefits of input/output modules?

8. (for electrical engineers) Explain the operation of AC input and output conditioning circuits.

9. What will happen if a DC output is switched by an AC output.

10. Explain why a stop button must be normally closed and a start button must be normally open.

11. For the circuit shown in the figure below, list the input and output addresses for the PLC. If switch A controls the light, switch B the motor, and C the solenoid, write a simple ladder logic program.

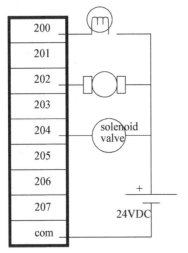
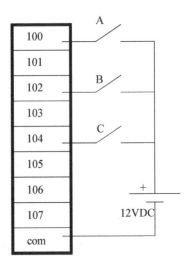

12. We have a PLC rack with a 24 VDC input card in slot 3, and a 120VAC output card in slot 2. The inputs are to be connected to 4 push buttons. The outputs are to drive a 120VAC light bulb, a 240VAC motor, and a 24VDC operated hydraulic valve. Draw the electrical connections for the inputs and outputs. Show all other power supplies and other equipment/components required.

13. You are planning a project that will be controlled by a PLC. Before ordering parts you decide to plan the basic wiring and select appropriate input and output cards. The devices that we will use for inputs are 2 limit switches, a push button and a thermal switch. The output will be for a 24Vdc solenoid valve, a 110Vac light bulb, and a 220Vac 50HP motor. Sketch the basic wiring below including PLC cards.

14. Add three push buttons as inputs to the figure below. You must also select a power supply, and show all neces-

sary wiring.

1
com
2
com
3
com
4
com
5
com

15. Three 120Vac outputs are to be connected to the output card below. Show the 120Vac source, and all wiring.

V
00
01
02
03
04
05
06
07

16. Sketch the wiring for PLC outputs that are listed below.
    - a double acting hydraulic solenoid valve (with two coils)
    - a 24Vdc lamp
    - a 120 Vac high current lamp
    - a low current 12Vdc motor

17. Draw a ladder wiring diagram for a system that has 2 PNP inputs, and 2 solenoid outputs. All inputs and out-puts are 24Vdc. Include ALL safety circuitry.

2.7 ASSIGNMENT PROBLEMS

1. Describe what could happen if a normally closed start button was used on a system, and the wires to the button were cut.

2. Describe what could happen if a normally open stop button was used on a system and the wires to the button were cut.

3. a) For the input ('in') and output ('out') cards below, add three output lights and three normally open push button

inputs. b) Redraw the outputs so that it uses a relay output card.

in:0.I.Data.x

0
1
2
3
4
5
6
7
com

out:1.O.Data.x

V
0
1
2
3
4
5
6
7

4. Draw an electrical wiring (ladder) diagram for PLC outputs that are listed below.
   - a solenoid controlled hydraulic valve
   - a 24Vdc lamp
   - a 120 Vac high current lamp
   - a low current 12Vdc motor

5. Draw an electrical ladder diagram for a PLC that has a PNP and an NPN sensor for inputs. The outputs are two small indicator lights. You should use proper symbols for all components. You must also include all safety devices including fuses, disconnects, MCRs, etc...

6. Draw an electrical wiring diagram for a PLC controlling a system with both NPN and PNP input sensors. The outputs include an indicator light and a relay to control a 20A motor load. Include ALL safety circuitry.

7. Develop a wiring diagram for a system that has the following elements. Include all safety circuitry.
   2 NPN proximity sensors
   2 N.O. pushbuttons
   3 solenoid outputs
   A 440Vac 3ph. 20HP (i.e., large) motor

8. Develop a ladder wiring diagram, including all safety circuitry that uses an PNP and an NPN input sensors. The outputs is a relay controlled AC light.

9. Draw a complete ladder wiring diagram for a PLC based control system with the following components. Include all necessary safety circuitry.
   1 large 3 phase (AC) motor
   2 PNP sensors
   1 NO pushbutton
   1 NC pushbutton
   1 solenoid output

2.1 PRACTICE PROBLEM SOLUTIONS

1. no - a plc OUTPUT can switch a relay

2. input cards are connected to sensors to determine the state of the system. Output cards are connected to actuators that can drive the process.

3. sourcing outputs supply current that will pass through an electrical load to ground. Sinking inputs allow current to flow from the electrical load, to the common.

4. a motor starter typically has three phases

5. AC is easier, it has a zero crossing

6. it will lead to premature failure

7. by using separate modules, a PLC can be customized for different applications. If a single module fails, it can be replaced quickly, without having to replace the entire controller.

8. AC input conditioning circuits will rectify an AC input to a DC waveform with a ripple. This will be smoothed, and reduced to a reasonable voltage level to drive an optocoupler. An AC output circuit will switch an AC output with a triac, or a relay.

9. an AC output is a triac. When a triac output is turned off, it will not actually turn off until the AC voltage goes to 0V. Because DC voltages don't go to 0V, it will never turn off.

10. If a NC stop button is damaged, the machine will act as if the stop button was pushed and shut down safely. If a NO start button is damaged the machine will not be able to start.

11.

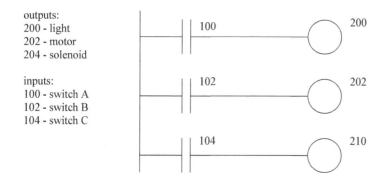

```
outputs:
200 - light
202 - motor
204 - solenoid

inputs:
100 - switch A
102 - switch B
104 - switch C
```

12.

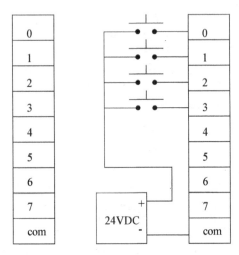

| 0 |
| 1 |
| 2 |
| 3 |
| 4 |
| 5 |
| 6 |
| 7 |
| com |

13.

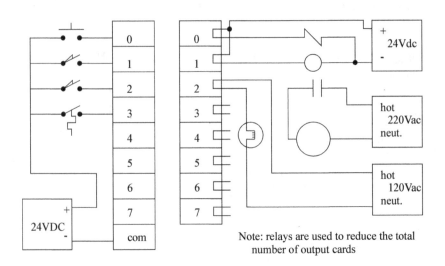

Note: relays are used to reduce the total
number of output cards

14.

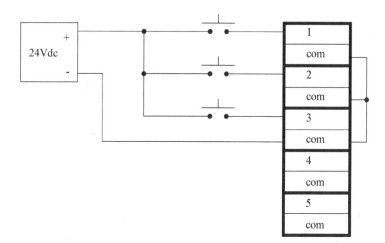

15.

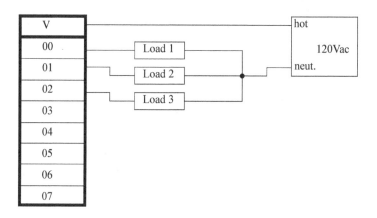

16.

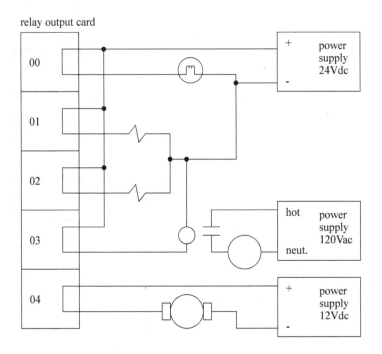

relay output card

17.

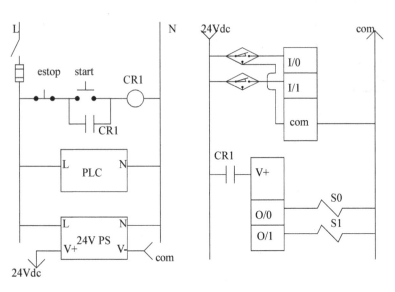

3LOGICAL SENSORS

Topics:

- Sensor wiring; switches, TTL, sourcing, sinking
- Proximity detection; contact switches, photo-optics, capacitive, inductive and ultrasonic

Objectives:

- Understand the different types of sensor outputs.
- Know the basic sensor types and understand application issues.

Sensors allow a PLC to detect the state of a process. Logical sensors can only detect a state that is either true or false. Examples of physical phenomena that are typically detected are listed below.

- inductive proximity - is a metal object nearby?
- capacitive proximity - is a dielectric object nearby?
- optical presence - is an object breaking a light beam or reflecting light?
- mechanical contact - is an object touching a switch?

Recently, the cost of sensors has dropped and they have become commodity items, typically between $50 and $100. They are available in many forms from multiple vendors such as Allen Bradley, Omron, Hyde Park and Turck. In applications sensors are interchangeable between PLC vendors, but each sensor will have specific interface requirements.

This chapter will begin by examining the various electrical wiring techniques for sensors, and conclude with an examination of many popular sensor types.

3.1 SENSOR WIRING

When a sensor detects a logical change it must signal that change to the PLC. This is typically done by switching a voltage or current on or off. In some cases the output of the sensor is used to switch a load directly, completely eliminating the PLC. Typical outputs from sensors (and inputs to PLCs) are listed below in relative popularity.

Sinking/Sourcing - Switches current on or off.
Plain Switches - Switches voltage on or off.
Solid State Relays - These switch AC outputs.
TTL (Transistor Transistor Logic) - Uses 0V and 5V to indicate logic levels.

3.1.1 Switches

The simplest example of sensor outputs are switches and relays. A simple example is shown in Figure 3.1.

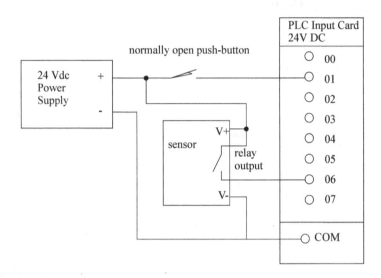

normally open push-button

24 Vdc +
Power
Supply -

sensor

V+

relay
output

V-

○ 00
○ 01
○ 02
○ 03
○ 04
○ 05
○ 06
○ 07

○ COM

Figure 3.1   *An Example of Switched Sensors*

In the figure a NO contact switch is connected to input *01*. A sensor with a relay output is also shown. The sensor must be powered separately, therefore the *V+* and *V-* terminals are connected to the power supply. The output of the sensor will become active when a phenomenon has been detected. This means the internal switch (probably a relay) will be closed allowing current to flow and the positive voltage will be applied to input *06*.

3.1.2 Transistor Transistor Logic (TTL)

Transistor-Transistor Logic (TTL) is based on two voltage levels, 0V for false and 5V for true. The voltages can actually be slightly larger than 0V, or lower than 5V and still be detected correctly. This method is very susceptible to electrical noise on the factory floor, and should only be used when necessary. TTL outputs are common on electronic devices and computers, and will be necessary sometimes. When connecting to other devices simple circuits can be used to improve the signal, such as the Schmitt trigger in Figure 3.1.

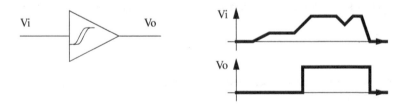

Vi   Vo   Vi

Vo

Figure 3.2   *A Schmitt Trigger*

A Schmitt trigger will receive an input voltage between 0-5V and convert it to 0V or 5V. If the voltage is in an ambiguous range, about 1.5-3.5V it will be ignored.

If a sensor has a TTL output the PLC must use a TTL input card to read the values. If the TTL sensor is being used for other applications it should be noted that the maximum current output is normally about 20mA.

3.1.3 Sinking/Sourcing

Sinking sensors allow current to flow into the sensor to the voltage common, while sourcing sensors allow current to flow out of the sensor from a positive source. For both of these methods the emphasis is on current flow, not voltage. By using current flow, instead of voltage, many of the electrical noise problems are reduced.

When discussing sourcing and sinking we are referring to the *output* of the sensor that is acting like a switch. In fact the output of the sensor is normally a transistor, that will act like a switch (with some voltage loss). A PNP transistor is used for the sourcing output, and an NPN transistor is used for the sinking input. When discussing these sensors the term sourcing is often interchanged with PNP, and sinking with NPN. A simplified example of a sinking output sensor is shown in Figure 3.1. The sensor will have some part that deals with detection, this is on the left. The sensor needs a voltage supply to operate, so a voltage supply is needed for the sensor. If the sensor has detected some phenomenon then it will trigger the active line. The active line is directly connected to an NPN transistor. (Note: for an NPN transistor the arrow always points away from the center.) If the voltage to the transistor on the *active line* is 0V, then the transistor will not allow current to flow into the sensor. If the voltage on the active line becomes larger (say 12V) then the transistor will switch on and allow current to flow into the sensor to the common.

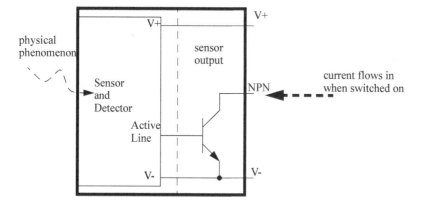

Aside: The sensor responds to a physical phenomenon. If the sensor is inactive (nothing detected) then the active line is low and the transistor is off, this is like an open switch. That means the NPN output will have no current in/out. When the sensor is active, it will make the active line high. This will turn on the transistor, and effectively close the switch. This will allow current to flow into the sensor to ground (hence sinking). The voltage on the NPN output will be pulled down to V-. Note: the voltage will always be 1-2V higher because of the transistor. When the sensor is off, the NPN output will float, and any digital circuitry needs to contain a pull-up resistor.

Figure 3.3   *A Simplified NPN/Sinking Sensor*

Sourcing sensors are the complement to sinking sensors. The sourcing sensors use a PNP transistor, as shown in Figure 3.1. (Note: PNP transistors are always drawn with the arrow pointing to the center.) When the sensor is inactive the active line stays at the V+ value, and the transistor stays switched off. When the sensor becomes active the active line will be made 0V, and the transistor will allow current to flow out of the sensor.

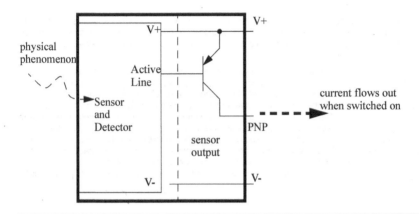

physical phenomenon

Active Line

Sensor and Detector

V+

V-

sensor output

PNP

V+

V-

current flows out when switched on

Aside: The sensor responds to the physical phenomenon. If the sensor is inactive (nothing detected) then the active line is high and the transistor is off, this is like an open switch. That means the PNP output will have no current in/out. When the sensor is active, it will make the active line high. This will turn on the transistor, and effectively close the switch. This will allow current to flow from V+ through the sensor to the output (hence sourcing). The voltage on the PNP output will be pulled up to V+. Note: the voltage will always be 1-2V lower because of the transistor. When off, the PNP output will float, if used with digital circuitry a pull-down resistor will be needed.

Figure 3.4   **A Simplified Sourcing/PNP Sensor**

Most NPN/PNP sensors are capable of handling currents up to a few amps, and they can be used to switch loads directly. (Note: always check the documentation for rated voltages and currents.) An example using sourcing and sinking sensors to control lights is shown in Figure 3.1. (Note: This example could be for a motion detector that turns on lights in dark hallways.)

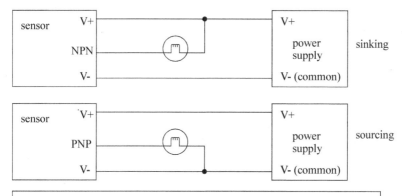

Note: remember to check the current and voltage ratings for the sensors.

Note: When marking power terminals, there will sometimes be two sets of markings. The more standard is V+ and COM, but sometimes you will see devices and power supplies without a COM (common), in this case assume the V- is the common.

Figure 3.5   ***Direct Control Using NPN/PNP Sensors***

In the sinking system in Figure 3.1 the light has V+ applied to one side. The other side is connected to the NPN *output* of the sensor. When the sensor turns on the current will be able to flow through the light, into the output to V- common. (Note: Yes, the current will be allowed to flow into the output for an NPN sensor.) In the sourcing arrangement the light will turn on when the output becomes active, allowing current to flow from the V+, thought the sensor, the light and to V- (the common).

At this point it is worth stating the obvious - The output of a sensor will be an input for a PLC. And, as we saw with the NPN sensor, this does not necessarily indicate where current is flowing. There are two viable approaches for connecting sensors to PLCs. The first is to always use PNP sensors and normal voltage input cards. The second option is to purchase input cards specifically designed for sourcing or sinking sensors. An example of a PLC card for sinking sensors is shown in Figure 3.1.

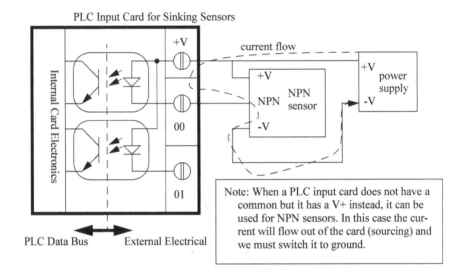

PLC Input Card for Sinking Sensors

Note: When a PLC input card does not have a common but it has a V+ instead, it can be used for NPN sensors. In this case the current will flow out of the card (sourcing) and we must switch it to ground.

ASIDE: This card is shown with 2 optocouplers (one for each output). Inside these devices the is an LED and a phototransistor, but no electrical connection. These devices are used to isolate two different electrical systems. In this case they protect the 5V digital levels of the PLC computer from the various external voltages and currents.

Figure 3.6   *A PLC Input Card for Sinking Sensors*

The dashed line in the figure represents the circuit, or current flow path when the sensor is active. This path enters the PLC input card first at a V+ terminal (Note: there is no common on this card) and flows through an optocoupler. This current will use light to turn on a phototransistor to tell the computer in the PLC the input current is flowing. The current then leaves the card at input *00* and passes through the sensor to V-. When the sensor is inactive the current will not flow, and the light in the optocoupler will be off. The optocoupler is used to help protect the PLC from electrical problems outside the PLC.

The input cards for PNP sensors are similar to the NPN cards, as shown in Figure 3.1.

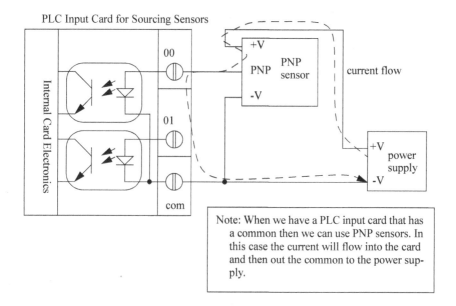

PLC Input Card for Sourcing Sensors

Note: When we have a PLC input card that has a common then we can use PNP sensors. In this case the current will flow into the card and then out the common to the power supply.

Figure 3.7   **PLC Input Card for Sourcing Sensors**

The current flow loop for an active sensor is shown with a dashed line. Following the path of the current we see that it begins at the *V+*, passes through the sensor, in the input *00*, through the optocoupler, out the common and to the *V-*.

Wiring is a major concern with PLC applications, so to reduce the total number of wires, two wire sensors have become popular. But, by integrating three wires worth of function into two, we now couple the power supply and sensing functions into one. Two wire sensors are shown in Figure 3.1.

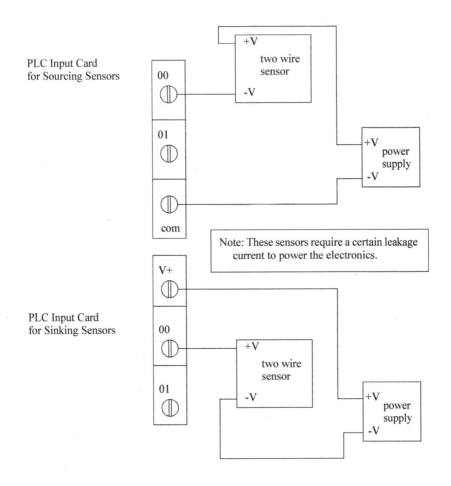

PLC Input Card
for Sourcing Sensors

00

+V

two wire
sensor

-V

01

+V

power
supply

-V

com

Note: These sensors require a certain leakage
current to power the electronics.

V+

PLC Input Card
for Sinking Sensors

00

+V

two wire
sensor

-V

01

+V

power
supply

-V

Figure 3.8   **Two Wire Sensors**

A two wire sensor can be used as either a sourcing or sinking input. In both of these arrangements the sensor will require a small amount of current to power the sensor, but when active it will allow more current to flow. This requires input cards that will allow a small amount of current to flow (called the leakage current), but also be able to detect when the current has exceeded a given value.

When purchasing sensors and input cards there are some important considerations. Most modern sensors have both PNP and NPN outputs, although if the choice is not available, PNP is the more popular choice. PLC cards can be confusing to buy, as each vendor refers to the cards differently. To avoid problems, look to see if the card is specifically for sinking or sourcing sensors, or look for a V+ (sinking) or COM (sourcing). Some vendors also sell cards that will allow you to have NPN and PNP inputs mixed on the same card.

When drawing wiring diagrams the symbols in Figure 3.1 are used for sinking and sourcing proximity sensors. Notice that in the sinking sensor when the switch closes (moves up to the terminal) it contacts the common. Closing the switch in the sourcing sensor connects the output to the V+. On the physical sensor the wires are color coded as indicated in the diagram. The brown wire is positive, the blue wire is negative and the output is white for sinking and black for sourcing. The outside shape of the sensor may change for other devices, such as photo sensors which are often shown as round circles.

NPN (sinking)

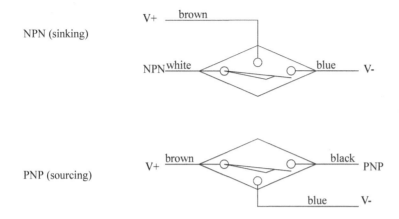

PNP (sourcing)

Figure 3.9   **Sourcing and Sinking Schematic Symbols**

3.1.4 Solid State Relays

Solid state relays switch AC currents. These are relatively inexpensive and are available for large loads. Some sensors and devices are available with these as outputs.

3.2 PRESENCE DETECTION

There are two basic ways to detect object presence; contact and proximity. Contact implies that there is mechanical contact and a resulting force between the sensor and the object. Proximity indicates that the object is near, but contact is not required. The following sections examine different types of sensors for detecting object presence. These sensors account for a majority of the sensors used in applications.

3.2.1 Contact Switches

Contact switches are available as normally open and normally closed. Their housings are reinforced so that they can take repeated mechanical forces. These often have rollers and wear pads for the point of contact. Lightweight contact switches can be purchased for less than a dollar, but heavy duty contact switches will have much higher costs. Examples of applications include motion limit switches and part present detectors.

3.2.2 Reed Switches

Reed switches are very similar to relays, except a permanent magnet is used instead of a wire coil. When the magnet is far away the switch is open, but when the magnet is brought near the switch is closed as shown in Figure 3.1. These are very inexpensive an can be purchased for a few dollars. They are commonly used for safety screens and doors because they are harder to *trick* than other sensors.

> Note: With this device the magnet is moved towards the reed switch. As it gets
> closer the switch will close. This allows proximity detection without contact, but
> requires that a separate magnet be attached to a moving part.

Figure 3.10   ***Reed Switch***

3.2.3 Optical (Photoelectric) Sensors

Light sensors have been used for almost a century - originally photocells were used for applications such as reading audio tracks on motion pictures. But modern optical sensors are much more sophisticated.

Optical sensors require both a light source (emitter) and detector. Emitters will produce light beams in the visible and invisible spectrums using LEDs and laser diodes. Detectors are typically built with photodiodes or phototransistors. The emitter and detector are positioned so that an object will block or reflect a beam when present. A basic optical sensor is shown in Figure 3.1.

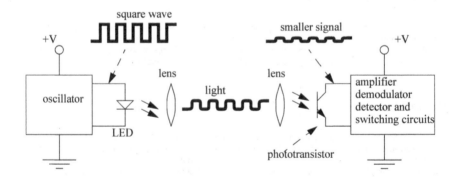

Figure 3.11   ***A Basic Optical Sensor***

In the figure the light beam is generated on the left, focused through a lens. At the detector side the beam is focused on the detector with a second lens. If the beam is broken the detector will indicate an object is present. The oscillating light wave is used so that the sensor can filter out normal light in the room. The light from the emitter is turned on and off at a set frequency. When the detector receives the light it checks to make sure that it is at the same frequency. If light is being received at the right frequency then the beam is not broken. The frequency of oscillation is in the KHz range, and too fast to be noticed. A side effect of the frequency method is that the sensors can be used with lower power at longer distances.

An emitter can be set up to point directly at a detector, this is known as opposed mode. When the beam is broken the part will be detected. This sensor needs two separate components, as shown in Figure 3.1. This arrangement works well with opaque and reflective objects with the emitter and detector separated by distances of up to hundreds of feet.

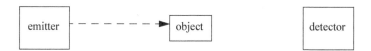

Figure 3.12   *Opposed Mode Optical Sensor*

Having the emitter and detector separate increases maintenance problems, and alignment is required. A preferred solution is to house the emitter and detector in one unit. But, this requires that light be reflected back as shown in Figure 3.1. These sensors are well suited to larger objects up to a few feet away.

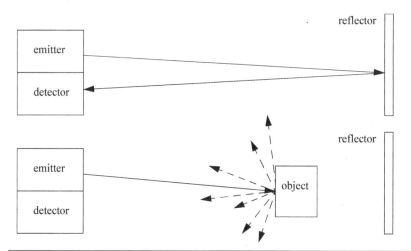

Note: the reflector is constructed with polarizing screens oriented at 90 deg. angles. If the light is reflected back directly the light does not pass through the screen in front of the detector. The reflector is designed to rotate the phase of the light by 90 deg., so it will now pass through the screen in front of the detector.

Figure 3.13   *Retroreflective Optical Sensor*

In the figure, the emitter sends out a beam of light. If the light is returned from the reflector most of the light beam is returned to the detector. When an object interrupts the beam between the emitter and the reflector the beam is no longer reflected back to the detector, and the sensor becomes active. A potential problem with this sensor is that reflective objects could return a good beam. This problem is overcome by polarizing the light at the emitter (with a filter), and then using a polarized filter at the detector. The reflector uses small cubic reflectors and when the light is reflected the polarity is rotated by 90 degrees. If the light is reflected off the object the light will not be rotated by 90 degrees. So the polarizing filters on the emitter and detector are rotated by 90 degrees, as shown in Figure 3.1. The reflector is very similar to reflectors used on bicycles.

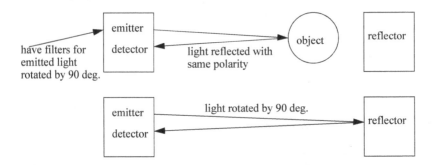

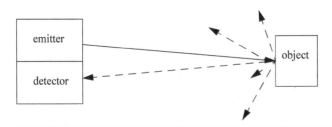

Note: with diffuse reflection the light is scattered. This reduces the quantity of light returned. As a result the light needs to be amplified using lenses.

Figure 3.14   ***Polarized Light in Retroreflective Sensors***

For retroreflectors the reflectors are quite easy to align, but this method still requires two mounted components. A diffuse sensors is a single unit that does not use a reflector, but uses focused light as shown in Figure 3.1.

Figure 3.15   ***Diffuse Optical Sensor***

Diffuse sensors use light focused over a given range, and a sensitivity adjustment is used to select a distance. These sensors are the easiest to set up, but they require well controlled conditions. For example if it is to pick up light and dark colored objects problems would result.

When using opposed mode sensors the emitter and detector must be aligned so that the emitter beam and detector window overlap, as shown in Figure 3.1. Emitter beams normally have a cone shape with a small angle of divergence (a few degrees of less). Detectors also have a cone shaped volume of detection. Therefore when aligning opposed mode sensor care is required not just to point the emitter at the detector, but also the detector at the emitter. Another factor that must be considered with this and other sensors is that the light intensity decreases over distance, so the sensors will have a limit to separation distance.

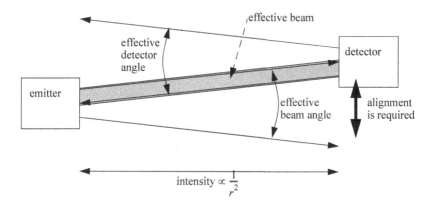

Figure 3.16   *Beam Divergence and Alignment*

If an object is smaller than the width of the light beam it will not be able to block the beam entirely when it is in front as shown in Figure 3.1. This will create difficulties in detection, or possibly stop detection altogether. Solutions to this problem are to use narrower beams, or wider objects. Fiber optic cables may be used with an opposed mode optical sensor to solve this problem, however the maximum effective distance is reduced to a couple feet.

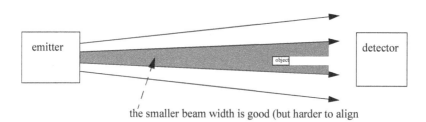

the smaller beam width is good (but harder to align

Figure 3.17   *The Relationship Between Beam Width and Object Size*

Separated sensors can detect reflective parts using reflection as shown in Figure 3.1. The emitter and detector are positioned so that when a reflective surface is in position the light is returned to the detector. When the surface is not present the light does not return.

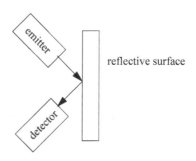

Figure 3.18   *Detecting Reflecting Parts*

Other types of optical sensors can also focus on a single point using beams that converge instead of diverge. The emitter beam is focused at a distance so that the light intensity is greatest at the focal distance. The detector can look at the point from another angle so that the two centerlines of the emitter and detector intersect at the point of interest. If an object is present before or after the focal point the detector will not see the reflected light. This technique can also be used to detect multiple points and ranges, as shown in Figure 3.1 where the net angle of refraction by the lens determines which detector is used. This type of approach, with many more detectors, is used for range sensing systems.

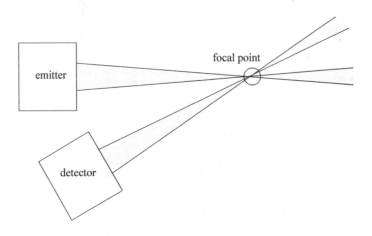

Figure 3.19   *Point Detection Using Focused Optics*

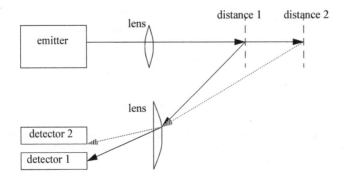

Figure 3.20   *Multiple Point Detection Using Optics*

Some applications do not permit full sized photooptic sensors to be used. Fiber optics can be used to separate the emitters and detectors from the application. Some vendors also sell photosensors that have the phototransistors and LEDs separated from the electronics.

Light curtains are an array of beams, set up as shown in Figure 3.1. If any of the beams are broken it indicates that somebody has entered a workcell and the machine needs to be shut down. This is an inexpensive replacement for some mechanical cages and barriers.

Figure 3.21   **A Light Curtain**

The optical reflectivity of objects varies from material to material as shown in Figure 3.1. These values show the percentage of incident light on a surface that is reflected. These values can be used for relative comparisons of materials and estimating changes in sensitivity settings for sensors.

		Reflectivity
nonshiny materials	Kodak white test card	90%
	white paper	80%
	kraft paper, cardboard	70%
	lumber (pine, dry, clean)	75%
	rough wood pallet	20%
	beer foam	70%
	opaque black nylon	14%
	black neoprene	4%
	black rubber tire wall	1.5%
shiny/transparent materials	clear plastic bottle	40%
	translucent brown plastic bottle	60%
	opaque white plastic	87%
	unfinished aluminum	140%
	straightened aluminum	105%
	unfinished black anodized aluminum	115%
	stainless steel microfinished	400%
	stainless steel brushed	120%

> Note: For shiny and transparent materials the reflectivity can be higher than 100% because of the return of ambient light.

Figure 3.22   **Table of Reflectivity Values for Different Materials [Banner Handbook of Photoelectric Sensing]**

3.2.4 Capacitive Sensors

Capacitive sensors are able to detect most materials at distances up to a few centimeters. Recall the basic relationship for capacitance.

$$C = \frac{Ak}{d}$$

where, C = capacitance (Farads)
k = dielectric constant
A = area of plates
d = distance between plates (electrodes)

In the sensor the area of the plates and distance between them is fixed. But, the dielectric constant of the space around them will vary as different materials are brought near the sensor. An illustration of a capacitive sensor is shown in Figure 3.1. an oscillating field is used to determine the capacitance of the plates. When this changes beyond a selected sensitivity the sensor output is activated.

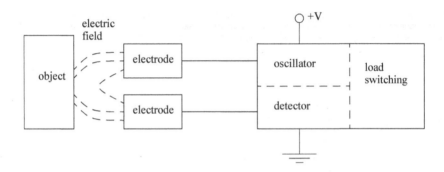

NOTE: For this sensor the proximity of any material near the electrodes will increase the capacitance. This will vary the magnitude of the oscillating signal and the detector will decide when this is great enough to determine proximity.

Figure 3.23   **A Capacitive Sensor**

These sensors work well for insulators (such as plastics) that tend to have high dielectric coefficients, thus increasing the capacitance. But, they also work well for metals because the conductive materials in the target appear as larger electrodes, thus increasing the capacitance as shown in Figure 3.1. In total the capacitance changes are normally in the order of pF.

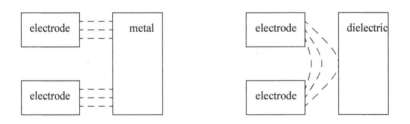

Figure 3.24   **Dielectrics and Metals Increase the Capacitance**

The sensors are normally made with rings (not plates) in the configuration shown in Figure 3.1. In the figure the two inner metal rings are the capacitor electrodes, but a third outer ring is added to compensate for variations. Without the compensator ring the sensor would be very sensitive to dirt, oil and other contaminants that might stick to the sensor.

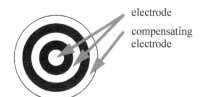

electrode

compensating
electrode

Note: the compensating electrode is used for negative feedback to make the sensor more resistant to variations, such as contaminations on the face of the sensor.

Figure 3.25   ***Electrode Arrangement for Capacitive Sensors***

A table of dielectric properties is given in Figure 3.1. This table can be used for estimating the relative size and sensitivity of sensors. Also, consider a case where a pipe would carry different fluids. If their dielectric constants are not very close, a second sensor may be desired for the second fluid.

Material	Constant	Material	Constant
ABS resin pellet	1.5-2.5	hexane	1.9
acetone	19.5	hydrogen cyanide	95.4
acetyl bromide	16.5	hydrogen peroxide	84.2
acrylic resin	2.7-4.5	isobutylamine	4.5
air	1.0	lime, shell	1.2
alcohol, industrial	16-31	marble	8.0-8.5
alcohol, isopropyl	18.3	melamine resin	4.7-10.2
ammonia	15-25	methane liquid	1.7
aniline	5.5-7.8	methanol	33.6
aqueous solutions	50-80	mica, white	4.5-9.6
ash (fly)	1.7	milk, powdered	3.5-4
bakelite	3.6	nitrobenzene	36
barley powder	3.0-4.0	neoprene	6-9
benzene	2.3	nylon	4-5
benzyl acetate	5	oil, for transformer	2.2-2.4
butane	1.4	oil, paraffin	2.2-4.8
cable sealing compound	2.5	oil, peanut	3.0
calcium carbonate	9.1	oil, petroleum	2.1
carbon tetrachloride	2.2	oil, soybean	2.9-3.5
celluloid	3.0	oil, turpentine	2.2
cellulose	3.2-7.5	paint	5-8
cement	1.5-2.1	paraffin	1.9-2.5
cement powder	5-10	paper	1.6-2.6
cereal	3-5	paper, hard	4.5
charcoal	1.2-1.8	paper, oil saturated	4.0
chlorine, liquid	2.0	perspex	3.2-3.5
coke	1.1-2.2	petroleum	2.0-2.2
corn	5-10	phenol	9.9-15
ebonite	2.7-2.9	phenol resin	4.9
epoxy resin	2.5-6	polyacetal (Delrin TM)	3.6
ethanol	24	polyamide (nylon)	2.5
ethyl bromide	4.9	polycarbonate	2.9
ethylene glycol	38.7	polyester resin	2.8-8.1
flour	2.5-3.0	polyethylene	2.3
FreonTM R22,R502 liq.	6.1	polypropylene	2.0-2.3
gasoline	2.2	polystyrene	3.0
glass	3.1-10	polyvinyl chloride resin	2.8-3.1
glass, raw material	2.0-2.5	porcelain	4.4-7
glycerine	47	press board	2-5

Material	Constant	Material	Constant
quartz glass	3.7	Teflon (TM), PCTFE	2.3-2.8
rubber	2.5-35	Teflon (TM), PTFE	2.0
salt	6.0	toluene	2.3
sand	3-5	trichloroethylene	3.4
shellac	2.0-3.8	urea resin	6.2-9.5
silicon dioxide	4.5	urethane	3.2
silicone rubber	3.2-9.8	vaseline	2.2-2.9
silicone varnish	2.8-3.3	water	48-88
styrene resin	2.3-3.4	wax	2.4-6.5
sugar	3.0	wood, dry	2-7
sugar, granulated	1.5-2.2	wood, pressed board	2.0-2.6
sulfur	3.4	wood, wet	10-30
sulfuric acid	84	xylene	2.4

Figure 3.26   *Dielectric Constants of Various Materials [Turck Proximity Sensors Guide]*

The range and accuracy of these sensors are determined mainly by their size. Larger sensors can have diameters of a few centimeters. Smaller ones can be less than a centimeter across, and have smaller ranges, but more accuracy.

3.2.5 Inductive Sensors

Inductive sensors use currents induced by magnetic fields to detect nearby metal objects. The inductive sensor uses a coil (an inductor) to generate a high frequency magnetic field as shown in Figure 3.1. If there is a metal object near the changing magnetic field, current will flow in the object. This resulting current flow sets up a new magnetic field that opposes the original magnetic field. The net effect is that it changes the inductance of the coil in the inductive sensor. By measuring the inductance the sensor can determine when a metal have been brought nearby.

These sensors will detect any metals, when detecting multiple types of metal multiple sensors are often used.

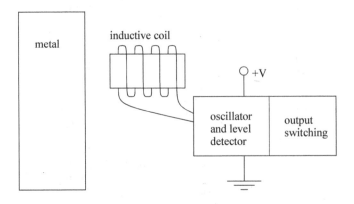

metal

inductive coil

+V

oscillator and level detector

output switching

Note: these work by setting up a high frequency field. If a target nears the field will induce eddy currents. These currents consume power because of resistance, so energy is in the field is lost, and the signal amplitude decreases. The detector examines filed magnitude to determine when it has decreased enough to switch.

Figure 3.27   *Inductive Proximity Sensor*

The sensors can detect objects a few centimeters away from the end. But, the direction to the object can be arbitrary as shown in Figure 3.1. The magnetic field of the unshielded sensor covers a larger volume around the head of the coil. By adding a shield (a metal jacket around the sides of the coil) the magnetic field becomes smaller, but also more directed. Shields will often be available for inductive sensors to improve their directionality and accuracy.

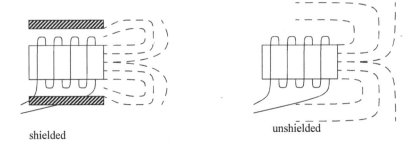

shielded

unshielded

Figure 3.28   *Shielded and Unshielded Sensors*

3.2.6 Ultrasonic

An ultrasonic sensor emits a sound above the normal hearing threshold of 16KHz. The time that is required for the sound to travel to the target and reflect back is proportional to the distance to the target. The two common types of sensors are;

electrostatic - uses capacitive effects. It has longer ranges and wider bandwidth, but is more sensitive to factors such as humidity.
piezoelectric - based on charge displacement during strain in crystal lattices. These are rugged and inexpensive.

These sensors can be very effective for applications such as fluid levels in tanks and crude distance measurement.

3.2.7 Hall Effect

Hall effect switches are basically transistors that can be switched by magnetic fields. Their applications are very similar to reed switches, but because they are solid state they tend to be more rugged and resist vibration. Automated machines often use these to do initial calibration and detect end stops.

3.2.8 Fluid Flow

We can also build more complex sensors out of simpler sensors. The example in Figure 3.1 shows a metal float in a tapered channel. As the fluid flow rate increases the pressure forces the float upwards. The tapered shape of the float ensures an equilibrium position proportional to flowrate. An inductive proximity sensor can be positioned so that it will detect when the float has reached a certain height, and the system has reached a given flowrate.

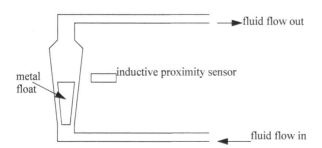

As the fluid flow increases the float is forced higher. A proximity sensor can be used to detect when the float reaches a certain height.

Figure 3.29   ***Flow Rate Detection With an Inductive Proximity Switch***

3.3 SUMMARY

- Sourcing sensors allow current to flow out from the V+ supply.
- Sinking sensors allow current to flow in to the V- supply.
- Photo-optical sensors can use reflected beams (retroreflective), an emitter and detector (opposed mode) and reflected light (diffuse) to detect a part.
- Capacitive sensors can detect metals and other materials.
- Inductive sensors can detect metals.
- Hall effect and reed switches can detect magnets.
- Ultrasonic sensors use sound waves to detect parts up to meters away.

3.4 PRACTICE PROBLEMS

(Note: Problem solutions are available at http://sites.google.com/site/automatedmanufacturingsystems/)

1. Given a clear plastic bottle, list 3 different types of sensors that could be used to detect it.

2. List 3 significant trade-offs between inductive, capacitive and photooptic sensors.

3. Why is a sinking output on a sensor not like a normal switch?

4. a) Sketch the connections needed for the PLC inputs and outputs below. The outputs include a 24Vdc light and a 120Vac light. The inputs are from 2 NO push buttons, and also from an optical sensor that has both PNP and NPN outputs.

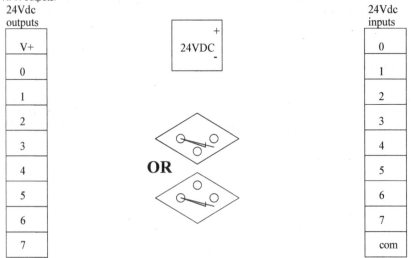

b) State why you used either the NPN or PNP output on the sensor.

5. Select a sensor to pick up a transparent plastic bottle from a manufacturer. Copy or print the specifications, and then draw a wiring diagram that shows how it will be wired to an appropriate PLC input card.

6. Sketch the wiring to connect a power supply and PNP sensor to the PLC input card shown below.

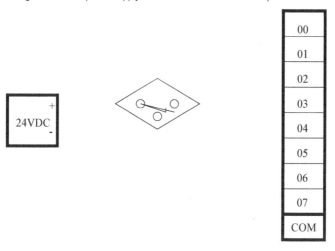

7. Sketch the wiring for inputs that include the following items.
   3 normally open push buttons
   1 thermal relay
   3 sinking sensors
   1 sourcing sensor

8. A PLC has eight 10-60Vdc inputs, and four relay outputs. It is to be connected to the following devices. Draw the required wiring.
   • Two inductive proximity sensors with sourcing and sinking outputs.
   • A NO run button and NC stop button.
   • A 120Vac light.
   • A 24Vdc solenoid.

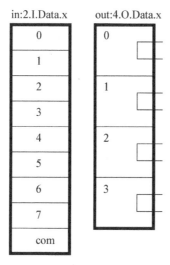

9. Draw a ladder wiring diagram (as done in the lab) for a system that has two push-buttons and a sourcing/sinking proximity sensors for 10-60Vdc inputs and two 120Vac output lights. Don't forget to include hard-wired start and

stop buttons with an MCR.

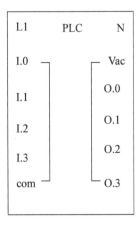

3.5 ASSIGNMENT PROBLEMS

1. What type of sensor should be used if it is to detect small cosmetic case mirrors as they pass along a belt. Explain your choice.

2. Summarize the trade-offs between capacitive, inductive and optical sensors in a table.

3. Clearly and concisely explain the difference between wiring PNP and NPN sensors.

4. a) Show the wiring for the following sensor, and circle the output that you are using, NPN or PNP. Redraw the

sensor using the correct symbol for the sourcing or sinking sensor chosen.

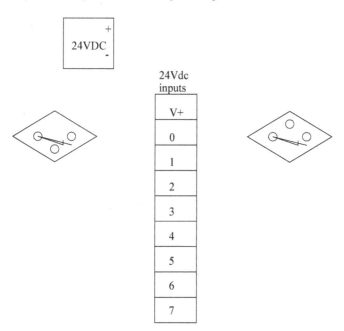

5. A PLC has three NPN and two PNP sensors as inputs, and outputs to control a 24Vdc solenoid and a small 115Vac motor. Develop the required wiring for the inputs and outputs.

3.1 PRACTICE PROBLEM SOLUTIONS

1. capacitive proximity, contact switch, photo-optic retroreflective/diffuse, ultrasonic

2. materials that can be sensed, environmental factors such as dirt, distance to object

3. the sinking output will pass only DC in a single direction, whereas a switch can pass AC and DC.

4.

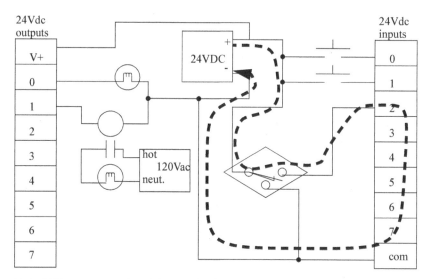

b) the PNP output was selected. because it will supply current, while the input card requires it. The dashed line indicates the current flow through the sensor and input card.

5.

A transparent bottle can be picked up with a capacitive, ultrasonic, diffuse optical sensor. A particular model can be selected at a manufacturers web site (eg., www.banner.com, www.hydepark.com, www.ab.com, etc.) The figure below shows the sensor connected to a sourcing PLC input card - therefore the sensor must be sinking, NPN.

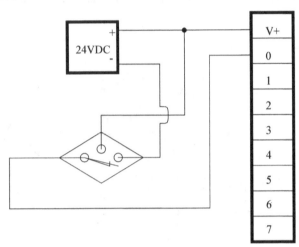

6.

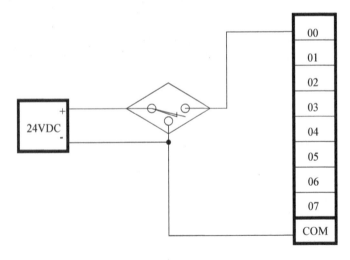

7.

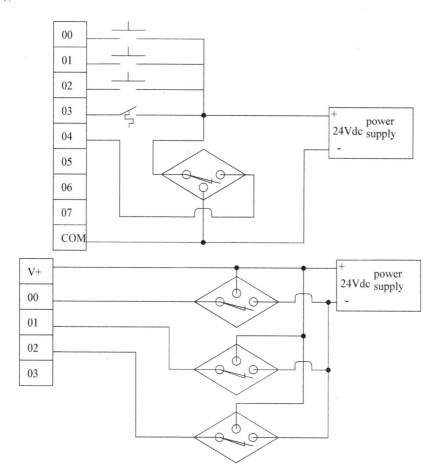

8.

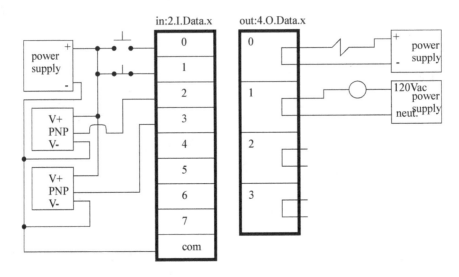

9.

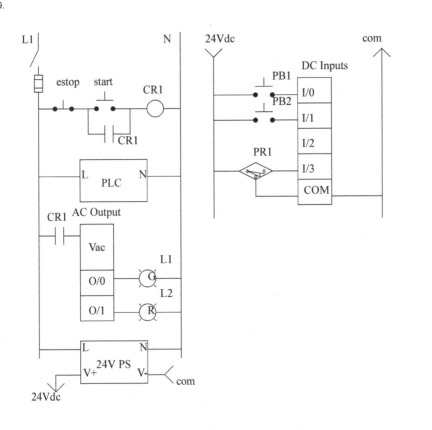

4LOGICAL ACTUATORS

Topics:

- Solenoids, valves and cylinders
- Hydraulics and pneumatics
- Other actuators

Objectives:

- Be aware of various actuators available.

Actuators Drive motions in mechanical systems. Most often this is by converting electrical energy into some form of mechanical motion.

4.1 SOLENOIDS

Solenoids are the most common actuator components. The basic principle of operation is there is a moving ferrous core (a piston) that will move inside wire coil as shown in Figure 4.1. Normally the piston is held outside the coil by a spring. When a voltage is applied to the coil and current flows, the coil builds up a magnetic field that attracts the piston and pulls it into the center of the coil. The piston can be used to supply a linear force. Well known applications of these include pneumatic values and car door openers.

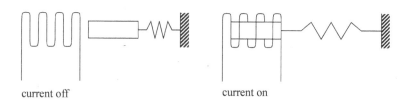

current off   current on

Figure 4.1   *A Solenoid*

As mentioned before, inductive devices can create voltage spikes and may need snubbers, although most industrial applications have low enough voltage and current ratings they can be connected directly to the PLC outputs. Most industrial solenoids will be powered by 24Vdc and draw a few hundred mA.

4.2 VALVES

The flow of fluids and air can be controlled with solenoid controlled valves. An example of a solenoid controlled valve is shown in Figure 4.1. The solenoid is mounted on the side. When actuated it will drive the central spool left. The top of the valve body has two ports that will be connected to a device such as a hydraulic cylinder. The bottom of the valve body has a single pressure line in the center with two exhausts to the side. In the top drawing the power flows in through the center to the right hand cylinder port. The left hand cylinder port is allowed to exit through an exhaust port. In the bottom drawing the solenoid is in a new position and the pressure is now applied to the left hand port on the top, and the right hand port can exhaust. The symbols to the left of the figure show the schematic equivalent of the actual valve positions. Valves are also available that allow the valves to be blocked when unused.

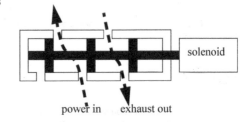

exhaust out   power in

The solenoid has two positions and when actuated will change the direction that fluid flows to the device. The symbols shown here are commonly used to represent this type of valve.

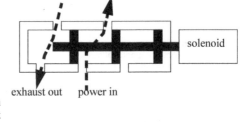

power in   exhaust out

Figure 4.2   *A Solenoid Controlled 5 Ported, 4 Way 2 Position Valve*

Valve types are listed below. In the standard terminology, the 'n-way' designates the number of connections for inlets and outlets. In some cases there are redundant ports for exhausts. The normally open/closed designation indicates the valve condition when power is off. All of the valves listed are two position valve, but three position valves are also available.

2-way normally closed - these have one inlet, and one outlet. When unenergized, the valve is closed. When energized, the valve will open, allowing flow. These are used to permit flows.

2-way normally open - these have one inlet, and one outlet. When unenergized, the valve is open, allowing flow. When energized, the valve will close. These are used to stop flows. When system power is off, flow will be allowed.

3-way normally closed - these have inlet, outlet, and exhaust ports. When unenergized, the outlet port is connected to the exhaust port. When energized, the inlet is connected to the outlet port. These are used for single acting cylinders.

3-way normally open - these have inlet, outlet and exhaust ports. When unenergized, the inlet is connected to the outlet. Energizing the valve connects the outlet to the exhaust. These are used for single acting cylinders

3-way universal - these have three ports. One of the ports acts as an inlet or outlet, and is connected to one of the other two, when energized/unenergized. These can be used to divert flows, or select alternating sources.

4-way - These valves have four ports, two inlets and two outlets. Energizing the valve causes connection between the inlets and outlets to be reversed. These are used for double acting cylinders.

Some of the ISO symbols for valves are shown in Figure 4.1. When using the symbols in drawings the connections are shown for the unenergized state. The arrows show the flow paths in different positions. The small triangles indicate an exhaust port.

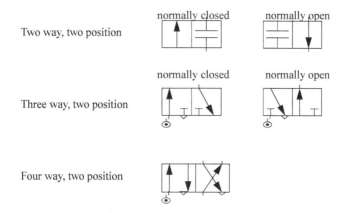

Two way, two position

normally closed   normally open

Three way, two position

normally closed   normally open

Four way, two position

Figure 4.3   ***ISO Valve Symbols***

When selecting valves there are a number of details that should be considered, as listed below.

pipe size - inlets and outlets are typically threaded to accept NPT (national pipe thread).
flow rate - the maximum flow rate is often provided to hydraulic valves.
operating pressure - a maximum operating pressure will be indicated. Some valves will also
    require a minimum pressure to operate.
electrical - the solenoid coil will have a fixed supply voltage (AC or DC) and current.
response time - this is the time for the valve to fully open/close. Typical times for valves range from
    5ms to 150ms.
enclosure - the housing for the valve will be rated as,
        type 1 or 2 - for indoor use, requires protection against splashes
        type 3 - for outdoor use, will resists some dirt and weathering
        type 3R or 3S or 4 - water and dirt tight
        type 4X - water and dirt tight, corrosion resistant

4.3 CYLINDERS

A cylinder uses pressurized fluid or air to create a linear force/motion as shown in Figure 4.1. In the figure a fluid is pumped into one side of the cylinder under pressure, causing that side of the cylinder to expand, and advancing the piston. The fluid on the other side of the piston must be allowed to escape freely - if the incompressible fluid was trapped the cylinder could not advance. The force the cylinder can exert is proportional to the cross sectional area of the cylinder.

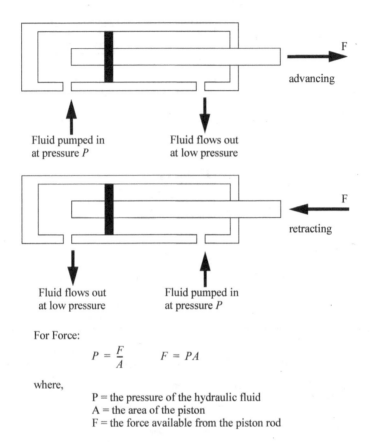

Fluid pumped in
at pressure P Fluid flows out
at low pressure

Fluid flows out
at low pressure   Fluid pumped in
at pressure P

For Force:

$$P = \frac{F}{A} \qquad F = PA$$

where,

P = the pressure of the hydraulic fluid
A = the area of the piston
F = the force available from the piston rod

Figure 4.4   *A Cross Section of a Hydraulic Cylinder*

Single acting cylinders apply force when extending and typically use a spring to retract the cylinder. Double acting cylinders apply force in both direction.

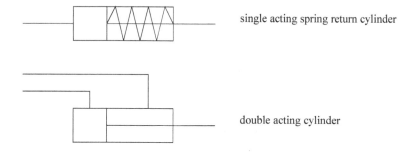

single acting spring return cylinder

double acting cylinder

Figure 4.5　　　*Schematic Symbols for Cylinders*

Magnetic cylinders are often used that have a magnet on the piston head. When it moves to the limits of motion, reed switches will detect it.

4.4 HYDRAULICS

Hydraulics use incompressible fluids to supply very large forces at slower speeds and limited ranges of motion. If the fluid flow rate is kept low enough, many of the effects predicted by Bernoulli's equation can be avoided. The system uses hydraulic fluid (normally an oil) pressurized by a pump and passed through hoses and valves to drive cylinders. At the heart of the system is a pump that will give pressures up to hundreds or thousands of psi. These are delivered to a cylinder that converts it to a linear force and displacement.

Hydraulic systems normally contain the following components;

      1. Hydraulic Fluid
      2. An Oil Reservoir
      3. A Pump to Move Oil, and Apply Pressure
      4. Pressure Lines
      5. Control Valves - to regulate fluid flow
      6. Piston and Cylinder - to actuate external mechanisms

The hydraulic fluid is often a noncorrosive oil chosen so that it lubricates the components. This is normally stored in a reservoir as shown in Figure 4.1. Fluid is drawn from the reservoir to a pump where it is pressurized. This is normally a geared pump so that it may deliver fluid at a high pressure at a constant flow rate. A flow regulator is normally placed at the high pressure outlet from the pump. If fluid is not flowing in other parts of the system this will allow fluid to recirculate back to the reservoir to reduce wear on the pump. The high pressure fluid is delivered to solenoid controlled vales that can switch fluid flow on or off. From the vales fluid will be delivered to the hydraulics at high pressure, or exhausted back to the reservoir.

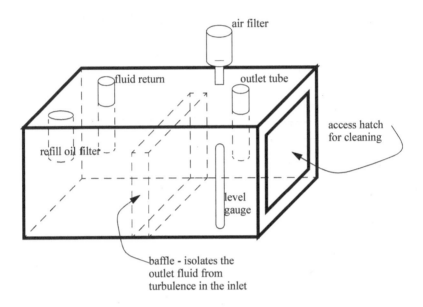

Figure 4.6   *A Hydraulic Fluid Reservoir*

Hydraulic systems can be very effective for high power applications, but the use of fluids, and high pressures can make this method awkward, messy, and noisy for other applications.

4.5 PNEUMATICS

Pneumatic systems are very common, and have much in common with hydraulic systems with a few key differences. The reservoir is eliminated as there is no need to collect and store the air between uses in the system. Also because air is a gas it is compressible and regulators are not needed to recirculate flow. But, the compressibility also means that the systems are not as stiff or strong. Pneumatic systems respond very quickly, and are commonly used for low force applications in many locations on the factory floor.

Some basic characteristics of pneumatic systems are,

- stroke from a few millimeters to meters in length (longer strokes have more springiness
- the actuators will give a bit - they are springy
- pressures are typically up to 85psi above normal atmosphere
- the weight of cylinders can be quite low
- additional equipment is required for a pressurized air supply- linear and rotary actuators are available.
- dampers can be used to cushion impact at ends of cylinder travel.

When designing pneumatic systems care must be taken to verify the operating location. In particular the elevation above sea level will result in a dramatically different air pressure. For example, at sea level the air pressure is about 14.7 psi, but at a height of 7,800 ft (Mexico City) the air pressure is 11.1 psi. Other operating environments, such as in submersibles, the air pressure might be higher than at sea level.

Some symbols for pneumatic systems are shown in Figure 4.1. The flow control valve is used to restrict the flow, typically to slow motions. The shuttle valve allows flow in one direction, but blocks it in the other. The receiver tank allows pressurized air to be accumulated. The dryer and filter help remove dust and moisture from the air, prolonging the life of the valves and cylinders.

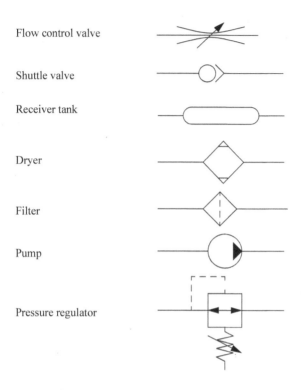

Flow control valve

Shuttle valve

Receiver tank

Dryer

Filter

Pump

Pressure regulator

Figure 4.7   ***Pneumatics Components***

4.6 MOTORS

Motors are common actuators, but for logical control applications their properties are not that important. Typically logical control of motors consists of switching low current motors directly with a PLC, or for more powerful motors using a relay or motor starter. Motors will be discussed in greater detail in the chapter on continuous actuators.

4.7 OTHERS

There are many other types of actuators including those on the brief list below.

Heaters - The are often controlled with a relay and turned on and off to maintain a temperature within a range.
Lights - Lights are used on almost all machines to indicate the machine state and provide feedback to the operator. most lights are low current and are connected directly to the PLC.
Sirens/Horns - Sirens or horns can be useful for unattended or dangerous machines to make con-

ditions well known. These can often be connected directly to the PLC.

Computers - some computer based devices may use TTL 0/5V logic levels to trigger actions. Generally these are prone to electrical noise and should be avoided if possible.

4.8 SUMMARY

- Solenoids can be used to convert an electric current to a limited linear motion.
- Hydraulics and pneumatics use cylinders to convert fluid and gas flows to limited linear motions.
- Solenoid valves can be used to redirect fluid and gas flows.
- Pneumatics provides smaller forces at higher speeds, but is not *stiff*. Hydraulics provides large forces and is rigid, but at lower speeds.
- Many other types of actuators can be used.

4.9 PRACTICE PROBLEMS

(Note: Problem solutions are available at http://sites.google.com/site/automatedmanufacturingsystems/)

1. A piston is to be designed to exert an actuation force of 120 lbs on its extension stroke. The inside diameter of the cylinder is 2.0" and the ram diameter is 0.375". What shop air pressure will be required to provide this actuation force? Use a safety factor of 1.3.

2. Draw a simple hydraulic system that will advance and retract a cylinder using PLC outputs. Sketches should include details from the PLC output card to the hydraulic cylinder.

3. Develop an electrical ladder diagram and pneumatic diagram for a PLC controlled system. The system includes the components listed below. The system should include all required safety and wiring considerations.
   a 3 phase 50 HP motor
   1 NPN sensor
   1 NO push button
   1 NC limit switch
   1 indicator light
   a doubly acting pneumatic cylinder

4. What are the trade-offs between 3-phase and single-phase AC power.

4.10 ASSIGNMENT PROBLEMS

1. Draw a schematic symbol for a solenoid controlled pneumatic valve and explain how the valve operates.

2. A PLC based system has 3 proximity sensors, a start button, and an E-stop as inputs. The system controls a pneumatic system with a solenoid controlled valve. It also controls a robot with a TTL output. Develop a complete wiring diagram including all safety elements.

3. A system contains a pneumatic cylinder with two inductive proximity sensors that will detect when the cylinder is fully advanced or retracted. The cylinder is controlled by a solenoid controlled valve. Draw electrical and pneumatic schematics for a system.

4. Draw an electrical ladder wiring diagram for a PLC controlled system that contains 2 PNP sensors, a NO push button, a NC limit switch, a contactor controlled AC motor and an indicator light. Include all safety circuitry.

5. We are to connect a PLC to detect boxes moving down an assembly line and divert larger boxes. The line is 12 inches wide and slanted so the boxes fall to one side as they travel by. One sensor will be mounted on the lower side of the conveyor to detect when a box is present. A second sensor will be mounted on the upper side of the conveyor to determine when a larger box is present. If the box is present, an output to a pneumatic solenoid will be actuated to divert the box. Your job is to select a specific PLC, sensors, and solenoid valve. Details (the absolute minimum being model numbers) are expected with a ladder wiring diagram. (Note: take advantage of manufacturers web sites.)

6. Develop a wiring diagram for a system that has the following elements. Include all safety circuitry.

2 NPN proximity sensors

2 N.O. pushbuttons

3 solenoid outputs

A 440Vac 3ph. 20HP (i.e., large) motor

7. Draw an electrical ladder wiring diagram for a PLC controlled system that has the following inputs; 2 PNP sensors, 2 NPN sensors, and a NC limit switch. The outputs include a 24Vdc solenoid valve and a very large 3 phase AC motor.

4.1 PRACTICE PROBLEM SOLUTIONS

1. A = pi*r^2 = 3.14159in^2, P=FS*(F/A)=1.3(120/3.14159)=49.7psi. Note, if the cylinder were retracting we would need to subtract the rod area from the piston area. Note: this air pressure is much higher than normally found in a shop, so it would not be practical, and a redesign would be needed.

2.

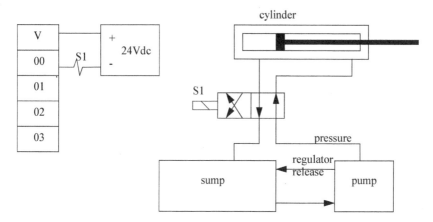

3.

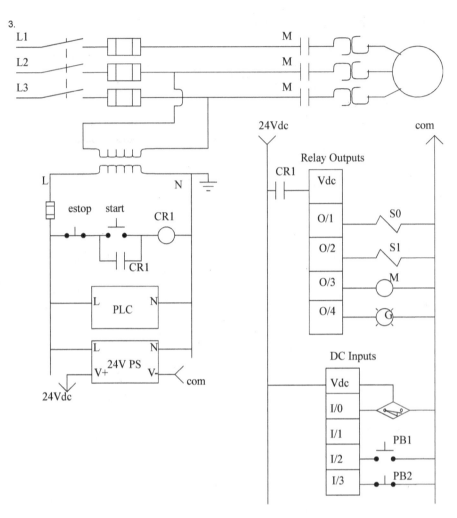

4. 3-phase power is ideal for large loads such as motors. Single phase power is suited to small loads, and the power usage on each phase must be balanced someplace on the electrical grid.

78

5BOOLEAN LOGIC DESIGN

Topics:

- Boolean algebra
- Converting between Boolean algebra and logic gates and ladder logic
- Logic examples

Objectives:

- Be able to simplify designs with Boolean algebra

The process of converting control objectives into a ladder logic program requires structured thought. Boolean algebra provides the tools needed to analyze and design these systems.

Boolean algebra was developed in the 1800's by James Boole, an Irish mathematician. It was found to be extremely useful for designing digital circuits, and it is still heavily used by electrical engineers and computer scientists. The techniques can model a logical system with a single equation. The equation can then be simplified and/or manipulated into new forms. The same techniques developed for circuit designers adapt very well to ladder logic programming.

Boolean equations consist of variables and operations and look very similar to normal algebraic equations. The three basic operators are AND, OR and NOT; more complex operators include exclusive or (EOR), not and (NAND), not or (NOR). Small truth tables for these functions are shown in Figure 5.1. Each operator is shown in a simple equation with the variables A and B being used to calculate a value for X. Truth tables are a simple (but bulky) method for showing all of the possible combinations that will turn an output on or off.

Note: By convention a false state is also called off or 0 (zero). A true state is also called on or 1.

AND

A
B ─ D ─ X

$X = A \cdot B$

A	B	X
0	0	0
0	1	0
1	0	0
1	1	1

OR

A
B ─ D ─ X

$X = A + B$

A	B	X
0	0	0
0	1	1
1	0	1
1	1	1

NOT

A ─ Do ─ X

$X = \overline{A}$

A	X
0	1
1	0

NAND

A
B ─ D o ─ X

$X = \overline{A \cdot B}$

A	B	X
0	0	1
0	1	1
1	0	1
1	1	0

NOR

A
B ─ D o ─ X

$X = \overline{A + B}$

A	B	X
0	0	1
0	1	0
1	0	0
1	1	0

EOR

A
B ─ X ─ X

$X = A \oplus B$

A	B	X
0	0	0
0	1	1
1	0	1
1	1	0

Note: The symbols used in these equations, such as + for OR are not universal standards and some authors will use different notations.

Note: The EOR function is available in gate form, but it is more often converted to its equivalent, as shown below.

$$X = A \oplus B = A \cdot \overline{B} + \overline{A} \cdot B$$

Figure 5.1   *Boolean Operations with Truth Tables and Gates*

In a Boolean equation the operators will be put in a more complex form as shown in Figure 5.1. The variable for these equations can only have a value of 0 for false, or 1 for true. The solution of the equation follows rules similar to normal algebra. Parts of the equation inside parenthesis are to be solved first. Operations are to be done in the sequence NOT, AND, OR. In the example the NOT function for C is done first, but the NOT over the first set of parentheses must wait until a single value is available. When there is a choice the AND operations are done before the OR operations. For the given set of variable values the result of the calculation is false.

given
$$X = \overline{(A + B \cdot C)} + A \cdot (B + \overline{C})$$

assuming A=1, B=0, C=1

$$X = \overline{(1 + 0 \cdot 1)} + 1 \cdot (0 + \overline{1})$$

$$X = \overline{(1 + 0)} + 1 \cdot (0 + 0)$$

$$X = \overline{(1)} + 1 \cdot (0)$$

$$X = 0 + 0$$

$$X = 0$$

Figure 5.2   *A Boolean Equation*

The equations can be manipulated using the basic axioms of Boolean shown in Figure 5.1. A few of the axioms (associative, distributive, commutative) behave like normal algebra, but the other axioms have subtle differences that must not be ignored.

Idempotent

$$A + A = A \qquad\qquad\qquad A \cdot A = A$$

Associative

$$(A + B) + C = A + (B + C) \qquad\qquad (A \cdot B) \cdot C = A \cdot (B \cdot C)$$

Commutative

$$A + B = B + A \qquad\qquad\qquad A \cdot B = B \cdot A$$

Distributive

$$A + (B \cdot C) = (A + B) \cdot (A + C) \qquad A \cdot (B + C) = (A \cdot B) + (A \cdot C)$$

Identity

$$A + 0 = A \qquad\qquad\qquad A + 1 = 1$$

$$A \cdot 0 = 0 \qquad\qquad\qquad A \cdot 1 = A$$

Complement

$$A + \bar{A} = 1 \qquad\qquad\qquad \overline{(\bar{A})} = A$$

$$A \cdot \bar{A} = 0 \qquad\qquad\qquad \bar{1} = 0$$

DeMorgan's

$$\overline{(A + B)} = \bar{A} \cdot \bar{B} \qquad\qquad \overline{(A \cdot B)} = \bar{A} + \bar{B}$$

Duality

interchange AND and OR operators, as well as all Universal, and Null
sets. The resulting equation is equivalent to the original.

Figure 5.3   ***The Basic Axioms of Boolean Algebra***

An example of equation manipulation is shown in Figure 5.1. The distributive axiom is applied to get equation (1). The idempotent axiom is used to get equation (2). Equation (3) is obtained by using the distributive axiom to move C outside the parentheses, but the identity axiom is used to deal with the lone C. The identity axiom is then used to simplify the contents of the parentheses to get equation (4). Finally the Identity axiom is used to get the final, simplified equation. Notice that using Boolean algebra has shown that 3 of the variables are entirely unneeded.

$$A = \bar{B} \cdot (C \cdot (\bar{D} + E + C) + \bar{F} \cdot C)$$

$$A = \bar{B} \cdot (\bar{D} \cdot C + E \cdot C + C \cdot C + \bar{F} \cdot C) \qquad (1)$$

$$A = \bar{B} \cdot (\bar{D} \cdot C + E \cdot C + C + \bar{F} \cdot C) \qquad (2)$$

$$A = \bar{B} \cdot C \cdot (\bar{D} + E + 1 + \bar{F}) \qquad (3)$$

$$A = \bar{B} \cdot C \cdot (1) \qquad (4)$$

$$A = \bar{B} \cdot C \qquad (5)$$

Figure 5.4 **Simplification of a Boolean Equation**

Note: When simplifying Boolean algebra, OR operators have a lower priority, so they should be manipulated first. NOT operators have the highest priority, so they should be simplified last. Consider the example from before.

$$X = \overline{(A + B \cdot C)} + A \cdot (B + \bar{C})$$
The higher priority operators are put in parentheses

$$X = \overline{(A) + (B \cdot C)} + A \cdot (B + \bar{C})$$
DeMorgan's theorem is applied

$$X = \overline{(A)} \cdot \overline{(B \cdot C)} + A \cdot (B + \bar{C})$$
DeMorgan's theorem is applied again

$$X = \bar{A} \cdot (\bar{B} + \bar{C}) + A \cdot (B + \bar{C})$$
The equation is expanded

$$X = \bar{A} \cdot \bar{B} + \bar{A} \cdot \bar{C} + A \cdot B + A \cdot \bar{C}$$
Terms with common terms are collected, here it is only NOT C

$$X = \bar{A} \cdot \bar{B} + (\bar{A} \cdot \bar{C} + A \cdot \bar{C}) + A \cdot B$$
The redundant term is eliminated

$$X = \bar{A} \cdot \bar{B} + \bar{C} \cdot (\bar{A} + A) + A \cdot B$$
A Boolean axiom is applied to simplify the equation further

$$X = \bar{A} \cdot \bar{B} + \bar{C} + A \cdot B$$

5.1 LOGIC DESIGN

Design ideas can be converted to Boolean equations directly, or with other techniques discussed later. The

Boolean equation form can then be simplified or rearranges, and then converted into ladder logic, or a circuit.

Aside: The logic for a seal-in circuit can be analyzed using a Boolean equation as shown below. Recall that the START is NO and the STOP is NC.

$$ON' = (START + ON) \cdot STOP$$

ON	STOP	START	ON'	
0	0	0	0	stop pushed, not active
0	0	1	0	stop pushed, not active
0	1	0	0	not active
0	1	1	1	start pushed, becomes active
1	0	0	0	stop pushed, not active
1	0	1	0	stop pushed, not active
1	1	0	1	active, start no longer pushed
1	1	1	1	becomes active and start pushed

If we can describe how a controller should work in words, we can often convert it directly to a Boolean equation, as shown in Figure 5.1. In the example a process description is given first. In actual applications this is obtained by talking to the designer of the mechanical part of the system. In many cases the system does not exist yet, making this a challenging task. The next step is to determine how the controller should work. In this case it is written out in a sentence first, and then converted to a Boolean expression. The Boolean expression may then be converted to a desired form. The first equation contains an EOR, which is not available in ladder logic, so the next line converts this to an equivalent expression (2) using ANDs, ORs and NOTs. The ladder logic developed is for the second equation. In the conversion the terms that are ANDed are in series. The terms that are ORed are in parallel branches, and terms that are NOTed use normally closed contacts. The last equation (3) is fully expanded and ladder logic for it is shown in Figure 5.1. This illustrates the same logical control function can be achieved with different, yet equivalent, ladder logic.

Process Description:
> A heating oven with two bays can heat one ingot in each bay. When the heater is on it provides enough heat for two ingots. But, if only one ingot is present the oven may become too hot, so a fan is used to cool the oven when it passes a set temperature.

Control Description:
> If the temperature is too high and there is an ingot in only one bay then turn on fan.

Define Inputs and Outputs:
> B1 = bay 1 ingot present
> B2 = bay 2 ingot present
> F = fan
> T = temperature overheat sensor

Boolean Equation:

$$F = T \cdot (B_1 \oplus B_2)$$

$$F = T \cdot (B_1 \cdot \overline{B_2} + \overline{B_1} \cdot B_2) \qquad (2)$$

$$F = B_1 \cdot \overline{B_2} \cdot T + \overline{B_1} \cdot B_2 \cdot T \qquad (3)$$

Ladder Logic for Equation (2):

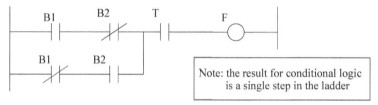

Note: the result for conditional logic is a single step in the ladder

Warning: in spoken and written english OR and EOR are often not clearly defined. Consider the traffic directions "Go to main street then turn left or right." Does this *or* mean that you can drive either way, or that the person isn't sure which way to go? Consider the expression "The cars are red or blue.", Does this mean that the cars can be either red or blue, or all of the cars are red, or all of the cars are blue. A good literal way to describe this condition is "one or the other, but not both".

Figure 5.5   *Boolean Algebra Based Design of Ladder Logic*

Ladder Logic for Equation (3):

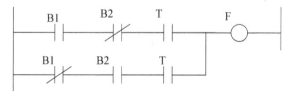

Figure 5.6   *Alternate Ladder Logic*

Boolean algebra is often used in the design of digital circuits. Consider the example in Figure 5.1. In this case we are presented with a circuit that is built with inverters, nand, nor and, and gates. This figure can be converted into a boolean equation by starting at the left hand side and working right. Gates on the left hand side are *solved* first, so they are put inside parentheses to indicate priority. Inverters are represented by putting a NOT operator on a variable in the equation. This circuit can't be directly converted to ladder logic because there are no equivalents to NAND and NOR gates. After the circuit is converted to a Boolean equation it is simplified, and then converted back into a (much simpler) circuit diagram and ladder logic.

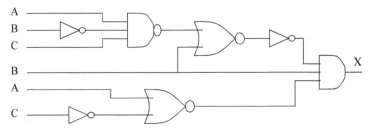

The circuit is converted to a Boolean equation and simplified. The most nested terms in the equation are on the left hand side of the diagram.

$$X = (\overline{(\overline{A \cdot \overline{B} \cdot C}) + B}) \cdot B \cdot \overline{(A + \overline{C})}$$

$$X = (\overline{A} + B + \overline{C} + B) \cdot B \cdot (\overline{A} \cdot C)$$

$$X = \overline{A} \cdot B \cdot \overline{A} \cdot C + B \cdot B \cdot \overline{A} \cdot C + \overline{C} \cdot B \cdot \overline{A} \cdot C + B \cdot B \cdot \overline{A} \cdot C$$
$$X = B \cdot \overline{A} \cdot C + B \cdot \overline{A} \cdot C + 0 + B \cdot \overline{A} \cdot C$$
$$X = B \cdot \overline{A} \cdot C$$

This simplified equation is converted back into a circuit and equivalent ladder logic.

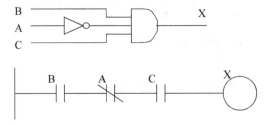

Figure 5.7   *Reverse Engineering of a Digital Circuit*

To summarize, we will obtain Boolean equations from a verbal description or existing circuit or ladder diagram. The equation can be manipulated using the axioms of Boolean algebra. after simplification the equation can be converted back into ladder logic or a circuit diagram. Ladder logic (and circuits) can behave the same even though they are in different forms. When simplifying Boolean equations that are to be implemented in ladder logic there are a few basic rules.

1. Eliminate NOTs that are for more than one variable. This normally includes replacing NAND and

NOR functions with simpler ones using DeMorgan's theorem.
2. Eliminate complex functions such as EORs with their equivalent.

These principles are reinforced with another design that begins in Figure 5.1. Assume that the Boolean equation that describes the controller is already known. This equation can be converted into both a circuit diagram and ladder logic. The circuit diagram contains about two dollars worth of integrated circuits. If the design was mass produced the final cost for the entire controller would be under $50. The prototype of the controller would cost thousands of dollars. If implemented in ladder logic the cost for each controller would be approximately $500. Therefore a large number of circuit based controllers need to be produced before the break even occurs. This number is normally in the range of hundreds of units. There are some particular advantages of a PLC over digital circuits for the factory and some other applications.

- the PLC will be more rugged,
- the program can be changed easily
- less skill is needed to maintain the equipment

Given the controller equation;

$$A = \bar{B} \cdot (\overline{C \cdot (\bar{D} + E + \bar{C})} + \bar{F} \cdot C)$$

The circuit is given below, and equivalent ladder logic is shown.

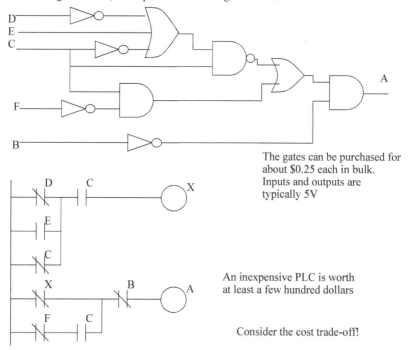

The gates can be purchased for about $0.25 each in bulk. Inputs and outputs are typically 5V

An inexpensive PLC is worth at least a few hundred dollars

Consider the cost trade-off!

Figure 5.8   *A Boolean Equation and Derived Circuit and Ladder Logic*

The initial equation is not the simplest. It is possible to simplify the equation to the form seen in Figure 5.1. If you are a visual learner you may want to notice that some simplifications are obvious with ladder logic - consider the C on both branches of the ladder logic in Figure 5.1.

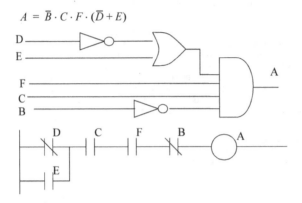

$$A = \bar{B} \cdot C \cdot F \cdot (\bar{D} + E)$$

Figure 5.9 *The Simplified Form of the Example*

The equation can also be manipulated to other forms that are more routine but less efficient as shown in Figure 5.1. The equation shown is in disjunctive normal form - in simpler words this is ANDed terms ORed together. This is also an example of a canonical form - in simpler terms this means a standard form. This form is more important for digital logic, but it can also make some PLC programming issues easier. For example, when an equation is simplified, it may not look like the original design intention, and therefore becomes harder to rework without starting from the beginning.

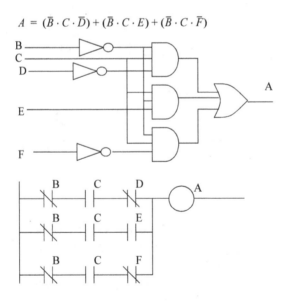

$$A = (\bar{B} \cdot C \cdot \bar{D}) + (\bar{B} \cdot C \cdot E) + (\bar{B} \cdot C \cdot \bar{F})$$

Figure 5.10 *A Canonical Logic Form*

5.1.1 Boolean Algebra Techniques

There are some common Boolean algebra techniques that are used when simplifying equations. Recognizing these forms are important to simplifying Boolean Algebra with ease. These are itemized, with proofs in Figure 5.1.

$$A + C\bar{A} = A + C$$ proof: $A + C\bar{A}$
$(A + C)(A + \bar{A})$
$(A + C)(1)$
$A + C$

$$AB + A = A$$ proof: $AB + A$
$AB + A1$
$A(B + 1)$
$A(1)$
A

$$\overline{A + B + C} = \bar{A}\bar{B}\bar{C}$$ proof: $\overline{A + B + C}$
$\overline{(A + B) + C}$
$\overline{(A + B)}\bar{C}$
$(\bar{A}\bar{B})\bar{C}$
$\bar{A}\bar{B}\bar{C}$

Figure 5.11   **_Common Boolean Algebra Techniques_**

5.2 COMMON LOGIC FORMS

Knowing a simple set of logic forms will support a designer when categorizing control problems. The following forms are provided to be used directly, or provide ideas when designing.

5.2.1 Complex Gate Forms

In total there are 16 different possible types of 2-input logic gates. The simplest are AND and OR, the other gates we will refer to as _complex_ to differentiate. The three popular complex gates that have been discussed before are NAND, NOR and EOR. All of these can be reduced to simpler forms with only ANDs and ORs that are suitable for ladder logic, as shown in Figure 5.1.

NAND
$$X = \overline{A \cdot B}$$

$$X = \bar{A} + \bar{B}$$

NOR
$$X = \overline{A + B}$$

$$X = \bar{A} \cdot \bar{B}$$

EOR
$$X = A \oplus B$$

$$X = A \cdot \bar{B} + \bar{A} \cdot B$$

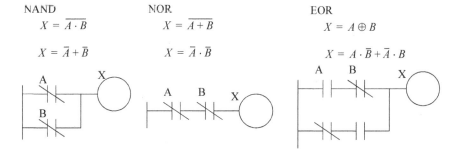

Figure 5.12   ***Conversion of Complex Logic Functions***

5.2.2 Multiplexers

Multiplexers allow multiple devices to be connected to a single device. These are very popular for telephone systems. A telephone *switch* is used to determine which telephone will be connected to a limited number of
lines to other telephone switches. This allows telephone calls to be made to somebody far away without a dedicated wire to the other telephone. In older telephone switch boards, operators physically connected wires by plug-
ging them in. In modern computerized telephone switches the same thing is done, but to digital voice signals.

In Figure 5.1 a multiplexer is shown that will take one of four inputs bits D1, D2, D3 or D4 and make it the
output X, depending upon the values of the address bits, A1 and A2.

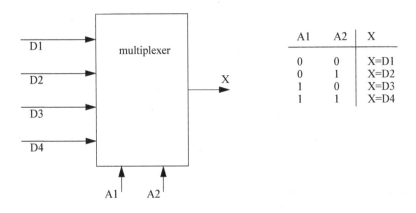

A1	A2	X
0	0	X=D1
0	1	X=D2
1	0	X=D3
1	1	X=D4

Figure 5.13   **A Multiplexer**

Ladder logic form the multiplexer can be seen in Figure 5.1.

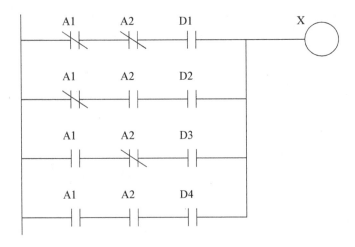

Figure 5.14   A Multiplexer in Ladder Logic

5.3 SIMPLE DESIGN CASES

The following cases are presented to illustrate various combinatorial logic problems, and possible solutions. It is recommended that you try to satisfy the description before looking at the solution.

5.3.1 Basic Logic Functions

Problem: Develop a program that will cause output D to go true when switch A and switch B are closed or when switch C is closed.

Solution:

$$D = (A \cdot B) + C$$

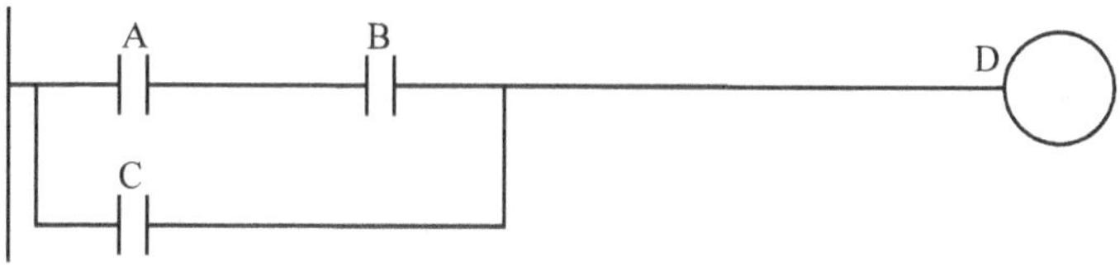

Figure 5.15   *Sample Solution for Logic Case Study A*

Problem: Develop a program that will cause output D to be on when push button A is on, or either B or C are on.

Solution:

$$D = A + (B \oplus C)$$

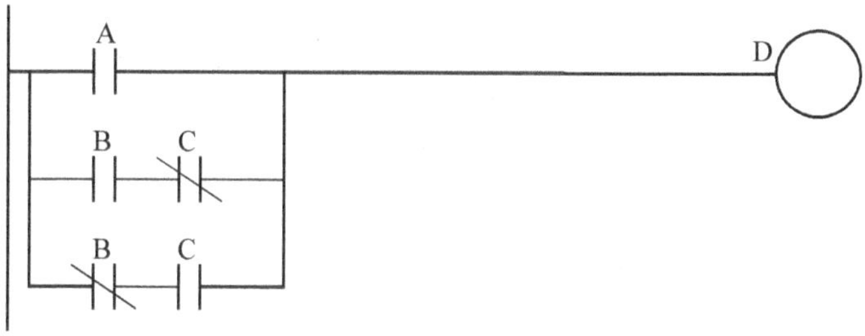

Figure 5.16   *Sample Solution for Logic Case Study B*

5.3.2 Car Safety System

Problem: Develop Ladder Logic for a car door/seat belt safety system. When the car door is open, and the seatbelt is not done up, the ignition power must not be applied. If all is safe then the key will start the engine.

Solution:

Door Open   Seat Belt   Key

                                              Ignition

Figure 5.17   *Solution to Car Safety System Case*

5.3.3 Motor Forward/Reverse

Problem: Design a motor controller that has a forward and a reverse button. The motor forward and reverse outputs will only be on when one of the buttons is pushed. When both buttons are pushed the motor will not work.

Solution:

$$F = BF \cdot \overline{BR}$$ where,

$$R = \overline{BF} \cdot BR$$

F = motor forward
R = motor reverse
BF = forward button
BR = reverse button

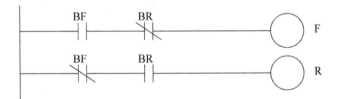

BF   BR   F

BF   BR   R

Figure 5.18   *Motor Forward, Reverse Case Study*

5.3.4 A Burglar Alarm

Consider the design of a burglar alarm for a house. When activated an alarm and lights will be activated to encourage the unwanted guest to leave. This alarm be activated if an unauthorized intruder is detected by window sensor and a motion detector. The window sensor is effectively a loop of wire that is a piece of thin metal foil that encircles the window. If the window is broken, the foil breaks breaking the conductor. This behaves like a normally closed switch. The motion sensor is designed so that when a person is detected the output will go on. As with any alarm an activate/deactivate switch is also needed. The basic operation of the alarm system, and the inputs and outputs of the controller are itemized in Figure 5.1.

The inputs and outputs are chosen to be;

A = Alarm and lights switch (1 = on)
W = Window/Door sensor (1 = OK)
M = Motion Sensor (0 = OK)
S = Alarm Active switch (1 = on)

The basic operation of the alarm can be described with rules.

1. If alarm is on, check sensors.
2. If window/door sensor is broken (turns off), sound alarm and turn on lights
3. If motion sensor goes on (detects thief) sound alarm and turn on lights.

Note: As the engineer, it is your responsibility to define these items before starting the work. If you do not do this first you are guaranteed to produce a poor design. It is important to develop a good list of inputs and outputs, and give them simple names so that they are easy to refer to. Most companies will use wire numbering schemes on their diagrams.

Figure 5.19   ***Controller Requirements List for Alarm***

The next step is to define the controller equation. In this case the controller has 3 different inputs, and a single output, so a truth table is a reasonable approach to formalizing the system. A Boolean equation can then be written using the truth table in Figure 5.1. Of the eight possible combinations of alarm inputs, only three lead to alarm conditions.

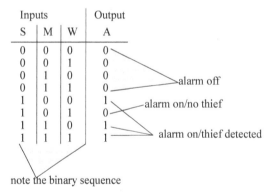

Figure 5.20   ***Truth Table for the Alarm***

The Boolean equation in Figure 5.1 is written by examining the truth table in Figure 5.1. There are three possible alarm conditions that can be represented by the conditions of all three inputs. For example take the last line in the truth table where when all three inputs are on the alarm should be one. This leads to the last term in the equation. The other two terms are developed the same way. After the equation has been written, it is simplified.

$$A = (S \cdot \overline{M} \cdot \overline{W}) + (S \cdot M \cdot \overline{W}) + (S \cdot M \cdot W)$$
$$\therefore A = S \cdot (\overline{M} \cdot \overline{W} + M \cdot \overline{W} + M \cdot W)$$
$$\therefore A = S \cdot ((\overline{M} \cdot \overline{W} + M \cdot \overline{W}) + (M \cdot \overline{W} + M \cdot W))$$
$$\therefore A = (S \cdot \overline{W}) + (S \cdot M) = S \cdot (\overline{W} + M)$$

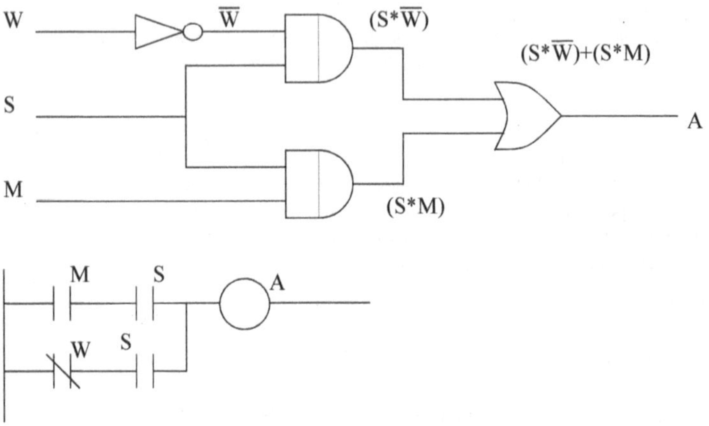

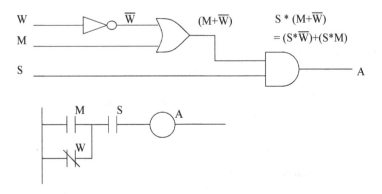

Figure 5.21   *A Boolean Equation and Implementation for the Alarm*

The equation and circuits shown in Figure can also be further simplified, as shown in Figure 5.1.

Figure 5.22   *The Simplest Circuit and Ladder Diagram*

Figure 5.23   *Alarm Implementation Using A High Level Programming Language*

5.4 SUMMARY

- Logic can be represented with Boolean equations.
- Boolean equations can be converted to (and from) ladder logic or digital circuits.
- Boolean equations can be simplified.
- Different controllers can behave the same way.
- Common logic forms exist and can be used to understand logic.
- Truth tables can represent all of the possible state of a system.

5.5 PRACTICE PROBLEMS

(Note: Problem solutions are available at http://sites.google.com/site/automatedmanufacturingsystems/)

1. Is the ladder logic in the figure below for an AND or an OR gate?

2. Draw a ladder diagram that will cause output D to go true when switch A and switch B are closed or when switch C is closed.

3. Draw a ladder diagram that will cause output D to be on when push button A is on, or either B or C are on.

4. Design ladder logic for a car that considers the variables below to control the motor *M*. Also add a second output that uses any outputs not used for motor control.

   - doors opened/closed (D)
   - keys in ignition (K)
   - motor running (M)
   - transmission in park (P)
   - ignition start (I)

5. a) Explain why a stop button must be normally closed and a start button must be normally open.

b) Consider a case where an input to a PLC is a normally closed stop button. The contact used in the ladder logic is normally open, as shown below. Why are they both not the same? (i.e., NC or NO)

6. Make a simple ladder logic program that will turn on the outputs with the binary patterns when the corresponding buttons are pushed.

OUTPUTS								INPUTS
H	G	F E	D C	B	A	
1 1	0 1	0 1	0 1	Input X on
1 0	1 0	0 0	0 1	Input Y on
1 0	0 1	0 1	1 1	Input Z on

7. Convert the following Boolean equation to the simplest possible ladder logic.

$$X = A \cdot (\overline{\overline{A} + \overline{A} \cdot B})$$

8. Simplify the following boolean equations.

a) $A(B + AB)$ b) $\overline{A(B + AB)}$

c) $\overline{A}(B + AB)$ d) $\overline{\overline{A}(B + AB)}$

9. Simplify the following Boolean equations,
a) $\overline{(A + B)} \cdot \overline{(A + \overline{B})}$

b) $ABCD + \overline{A}BCD + ABC\overline{D} + AB\overline{C}\overline{D}$

10. Simplify the Boolean expression below.

$$((A \cdot \overline{B}) + \overline{(\overline{B} + A)}) \cdot C + (\overline{B} \cdot C + B \cdot C)$$

11. Given the Boolean expression a) draw a digital circuit and b) a ladder diagram (do not simplify), c) simplify the expression.

$$X = A \cdot \overline{B} \cdot C + \overline{(C + B)}$$

12. Simplify the following Boolean equation and write corresponding ladder logic.

$$Y = \overline{(AB\overline{C}D + AB\overline{C}\overline{D} + \overline{A}BCD + \overline{A}\overline{B}CD)} + D$$

13. For the following Boolean equation,

$$X = A + B(A + C\overline{B} + D\overline{A}C) + ABCD$$

   a) Write out the logic for the unsimplified equation.
   b) Simplify the equation.
   c) Write out the ladder logic for the simplified equation.

14. a) Write a Boolean equation for the following truth table. (Hint: do this by writing an expression for each line with

96

a true output, and then ORing them together.)

A	B	C	D	Result
0	0	0	0	1
0	0	0	1	0
0	0	1	0	0
0	0	1	1	1
0	1	0	0	0
0	1	0	1	1
0	1	1	0	0
0	1	1	1	1
1	0	0	0	1
1	0	0	1	0
1	0	1	0	0
1	0	1	1	1
1	1	0	0	0
1	1	0	1	0
1	1	1	0	1
1	1	1	1	1

b) Write the results in a) in a Boolean equation.
c) Simplify the Boolean equation in b)

15. Simplify the following Boolean equation, and create the simplest ladder logic.

$$Y = \overline{\overline{C}\left[\overline{A} + \left(\overline{A} + (\overline{B}\overline{C}(\overline{A + B\overline{C}}))\right)\right]}$$

16. Simplify the following boolean equation with Boolean algebra and write the corresponding ladder logic.

$$X = \overline{(A + B \cdot \overline{A}) + \overline{(C + D + E\overline{C})}}$$

17. Convert the following ladder logic to a Boolean equation. Then simplify it, and convert it back to simpler ladder logic.

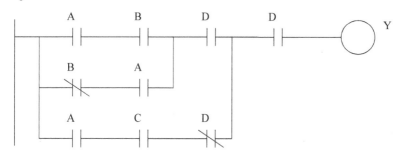

18. a) Develop the Boolean expression for the circuit below.
b) Simplify the Boolean expression.

c) Draw a simpler circuit for the equation in b).

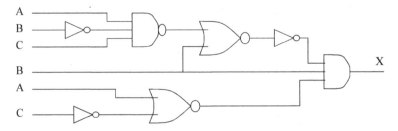

19. Given a system that is described with the following equation,

$$X = A + (B \cdot (\bar{A} + C) + C) + A \cdot B \cdot (\bar{D} + \bar{E})$$

   a) Simplify the equation using Boolean Algebra.
   b) Implement the original and then the simplified equation with a digital circuit.
   c) Implement the original and then the simplified equation in ladder logic.

20. Simplify the following and implement the original and simplified equations with gates and ladder logic.

$$A + (\bar{B} + \bar{C} + \bar{D}) \cdot (B + \bar{C}) + A \cdot B \cdot (\bar{C} + \bar{D})$$

21. Convert the following ladder logic to a Boolean equation. Simplify the equation and convert it back to ladder logic.

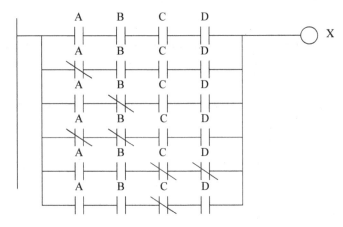

22. Use Boolean equations to develop simplified ladder logic for the following truth table where A, B, C and D are

inputs, and X and Y are outputs.

A	B	C	D	X	Y
0	0	0	0	0	0
0	0	0	1	1	0
0	0	1	0	0	0
0	0	1	1	1	0
0	1	0	0	0	0
0	1	0	1	0	1
0	1	1	0	0	1

5.6 ASSIGNMENT PROBLEMS

1. Simplify the following Boolean equation and implement it in ladder logic.

$$X = A + BA + B\bar{C} + \overline{D + C}$$

2. Simplify the following Boolean equation and write a ladder logic program to implement it.

$$X = (A\bar{B}C + \bar{A}BC + A\bar{B}\bar{C} + AB\bar{C} + ABC)$$

3. Convert the following ladder logic to a Boolean equation. Simplify the equation using Boolean algebra, and then convert the simplified equation back to ladder logic.

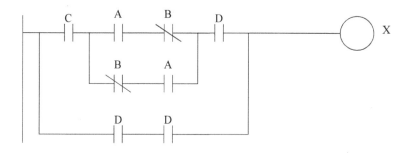

4. Convert the truth table below to a Boolean equation, and then simplify it. The output is X and the inputs

are A, B, C and D.

A	B	C	D	X
0	0	0	0	0
0	0	0	1	0
0	0	1	0	0
0	0	1	1	1
0	1	0	0	0
0	1	0	1	0
0	1	1	0	0
0	1	1	1	1
1	0	0	0	0
1	0	0	1	0
1	0	1	0	0
1	0	1	1	1
1	1	0	0	1
1	1	0	1	1
1	1	1	0	1
1	1	1	1	1

5. Simplify the following Boolean equation. Convert both the unsimplified and simplified equations to ladder logic.

$$X = \overline{(AB\overline{C})}(A + BC)$$

6. Convert the following ladder logic to a Boolean equation. Simplify the equation and convert it back to ladder logic.

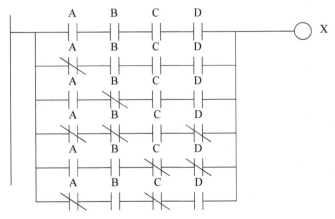

5.1 PRACTICE PROBLEM SOLUTIONS

1. AND

2.

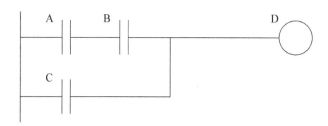

3.

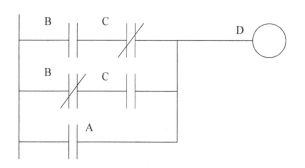

4.

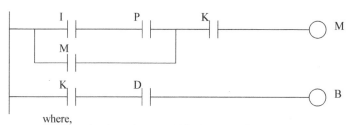

where,
B = the alarm that goes "Bing" to warn that the keys are still in the car.

5. a) If a NC stop button is damaged, the machine will not ac if the stop button was pushed and shut down safely. If a NO start button is damaged the machine will not be able to start.)

b) For the actual estop which is NC, when all is ok the power to the input is on, when there is a problem the power to the input is off. In the ladder logic an input that is on (indicating all is ok) will allow the rung to turn on the motor, otherwise an input that is off (indicating a stop) will break the rung and cut the power.)

6.

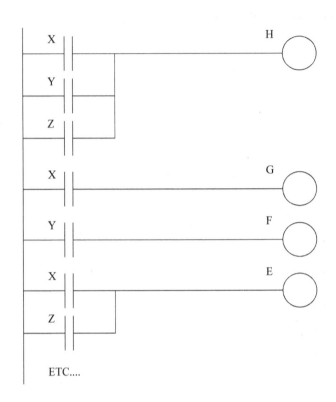

7.

8.

    a) AB b) $\overline{A} + B$ c) $\overline{A}B$ d) $A + B$

9.

  a) $$\overline{(A + B)} \cdot \overline{(A + \overline{B})} = (\overline{A}\overline{B})(\overline{A}B) = 0$$

  b) $$ABCD + \overline{A}BCD + ABC\overline{D} + AB\overline{C}\overline{D} = BCD + AB\overline{D} = B(CD + A\overline{D})$$

10. C

11.

$$X = \overline{B} \cdot (A \cdot C + \overline{C})$$

12.

$$Y = \overline{(AB\overline{C}D + AB\overline{C}\overline{D} + \overline{A}BCD + \overline{A}\overline{B}CD) + D}$$

$$Y = (AB\overline{C}D + AB\overline{C}\overline{D} + \overline{A}BCD + \overline{A}\overline{B}CD)\overline{D}$$

$$Y = (0 + AB\overline{C}\overline{D} + 0 + 0)\overline{D}$$

$$Y = AB\overline{C}\overline{D}$$

13.

a)

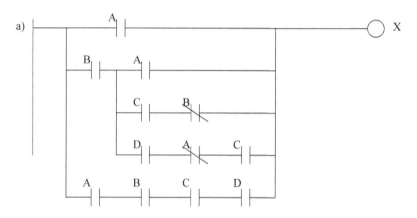

b) $A + DCB$

c)

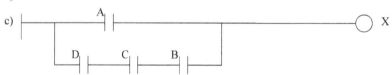

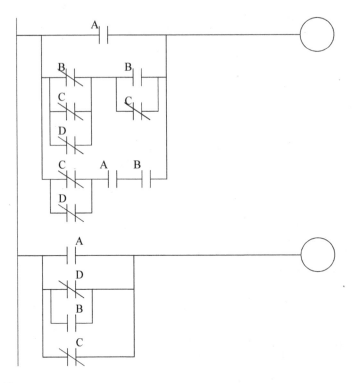

14.

$$\overline{A}\,\overline{B}\,\overline{C}\,\overline{D} + \overline{A}\,\overline{B}CD + \overline{A}B\overline{C}D + \overline{A}BCD + A\overline{B}\,\overline{C}\,\overline{D} + A\overline{B}CD + ABC\overline{D} + ABCD$$

$$\overline{B}\,\overline{C}\,\overline{D} + \overline{A}CD + \overline{B}CD + \overline{A}BD + BCD + ACD + ABC$$

$$\overline{B}\,\overline{C}\,\overline{D} + CD(\overline{A} + A) + CD(\overline{B} + B) + \overline{A}BD + ABC$$

$$\overline{B}\,\overline{C}\,\overline{D} + D(C + \overline{A}B) + ABC$$

104

15.

$$Y = \overline{C}\left(\overline{\overline{A} + \left(\overline{\overline{A} + (\overline{\overline{B}\,\overline{C}(A + B\overline{C})})}\right)}\right)$$

$$Y = \overline{C}\left(\overline{\overline{A} + \left(\overline{\overline{A} + (\overline{\overline{B}\,\overline{C}(A + \overline{B} + C)})}\right)}\right)$$

$$Y = \overline{C}\left(\overline{\overline{A} + (\overline{\overline{A} + (\overline{\overline{B}\,\overline{C}\,\overline{A}B\overline{C}})})}\right)$$

$$Y = \overline{C}\left(\overline{\overline{A} + (\overline{\overline{A} + \overline{0}})}\right)$$

$$Y = \overline{C}(\overline{\overline{A} + (\overline{\overline{A} + 1})})$$

$$Y = \overline{C}(\overline{\overline{A} + (\overline{1})})$$

$$Y = \overline{C}(\overline{\overline{A} + 0})$$

$$Y = \overline{C}\,\overline{\overline{A}}$$

$$Y = C + \overline{A}$$

16.

$$X = \overline{\overline{(A + B \cdot \overline{A})} + \overline{(C + D + E\overline{C})}}$$

OR

$$X = \overline{\overline{(A + B \cdot \overline{A})} + \overline{(C + D + E\overline{C})}}$$

$$X = \overline{(A + B \cdot \overline{A})}(C + D + E\overline{C})$$

$$X = \overline{(A)}(\overline{B \cdot \overline{A}})(C + D + E\overline{C})$$

$$X = (\overline{A})(\overline{B \cdot \overline{A}})(C + D + E\overline{C})$$

$$X = \overline{A}\,\overline{B}(C + D + E\overline{C})$$

$$X = \overline{A}\,\overline{B}(C + D + E)$$

$$X = \overline{A + B \cdot \overline{A}} + \overline{C}\,\overline{D}(\overline{E} + C)$$

$$X = \overline{A + B + \overline{C}\,\overline{D}\,\overline{E}}$$

$$X = \overline{A}\,\overline{B}(\overline{\overline{C}\,\overline{D}\,\overline{E}})$$

$$X = \overline{A}\,\overline{B}(C + D + E)$$

17.

18.

$C\bar{A}B$

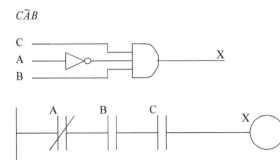

19.

a) $X = A + (B \cdot (\bar{A} + C) + C) + A \cdot B \cdot (\bar{D} + \bar{E})$

$X = A + (B \cdot \bar{A} + B \cdot C + C) + A \cdot B \cdot \bar{D} + A \cdot B \cdot \bar{E}$

$X = A \cdot (1 + B \cdot \bar{D} + B \cdot \bar{E}) + B \cdot \bar{A} + C \cdot (B + 1)$

$X = A + B \cdot \bar{A} + C$

b)

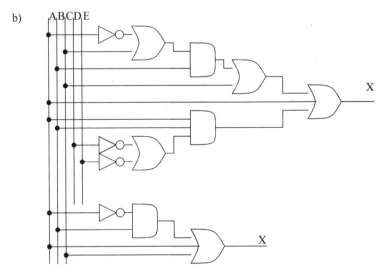

c)

20.

$$A + (\overline{B} + \overline{C} + \overline{D}) \cdot (B + \overline{C}) + A \cdot B \cdot (\overline{C} + \overline{D})$$

$$A \cdot (1 + B \cdot (\overline{C} + \overline{D})) + (\overline{B} + \overline{C} + \overline{D}) \cdot B + (\overline{B} + \overline{C} + \overline{D}) \cdot \overline{C}$$

$$A + (\overline{C} + \overline{D}) \cdot B + \overline{C}$$

$$A + \overline{C} \cdot B + \overline{D} \cdot B + \overline{C}$$

$$A + \overline{D} \cdot B + \overline{C}$$

$A + \overline{D} \cdot B + \overline{C}$

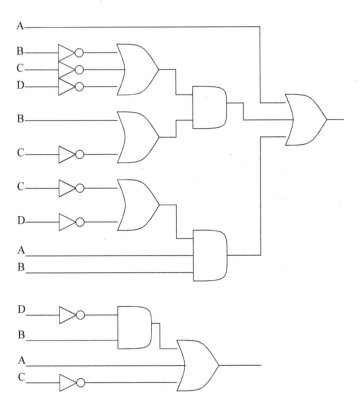

21.

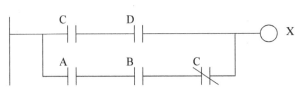

22.

(The equations.....) $\qquad X = D(\overline{B} + A) \qquad Y = B(D + C)$

6 KARNAUGH MAPS

Topics:

    • Truth tables and Karnaugh maps

Objectives:

    • Be able to simplify designs with Boolean algebra and Karnaugh maps

      Karnaugh maps allow us to convert a truth table to a simplified Boolean expression without using Boolean Algebra. The truth table in Figure 6.1 is an extension of the previous burglar alarm example, an alarm quiet input has been added.

Given

      A, W, M, S as before
      Q = Alarm Quiet (0 = quiet)

Step1: Draw the truth table

S	M	W	Q	A
0	0	0	0	0
0	0	0	1	0
0	0	1	0	0
0	0	1	1	0
0	1	0	0	0
0	1	0	1	0
0	1	1	0	0
0	1	1	1	0
1	0	0	0	0
1	0	0	1	1
1	0	1	0	0
1	0	1	1	0
1	1	0	0	0
1	1	0	1	1
1	1	1	0	0
1	1	1	1	1

Figure 6.1   *Truth Table for a Burglar Alarm*

      Instead of converting this directly to a Boolean equation, it is put into a tabular form as shown in Figure 6.1. The rows and columns are chosen from the input variables. The decision of which variables to use for rows or columns can be arbitrary - the table will look different, but you will still get a similar solution. For both the rows and columns the variables are ordered to show the values of the bits using NOTs. The sequence is not binary, but it is organized so that only one of the bits changes at a time, so the sequence of bits is 00, 01, 11, 10 - this step is very important. Next the values from the truth table that are true are entered into the Karnaugh map. Zeros can also be entered, but are not necessary. In the example the three true values from the truth table have been entered in the table.

Step 2: Divide the input variables up. I choose SQ and MW

Step 3: Draw a Karnaugh map based on the input variables

	$\overline{M}\,\overline{W}$ (=00)	$\overline{M}W$ (=01)	MW (=11)	M$\overline{W}$ (=10)
$\overline{S}\,\overline{Q}$ (=00)				
$\overline{S}Q$ (=01)				
SQ (=11)	1		1	1
S$\overline{Q}$ (=10)				

Added for clarity

Note: The inputs are arranged so that only one bit changes at a time for the Karnaugh map. In the example above notice that any adjacent location, even the top/bottom and left/right extremes follow this rule. This is done so that changes are visually grouped. If this pattern is not used then it is much more difficult to group the bits.

Figure 6.2   ***The Karnaugh Map***

When bits have been entered into the Karnaugh map there should be some obvious patterns. These patterns typically have some sort of symmetry. In Figure 6.1 there are two patterns that have been circled. In this case one of the patterns is because there are two bits beside each other. The second pattern is harder to see because the bits in the left and right hand side columns are beside each other. (Note: Even though the table has a left and right hand column, the sides and top/bottom wrap around.) Some of the bits are used more than once, this will lead to some redundancy in the final equation, but it will also give a simpler expression.

The patterns can then be converted into a Boolean equation. This is done by first observing that all of the patterns sit in the third row, therefore the expression will be ANDed with SQ. There are two patterns in the third row, one has M as the common term, the second has $\overline{W}$ as the common term. These can now be combined into the equation. Finally the equation is converted to ladder logic.

Step 4: Look for patterns in the map

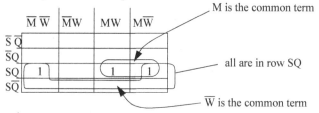

Step 5: Write the equation using the patterns

$$A = S \cdot Q \cdot (M + \overline{W})$$

Step 6: Convert the equation into ladder logic

Figure 6.3 *Recognition of the Boolean Equation from the Karnaugh Map*

Karnaugh maps are an alternative method to simplifying equations with Boolean algebra. It is well suited to visual learners, and is an excellent way to verify Boolean algebra calculations. The example shown was for four variables, thus giving two variables for the rows and two variables for the columns. More variables can also be used. If there were five input variables there could be three variables used for the rows or columns with the pattern 000, 001, 011, 010, 110, 111, 101, 100. If there is more than one output, a Karnaugh map is needed for each output.

Aside: A method developed by David Luque Sacaluga uses a circular format for the table. A brief example is shown below for comparison.

A	B	C	D	X
0	0	0	0	0
0	0	0	1	0
0	0	1	0	0
0	0	1	1	0
0	1	0	0	0
0	1	0	1	0
0	1	1	0	1
0	1	1	1	1
1	0	0	0	0
1	0	0	1	0
1	0	1	0	0
1	0	1	1	0
1	1	0	0	0
1	1	0	1	0
1	1	1	0	1
1	1	1	1	1

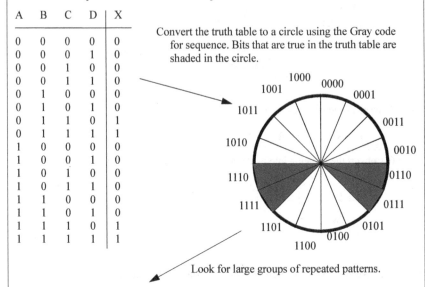

Convert the truth table to a circle using the Gray code for sequence. Bits that are true in the truth table are shaded in the circle.

Look for large groups of repeated patterns.

1. In this case 'B' is true in the bottom half of the circle, so the equation becomes,
$$X = B \cdot (\ldots)$$

2. There is left-right symmetry, with 'C' as the common term, so the equation becomes
$$X = B \cdot C \cdot (\ldots)$$

3. The equation covers all four values, so the final equation is,
$$X = B \cdot C$$

Figure 6.4   ***Aside: An Alternate Approach***

6.1 SUMMARY

• Karnaugh maps can be used to convert a truth table to a simplified Boolean equation.

6.2 PRACTICE PROBLEMS

(Note: Problem solutions are available at http://sites.google.com/site/automatedmanufacturingsystems/)

1. Setup the Karnaugh map for the truth table below.

A	B	C	D	Result
0	0	0	0	0
0	0	0	1	0
0	0	1	0	0
0	0	1	1	1
0	1	0	0	1
0	1	0	1	1
0	1	1	0	1
0	1	1	1	1
1	0	0	0	0
1	0	0	1	0
1	0	1	0	1
1	0	1	1	1
1	1	0	0	0
1	1	0	1	0
1	1	1	0	1
1	1	1	1	1

2. Use a Karnaugh map to simplify the following truth table, and implement it in ladder logic.

A	B	C	D	X
0	0	0	0	0
0	0	0	1	0
0	0	1	0	0
0	0	1	1	0
0	1	0	0	0
0	1	0	1	0
0	1	1	0	1
0	1	1	1	1
1	0	0	0	0
1	0	0	1	0
1	0	1	0	0
1	0	1	1	0
1	1	0	0	0
1	1	0	1	0
1	1	1	0	1
1	1	1	1	1

3. Write the simplest Boolean equation for the Karnaugh map below,

	CD	$C\bar{D}$	$\bar{C}\bar{D}$	$\bar{C}D$
AB	1	0	0	1
$A\bar{B}$	0	0	0	0
$\bar{A}\bar{B}$	0	0	0	0
$\bar{A}B$	0	1	1	0

4. Given the truth table below find the most efficient ladder logic to implement it. Use a structured technique such

A	B	C	D	X	Y
0	0	0	0	0	0
0	0	0	1	0	1
0	0	1	0	0	0
0	0	1	1	0	0
0	1	0	0	0	0
0	1	0	1	0	0
0	1	1	0	0	1
0	1	1	1	0	1
1	0	0	0	1	0
1	0	0	1	1	1
1	0	1	0	0	0
1	0	1	1	0	0
1	1	0	0	1	0
1	1	0	1	1	0
1	1	1	0	0	1
1	1	1	1	0	1

5. Examine the truth table below and design the simplest ladder logic using a Karnaugh map.

D	E	F	G	Y
0	0	0	0	0
0	0	0	1	0
0	0	1	0	0
0	0	1	1	0
0	1	0	0	0
0	1	0	1	1
0	1	1	0	0
0	1	1	1	1
1	0	0	0	0
1	0	0	1	1
1	0	1	0	0
1	0	1	1	1
1	1	0	0	0
1	1	0	1	1
1	1	1	0	0
1	1	1	1	1

6. Find the simplest Boolean equation for the Karnaugh map below without using Boolean algebra to simplify it. Draw the ladder logic.

	$\bar{A}\bar{B}\bar{C}$	$\bar{A}\bar{B}C$	$\bar{A}BC$	$\bar{A}B\bar{C}$	$AB\bar{C}$	ABC	$A\bar{B}C$	$A\bar{B}\bar{C}$
$\bar{D}\bar{E}$	1	1	0	1	0	0	0	0
$\bar{D}E$	1	1	0	0	0	0	0	0
DE	1	1	0	0	0	0	0	0
$D\bar{E}$	1	1	0	1	0	0	0	0

7. Given the following truth table for inputs A, B, C and D and output X. Convert it to simplified ladder logic using a Karnaugh map.

A	B	C	D	X
0	0	0	0	0
0	0	0	1	0
0	0	1	0	0
0	0	1	1	0
0	1	0	0	0
0	1	0	1	1
0	1	1	0	0
0	1	1	1	1
1	0	0	0	0
1	0	0	1	0
1	0	1	0	0
1	0	1	1	0
1	1	0	0	1
1	1	0	1	1
1	1	1	0	1
1	1	1	1	1

8. Consider the following truth table. Convert it to a Karnaugh map and develop a simplified Boolean equation

(without Boolean algebra). Draw the corresponding ladder logic.

	inputs				output
A	B	C	D	E	X
0	0	0	0	0	
0	0	0	0	1	
0	0	0	1	0	
0	0	0	1	1	
0	0	1	0	0	
0	0	1	0	1	
0	0	1	1	0	1
0	0	1	1	1	1
0	1	0	0	0	
0	1	0	0	1	
0	1	0	1	0	1
0	1	0	1	1	1
0	1	1	0	0	
0	1	1	0	1	
0	1	1	1	0	
0	1	1	1	1	1
1	0	0	0	0	
1	0	0	0	1	
1	0	0	1	0	
1	0	0	1	1	
1	0	1	0	0	
1	0	1	0	1	
1	0	1	1	0	
1	0	1	1	1	1
1	1	0	0	0	
1	1	0	0	1	
1	1	0	1	0	
1	1	0	1	1	
1	1	1	0	0	1
1	1	1	0	1	1
1	1	1	1	0	1
1	1	1	1	1	1

9. Given the truth table below

A	B	C	D	Z
0	0	0	0	0
0	0	0	1	0
0	0	1	0	0
0	0	1	1	0
0	1	0	0	0
0	1	0	1	1
0	1	1	0	1
0	1	1	1	1
1	0	0	0	0
1	0	0	1	1
1	0	1	0	0
1	0	1	1	0
1	1	0	0	0
1	1	0	1	1
1	1	1	0	1
1	1	1	1	1

a) find a Boolean algebra expression using a Karnaugh map.
b) draw a ladder diagram using the truth table (not the Boolean expression).

116

10. Convert the following ladder logic to a Karnaugh map.

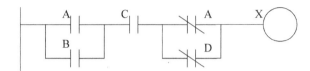

11. a) Construct a truth table for the following problem.
        i) there are three buttons A, B, C.
        ii) the output is on if any two buttons are pushed.
        iii) if C is pressed the output will always turn on.
   b) Develop a Boolean expression.
   c) Develop a Boolean expression using a Karnaugh map.

12. Develop the simplest Boolean expression for the Karnaugh map below,
   a) graphically.
   b) by Boolean Algebra

	$\overline{A}\,\overline{B}$	$A\overline{B}$	AB	$\overline{A}B$
CD	1			1
$C\overline{D}$		1	1	
$\overline{C}\,\overline{D}$				
$\overline{C}D$	1			1

13. Consider the following boolean equation.

$$X = \overline{(A + B\overline{A})\overline{A}} + \overline{(CD + \overline{C}D + C\overline{D})}$$

   a) Can this Boolean equation be converted directly ladder logic. Explain your answer, and if neces-
      sary, make any changes required so that it may be converted to ladder logic.
   b) Write out ladder logic, based on the result in step a).
   c) Simplify the equation using Boolean algebra and write out new ladder logic.
   d) Write a Karnaugh map for the Boolean equation, and show how it can be used to obtain a sim-
      plified Boolean equation.

6.3 ASSIGNMENT PROBLEMS

1. Use the Karnaugh map below to create a simplified Boolean equation. Then use the equation to create ladder
   logic.

	AB	$A\overline{B}$	$\overline{A}B$	$\overline{A}B$
CD	1	1	1	1
$C\overline{D}$	1	0	0	1
$\overline{C}\,\overline{D}$	0	0	0	1
$\overline{C}D$	0	0	0	1

2. Use a Karnaugh map to develop simplified ladder logic for the following truth table where A, B, C and D are
   inputs, and X and Y are outputs.

117

A	B	C	D	X	Y
0	0	0	0	0	0
0	0	0	1	1	0
0	0	1	0	0	0
0	0	1	1	1	0
0	1	0	0	0	0
0	1	0	1	0	1
0	1	1	0	0	1
0	1	1	1	0	1
1	0	0	0	0	0
1	0	0	1	1	0
1	0	1	0	0	0
1	0	1	1	1	0
1	1	0	0	0	0
1	1	0	1	1	1
1	1	1	0	0	1
1	1	1	1	1	1

3. You are planning the basic layout for a control system with the criteria provided below. You need to plan the wiring for the input and output cards, and then write the ladder logic for the controller. You decide to use a Boolean logic design technique to design the ladder logic. AND, your design will be laid out on the design sheets found later in this book.
- There are two inputs from PNP photoelectric sensors *part* and *busy*.
- There is a NO *cycle* button, and NC *stop* button.
- There are two outputs to indicator lights, the *running* light and the *stopped* light.
- There is an output to a conveyor, that will drive a high current 120Vac motor.
- The conveyor is to run when the *part* sensor is on and while the *cycle* button is pushed, but the *busy* sensor is off. If the *stop* button is pushed the conveyor will stop.
- While the conveyor is running the *running* light will be on, otherwise the *stopped* light will be on.

4. Convert the following truth table to simplified ladder logic using a Karnaugh map AND Boolean equations. The inputs are A, B, C and D and the output is X.

A	B	C	D	X
0	0	0	0	1
0	0	0	1	1
0	0	1	0	0
0	0	1	1	0
0	1	0	0	1
0	1	0	1	1
0	1	1	0	0
0	1	1	1	0
1	0	0	0	1
1	0	0	1	0
1	0	1	0	1
1	0	1	1	0
1	1	0	0	1
1	1	0	1	0
1	1	1	0	1
1	1	1	1	0

6.1 PRACTICE PROBLEM SOLUTIONS

1.

	AB	$A\overline{B}$	$\overline{A}\,\overline{B}$	$\overline{A}B$
CD	1	1	1	1
$C\overline{D}$	1	1	0	1
$\overline{C}\,\overline{D}$	0	0	0	1
$\overline{C}D$	0	0	0	1

2.

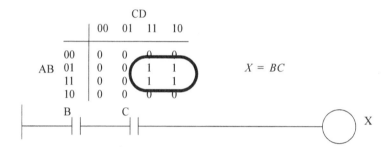

		CD			
		00	01	11	10
	00	0	0	0	0
AB	01	0	0	1	1
	11	0	0	1	1
	10	0	0	0	0

$X = BC$

3.

	CD	$C\overline{D}$	$\overline{C}\,\overline{D}$	$\overline{C}D$
AB	1	0	0	1
$A\overline{B}$	0	0	0	0
$\overline{A}\,\overline{B}$	0	0	0	0
$\overline{A}B$	0	1	1	0

-For all, B is true

$$B(AD + \overline{A}\overline{D})$$

4.

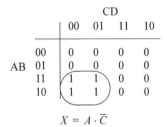

FOR X

CD

	00	01	11	10
AB 00	0	0	0	0
01	0	0	0	0
11	1	1	0	0
10	1	1	0	0

$$X = A \cdot \bar{C}$$

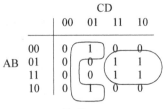

FOR Y

CD

	00	01	11	10
AB 00	0	1	0	0
01	0	0	1	1
11	0	0	1	1
10	0	1	0	0

$$Y = \bar{B} \cdot \bar{C} \cdot D + B \cdot C$$

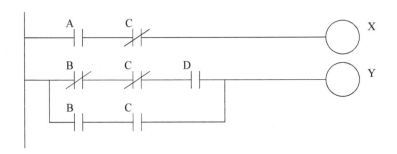

5.

FG

	00	01	11	10
DE 00	0	0	0	0
01	0	1	1	0
11	0	1	1	0
10	0	1	1	0

$$Y = G(E + D)$$

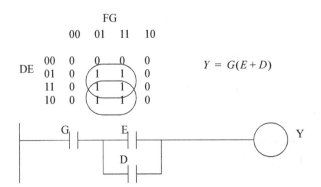

120

6.

	$\overline{A}\overline{B}\overline{C}$	$\overline{A}\overline{B}C$	$\overline{A}BC$	$\overline{A}B\overline{C}$	$AB\overline{C}$	ABC	$A\overline{B}C$	$A\overline{B}\overline{C}$
$\overline{D}\overline{E}$	1	1	0	1	0	0	0	0
$\overline{D}E$	1	1	0	0	0	0	0	0
DE	1	1	0	0	0	0	0	0
$D\overline{E}$	1	1	0	1	0	0	0	0

$\overline{A}\overline{B}$ $\overline{A}B\overline{C}\overline{E}$

$output = \overline{A}\overline{B} + \overline{A}B\overline{C}\overline{E}$

7.

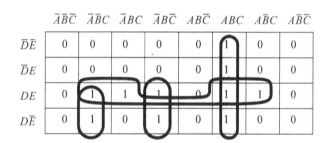

8.

	$\overline{A}\overline{B}\overline{C}$	$\overline{A}\overline{B}C$	$\overline{A}BC$	$\overline{A}B\overline{C}$	$AB\overline{C}$	ABC	$A\overline{B}C$	$A\overline{B}\overline{C}$
$\overline{D}\overline{E}$	0	0	0	0	0	1	0	0
$\overline{D}E$	0	0	0	0	0	1	0	0
DE	0	1	1	1	0	1	1	0
$D\overline{E}$	0	1	0	1	0	1	0	0

$$X = ABC + D(\overline{A}\overline{B}C + \overline{A}B\overline{C} + EC)$$

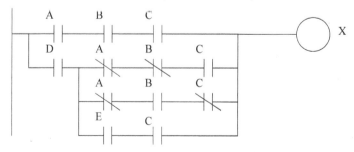

121

9.

	AB	A$\overline{B}$	$\overline{A}\,\overline{B}$	$\overline{A}$B
CD	1	0	0	1
C$\overline{D}$	1	0	0	1
$\overline{C}\,\overline{D}$	0	0	0	0
$\overline{C}$D	1	1	0	1

$Z=B*(C+D)+A\,\overline{B}\,\overline{C}\,D$

10.

A	B	C	D	X
0	0	0	0	0
0	0	0	1	0
0	0	1	0	0
0	0	1	1	0
0	1	0	0	0
0	1	0	1	0
0	1	1	0	1
0	1	1	1	1
1	0	0	0	0
1	0	0	1	0
1	0	1	0	1
1	0	1	1	0
1	1	0	0	0
1	1	0	1	0
1	1	1	0	1
1	1	1	1	0

	CD	$C\bar{D}$	$\bar{C}\bar{D}$	$\bar{C}D$
AB	0	1	0	0
$A\bar{B}$	0	1	0	0
$\bar{A}\bar{B}$	0	0	0	0
$\bar{A}B$	1	1	0	0

11.

A	B	C	out
0	0	0	0
0	0	1	1
0	1	0	0
0	1	1	1
1	0	0	0
1	0	1	1
1	1	0	1
1	1	1	1

$C + A \cdot B$

	AB	A$\bar{B}$	$\bar{A}\,\bar{B}$	$\bar{A}$B
C	1	1	1	1
$\bar{C}$	1	0	0	0

12.

$$D\bar{A} + AC\bar{D}$$

$$\bar{A}\bar{B}CD + \bar{A}BCD + A\bar{B}C\bar{D} + ABC\bar{D} + \bar{A}\bar{B}\bar{C}D + \bar{A}B\bar{C}D$$

$$\bar{A}CD + AC\bar{D} + \bar{A}\bar{C}D$$

$$\bar{A}D + AC\bar{D}$$

13.

a) $X = \bar{A}\bar{B} + A + (\bar{C} + \bar{D})(C + \bar{D})(\bar{C} + D)$

c) $X = A + \bar{B} + \bar{C}\bar{D}$

d)

	$\bar{C}\bar{D}$	$\bar{C}D$	CD	$C\bar{D}$
$\bar{A}\bar{B}$	1	1	1	1
$\bar{A}B$	1	0	0	0
AB	1	1	1	1
$A\bar{B}$	1	1	1	1

7PLC OPERATION

Topics:

- The computer structure of a PLC
- The sanity check, input, output and logic scans
- Status and memory types

Objectives:

- Understand the operation of a PLC.

For simple programming the relay model of the PLC is sufficient. As more complex functions are used the more complex vonNeumann model of the PLC must be used. A vonNeumann computer processes one instruction at a time. Most computers operate this way, although they appear to be doing many things at once. Consider the computer components shown in Figure 7.1.

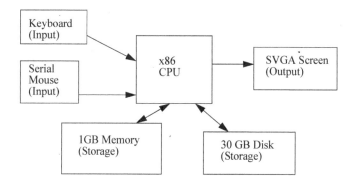

Figure 7.1   *Simplified Personal Computer Architecture*

Input is obtained from the keyboard and mouse, output is sent to the screen, and the disk and memory are used for both input and output for storage. (Note: the directions of these arrows are very important to engineers, always pay attention to indicate where information is flowing.) This figure can be redrawn as in Figure 7.1 to clarify the role of inputs and outputs.

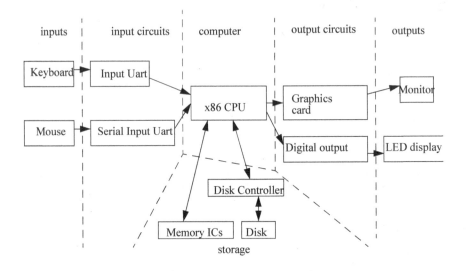

| inputs | input circuits | computer | output circuits | outputs |

Figure 7.2   *An Input-Output Oriented Architecture*

In this figure the data enters the left side through the inputs. (Note: most engineering diagrams have inputs on the left and outputs on the right.) It travels through buffering circuits before it enters the CPU. The CPU outputs data through other circuits. Memory and disks are used for storage of data that is not destined for output. If we look at a personal computer as a controller, it is controlling the user by outputting stimuli on the screen, and inputting responses from the mouse and the keyboard.

A PLC is also a computer controlling a process. When fully integrated into an application the analogies become;

> inputs - the keyboard is analogous to a proximity switch
> input circuits - the serial input uart is like a 24Vdc input card
> computer - the x86 CPU is like a PLC CPU unit
> output circuits - a graphics card is like a triac output card
> outputs - a monitor is like a light
> storage - memory in PLCs is similar to memories in personal computers

It is also possible to implement a PLC using a normal Personal Computer, although this is not advisable. In the case of a PLC the inputs and outputs are designed to be more reliable and rugged for harsh production environments.

7.1 OPERATION SEQUENCE

All PLCs have four basic stages of operations that are repeated many times per second. Initially when turned on the first time it will check it's own hardware and software for faults. If there are no problems it will copy all the input and copy their values into memory, this is called the input scan. Using only the memory copy of the inputs the ladder logic program will be solved once, this is called the logic scan. While solving the ladder logic the output values are only changed in temporary memory. When the ladder scan is done the outputs will updated using the temporary values in memory, this is called the output scan. The PLC now restarts the process by starting a self check for faults. This process typically repeats 10 to 100 times per second as is shown in Figure 7.1.

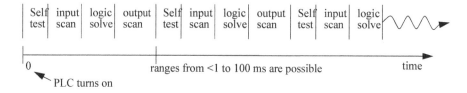

| Self test | input scan | logic solve | output scan | Self test | input scan | logic solve | output scan | Self test | input scan | logic solve |

0

ranges from <1 to 100 ms are possible   time

PLC turns on

SELF TEST - Checks to see if all cards error free, reset watch-dog timer, etc. (A watchdog timer will cause an error, and shut down the PLC if not reset within a short period of time - this would indicate that the ladder logic is not being scanned normally).

INPUT SCAN - Reads input values from the input cards, and copies their values to memory. This makes the PLC operation faster, and avoids cases where an input changes from the start to the end of the program (e.g., an emergency stop). There are special PLC functions that read the inputs directly, and avoid the input tables.

LOGIC SOLVE/SCAN - Based on the input table in memory, the program is executed 1 step at a time, and outputs are updated. This is the focus of the later sections.

OUTPUT SCAN - The output table is copied from memory to the outputs. These then drive the output devices.

Figure 7.3   **PLC Scan Cycle**

The input and output scans often confuse the beginner, but they are important. The input scan takes a *snapshot* of the inputs, and solves the logic. This prevents potential problems that might occur if an input that is used in multiple places in the ladder logic program changed while half way through a ladder scan. Thus changing the behaviors of half of the ladder logic program. This problem could have severe effects on complex programs that are developed later in the book. One side effect of the input scan is that if a change in input is too short in duration, it might fall between input scans and be missed.

When the PLC is initially turned on the normal outputs will be turned off. This does not affect the values of the inputs.

7.1.1 The Input and Output Scans

When the inputs to the PLC are scanned the physical input values are copied into memory. When the outputs to a PLC are scanned they are copied from memory to the physical outputs. When the ladder logic is scanned it uses the values in memory, not the actual input or output values. The primary reason for doing this is so that if a program uses an input value in multiple places, a change in the input value will not invalidate the logic. Also, if output bits were changed as each bit was changed, instead of all at once at the end of the scan the PLC would operate much slower.

7.1.2 The Logic Scan

Ladder logic programs are modelled after relay logic. In relay logic each element in the ladder will switch as quickly as possible. But in a program elements can only be examines one at a time in a fixed sequence. Consider the ladder logic in Figure 7.1, the ladder logic will be interpreted left-to-right, top-to-bottom. In the figure the ladder logic scan begins at the top rung. At the end of the rung it interprets the top output first, then the output branched below it. On the second rung it solves branches, before moving along the ladder logic rung.

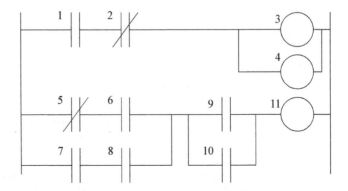

Figure 7.4 *Ladder Logic Execution Sequence*

The logic scan sequence become important when solving ladder logic programs which use outputs as inputs, as we will see in Chapter 8. It also becomes important when considering output usage. Consider Figure 7.1, the first line of ladder logic will examine input A and set output X to have the same value. The second line will examine input B and set the output X to have the opposite value. So the value of X was only equal to A until the second line of ladder logic was scanned. Recall that during the logic scan the outputs are only changed in memory, the actual outputs are only updated when the ladder logic scan is complete. Therefore the output scan would update the real outputs based upon the second line of ladder logic, and the first line of ladder logic would be ineffective.

Note: It is a common mistake for beginners to unintentionally repeat the same ladder logic output more than once. This will basically invalidate the first output, in this case the first line will never do anything.

Figure 7.5 *A Duplicated Output Error*

7.2 PLC STATUS

The lack of keyboard, and other input-output devices is very noticeable on a PLC. On the front of the PLC there are normally limited status lights. Common lights indicate;

> power on - this will be on whenever the PLC has power
> program running - this will often indicate if a program is running, or if no program is running
> fault - this will indicate when the PLC has experienced a major hardware or software problem

These lights are normally used for debugging. Limited buttons will also be provided for PLC hardware. The most common will be a run/program switch that will be switched to program when maintenance is being conducted, and back to run when in production. This switch normally requires a key to keep unauthorized personnel from altering the PLC program or stopping execution. A PLC will almost never have an on-off switch or reset button on the front. This needs to be designed into the remainder of the system.

The status of the PLC can be detected by ladder logic also. It is common for programs to check to see if they are being executed for the first time, as shown in Figure 7.1. The 'first scan' or 'first pass' input will be true the very first time the ladder logic is scanned, but false on every other scan. In this case the address for 'first pass' in ControlLogix is 'S:FS'. With the logic in the example the first scan will seal on 'light', until 'clear' is turned on. So the light will turn on after the PLC has been turned on, but it will turn off and stay off after 'clear' is turned on. The 'first scan' bit is also referred to at the 'first pass' bit.

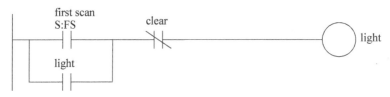

Figure 7.6   *An program that checks for the first scan of the PLC*

7.3 MEMORY TYPES

There are a few basic types of computer memory that are in use today.

        RAM (Random Access Memory) - this memory is fast, but it will lose its contents when power is lost, this is known as volatile memory. Every PLC uses this memory for the central CPU when running the PLC.

        ROM (Read Only Memory) - this memory is permanent and cannot be erased. It is often used for storing the operating system for the PLC.

        EPROM (Erasable Programmable Read Only Memory) - this is memory that can be programmed to behave like ROM, but it can be erased with ultraviolet light and reprogrammed.

        EEPROM (Electronically Erasable Programmable Read Only Memory) - This memory can store programs like ROM. It can be programmed and erased using a voltage, so it is becoming more popular than EPROMs.

        Hard Disk - Software based PLCs run on top of another operating system (such as Windows) that will read and save values to a hard drive, in case power is lost.

All PLCs use RAM for the CPU and ROM to store the basic operating system for the PLC. When the power is on the contents of the RAM will be kept, but the issue is what happens when power to the memory is lost. Originally PLC vendors used RAM with a battery so that the memory contents would not be lost if the power was lost. This method is still in use, but is losing favor. EPROMs have also been a popular choice for programming PLCs. The EPROM is programmed out of the PLC, and then placed in the PLC. When the PLC is turned on the ladder logic program on the EPROM is loaded into the PLC and run. This method can be very reliable, but the erasing and programming technique can be time consuming. EEPROM memories are a permanent part of the PLC, and programs can be stored in them like EPROM. Memory costs continue to drop, and newer types (such as flash memory) are becoming available, and these changes will continue to impact PLCs.

7.4 SOFTWARE BASED PLCS

The dropping cost of personal computers is increasing their use in control, including the replacement of PLCs. Software is installed that allows the personal computer to solve ladder logic, read inputs from sensors and update outputs to actuators. These are important to mention here because they don't obey the previous timing model. For example, if the computer is running a game it may slow or halt the computer. This issue and others are currently being investigated and good solutions should be expected soon.

7.5 SUMMARY

- A PLC and computer are similar with inputs, outputs, memory, etc.
- The PLC continuously goes through a cycle including a sanity check, input scan, logic scan, and output scan.
- While the logic is being scanned, changes in the inputs are not detected, and the outputs are not updated.
- PLCs use RAM, and sometime EPROMs are used for permanent programs.

7.6 PRACTICE PROBLEMS

(Note: Problem solutions are available at http://sites.google.com/site/automatedmanufacturingsystems/)

1. Does a PLC normally contain RAM, ROM, EPROM and/or batteries.

2. What are the indicator lights on a PLC used for?

3. A PLC can only go through the ladder logic a few times per second. Why?

4. What will happen if the scan time for a PLC is greater than the time for an input pulse? Why?

5. What is the difference between a PLC and a desktop computer?

6. Why do PLCs do a self check every scan?

7. Will the test time for a PLC be long compared to the time required for a simple program.

8. What is wrong with the following ladder logic? What will happen if it is used?

9. What is the address for a memory location that indicates when a PLC has just been turned on?

7.7 ASSIGNMENT PROBLEMS

1. Describe the basic steps of operation for a PLC after it is turned on.

2. Repeating a normal output in ladder logic should not be done normally. Discuss why.

3. Why does removing a battery from some older PLCs clear the memory?

7.1 PRACTICE PROBLEM SOLUTIONS

1. Every PLC contains RAM and ROM, but they may also contain EPROM or batteries.

2. Diagnostic and maintenance

3. Even if the program was empty the PLC would still need to scan inputs and outputs, and do a self check.

4. The pulse may be missed if it occurs between the input scans

5. Some key differences include inputs, outputs, and uses. A PLC has been designed for the factory floor, so it does not have inputs such as keyboards and mice (although some newer types can). They also do not have outputs such as a screen or sound. Instead they have inputs and outputs for voltages and current. The PLC runs user designed programs for specialized tasks, whereas on a personal computer it is uncommon for a user to program their system.

6. This helps detect faulty hardware or software. If an error were to occur, and the PLC continued operating, the controller might behave in an unpredictable way and become dangerous to people and equipment. The self check helps detect these types of faults, and shut the system down safely.

7. Yes, the self check is equivalent to about 1ms in many PLCs, but a single program instruction is about 1 micro second.

8. The normal output Y is repeated twice. In this example the value of Y would always match B, and the earlier rung with A would have no effect on Y.

9. S2:1/14 for micrologix, S2:1/15 for PLC-5, S:FS for ControlLogix processor

8LATCHES, TIMERS, COUNTERS AND MORE

Topics:

      • Latches, timers, counters and MCRs
      • Design examples
      • Internal memory locations are available, and act like outputs

Objectives:

      • Understand latches, timers, counters and MCRs.
      • To be able to select simple internal memory bits.

More complex systems cannot be controlled with combinatorial logic alone. The main reason for this is that we cannot, or choose not to add sensors to detect all conditions. In these cases we can use events to estimate the condition of the system. Typical events used by a PLC include;

      first scan of the PLC - indicating the PLC has just been turned on
      time since an input turned on/off - a delay
      count of events - to wait until set number of events have occurred
      latch on or unlatch - to lock something on or turn it off

The common theme for all of these events is that they are based upon one of two questions "How many?" or "How long?". An example of an event based device is shown in Figure 8.1. The input to the device is a push button. When the push button is pushed the input to the device turns on. If the push button is then released and the device turns off, it is a logical device. If when the push button is release the device stays on, is will be one type of event based device. To reiterate, the device is event based if it can respond to one or more things that have happened before. If the device responds only one way to the immediate set of inputs, it is logical.

e.g. A Start Push Button

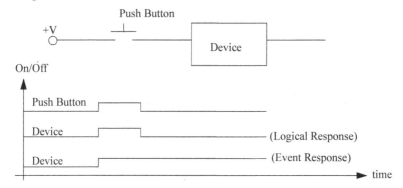

Figure 8.1   ***An Event Driven Device***

8.1 LATCHES

A latch is like a sticky switch - when pushed it will turn on, but stick in place, it must be pulled to release it and turn it off. A latch in ladder logic uses one instruction to latch, and a second instruction to unlatch, as shown in Figure 8.1. The output with an L inside will turn the output D on when the input A becomes true. D will stay on even if A turns off. Output D will turn off if input B becomes true and the output with a U inside becomes true (Note: this will seem a little backwards at first). If an output has been latched on, it will keep its value, even if the power has been turned off.

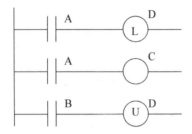

Figure 8.2   *A Ladder Logic Latch*

The operation of the ladder logic in Figure 8.1 is illustrated with a timing diagram in Figure 8.1. A timing diagram shows values of inputs and outputs over time. For example the value of input A starts low (false) and becomes high (true) for a short while, and then goes low again. Here when input A turns on both the outputs turn on. There is a slight delay between the change in inputs and the resulting changes in outputs, due to the program scan time. Here the dashed lines represent the output scan, sanity check and input scan (assuming they are very short.) The space between the dashed lines is the ladder logic scan. Consider that when A turns on initially it is not detected until the first dashed line. There is then a delay to the next dashed line while the ladder is scanned, and then the output at the next dashed line. When A eventually turns off, the normal output C turns off, but the latched output D stays on. Input B will unlatch the output D. Input B turns on twice, but the first time it is on is not long enough to be detected by an input scan, so it is ignored. The second time it is on it unlatches output D and output D turns off.

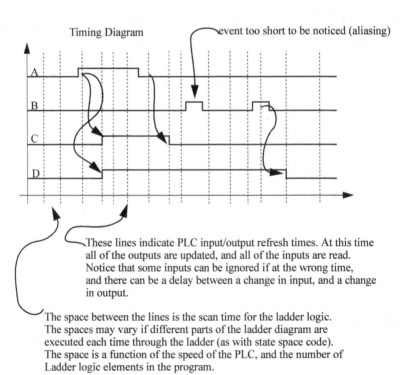

Figure 8.3   *A Timing Diagram for the Ladder Logic in Figure 8.1*

The timing diagram shown in Figure 8.1 has more details than are normal in a timing diagram as shown in Figure 8.1. The brief pulse would not normally be wanted, and would be designed out of a system either by extending the length of the pulse, or decreasing the scan time. An ideal system would run so fast that aliasing would not be possible.

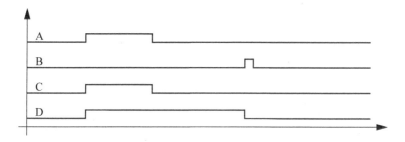

Figure 8.4   *A Typical Timing Diagram*

A more elaborate example of latches is shown in Figure 8.1. In this example the addresses are for an older Allen-Bradley Micrologix controller. The inputs begin with *I/*, followed by an input number. The outputs begin with *O/* , followed by an output number.

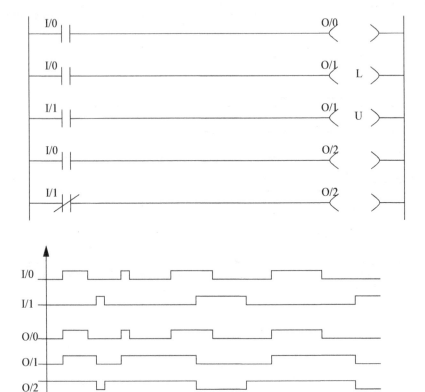

Figure 8.5   **A Latch Example**

A normal output should only appear once in ladder logic, but latch and unlatch instructions may appear multiple times. In Figure 8.1 a normal output O/2 is repeated twice. When the program runs it will examine the fourth line and change the value of O/2 in memory (remember the output scan does not occur until the ladder scan is done.) The last line is then interpreted and it overwrites the value of O/2. Basically, only the last line will change O/2.

Latches are not used universally by all PLC vendors, others such as Siemens use flip-flops. These have a similar behavior to latches, but a different notation as illustrated in Figure 8.1. Here the flip-flop is an output block that is connected to two different logic rungs. The first rung shown has an input A connected to the S setting terminal. When A goes true the output value Q will go true. The second rung has an input B connected to the R resetting terminal. When B goes true the output value Q will be turned off. The output Q will always be the inverse of $\overline{Q}$. Notice that the S and R values are equivalent to the L and U values from earlier examples.

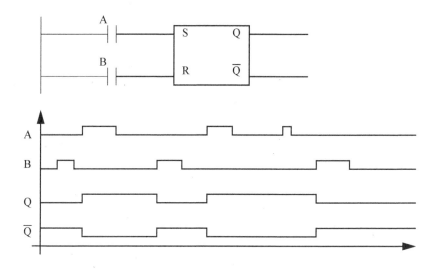

Figure 8.6   *Flip-Flops for Latching Values*

8.2 TIMERS

There are four fundamental types of timers shown in Figure 8.1. An on-delay timer will wait for a set time after a line of ladder logic has been true before turning on, but it will turn off immediately. An off-delay timer will turn on immediately when a line of ladder logic is true, but it will delay before turning off. Consider the example of an old car. If you turn the key in the ignition and the car does not start immediately, that is an on-delay. If you turn the key to stop the engine but the engine doesn't stop for a few seconds, that is an off delay. An on-delay timer can be used to allow an oven to reach temperature before starting production. An off delay timer can keep cooling fans on for a set time after the oven has been turned off.

	on-delay	off-delay
retentive	RTO	RTF
nonretentive	TON	TOF

TON - Timer ON
TOF - Timer OFf
RTO - Retentive Timer On
RTF - Retentive Timer oFf

Figure 8.7   *The Four Basic Timer Types*

A retentive timer will sum all of the on or off time for a timer, even if the timer never finished. A nonretentive timer will start timing the delay from zero each time. Typical applications for retentive timers include tracking the

137

time before maintenance is needed. A non retentive timer can be used for a start button to give a short delay before a conveyor begins moving.

An example of an Allen-Bradley TON timer is shown in Figure 8.1. The rung has a single input *A* and a function block for the *TON*. (Note: This timer block will look different for different PLCs, but it will contain the same information.) The information inside the timer block describes the timing parameters. The first item is the timer 'example'. This is a location in the PLC memory that will store the timer information. The preset is the millisecond delay for the timer, in this case it is 4s (4000ms). The accumulator value gives the current value of the timer as *0*. While the timer is running the accumulated value will increase until it reaches the preset value. Whenever the input *A* is true the *EN* output will be true. The *DN* output will be false until the accumulator has reached the preset value. The *EN* and *DN* outputs cannot be changed when programming, but these are important when debugging a ladder logic program. The second line of ladder logic uses the timer *DN* output to control another output *B*.

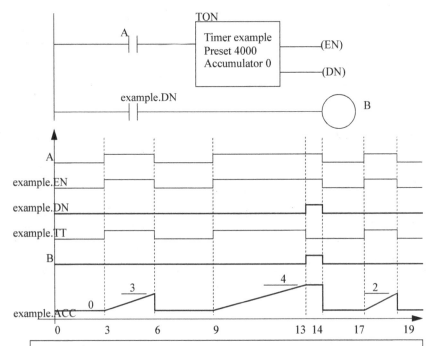

Note: For the older Allen-Bradley equipment the notations are similar, although the tag names are replaced with a more strict naming convention. The timers are kept in 'files' with names starting with 'T4:', followed by a timer number. The examples below show the older (PLC-5 and micrologix notations compared to the new RS-Logix (5000) notations. In the older PLCs the timer is given a unique number, in the RSLogix 5000 processors it is given a tag name (in this case 't') and type 'TIMER'.

Older	Newer
T4:0/DN	t.DN
T4:0/EN	t.EN
T4:0.PRE	t.PRE
T4:0.ACC	t.ACC
T4:0/TT	t.TT

Figure 8.8   **An Allen-Bradley TON Timer**

The timing diagram in Figure 8.1 illustrates the operation of the TON timer with a 4 second on-delay. *A* is the input to the timer, and whenever the timer input is true the *EN* enabled bit for the timer will also be true. If the accumulator value is equal to the preset value the *DN* bit will be set. Otherwise, the *TT* bit will be set and the accumulator value will begin increasing. The first time *A* is true, it is only true for 3 seconds before turning off, after this the value resets to zero. (Note: in a retentive time the value would remain at 3 seconds.) The second time *A* is true, it is on more than 4 seconds. After 4 seconds the *TT* bit turns off, and the *DN* bit turns on. But, when *A* is released the accumulator resets to zero, and the *DN* bit is turned off.

A value can be entered for the accumulator while programming. When the program is downloaded this value will be in the timer for the first scan. If the TON timer is not enabled the value will be set back to zero. Normally zero will be entered for the preset value.

The timer in Figure 8.1 is identical to that in Figure 8.1, except that it is retentive. The most significant difference is that when the input *A* is turned off the accumulator value does not reset to zero. As a result the timer turns on much sooner, and the timer does not turn off after it turns on. A reset instruction will be shown later that will allow the accumulator to be reset to zero.

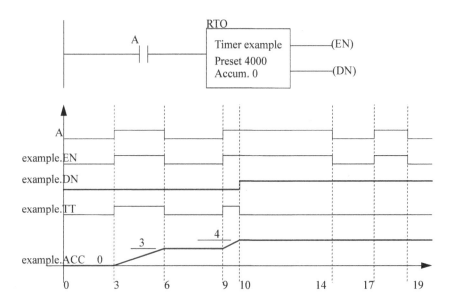

Figure 8.9   **An Allen Bradley Retentive On-Delay Timer**

An off delay timer is shown in Figure 8.1. This timer has a time base of 0.01s, with a preset value of 3500, giving a total delay of 3.5s. As before the *EN* enable for the timer matches the input. When the input *A* is true the *DN* bit is on. Is is also on when the input *A* has turned off and the accumulator is counting. The *DN* bit only turns off when the input *A* has been off long enough so that the accumulator value reaches the preset. This type of timer is not retentive, so when the input *A* becomes true, the accumulator resets. Off-delay timers are normally off (DN is false) until activated the first time.

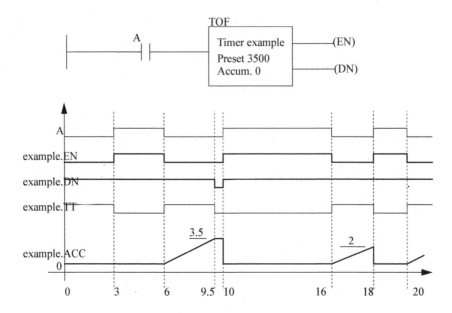

Figure 8.10   **An Allen Bradley Off-Delay Timer**

Retentive off-delay (RTF) timers have few applications and are rarely used, therefore many PLC vendors do not include them.

An example program is shown in Figure 8.1. In total there are four timers used in this example, t_1, t_2, t_3, and t_4. The timer instructions are shown with the accumulator values omitted, assuming that they start with a value of zero. All four different types of counters have the input *'go'*. Output *'done'* will turn on when the TON counter *t_1* is done. All four of the timers can be reset with input *'reset'*.

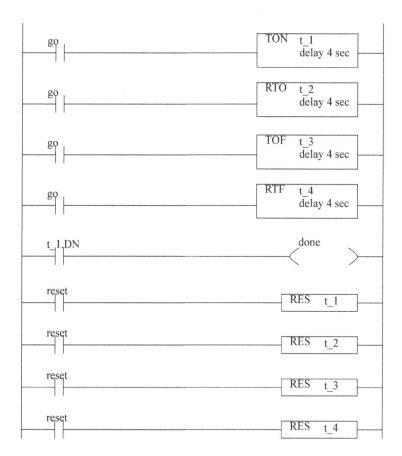

Figure 8.11   A Timer Example

A timing diagram for this example is shown in Figure 8.1. As input *go* is turned on the TON and RTO timers begin to count and reach 4s and turn on. When *reset* becomes true it resets both timers and they start to count for another second before *go* is turned off. After the input is turned off the TOF and RTF both start to count, but neither reaches the 4s preset. The input *go* is turned on again and the TON and RTO both start counting. The RTO turns on one second sooner because it had 1s stored from the 7-8s time period. After *go* turns off again both the off delay timers count down, and reach the 4 second delay, and turn on. These patterns continue across the diagram.

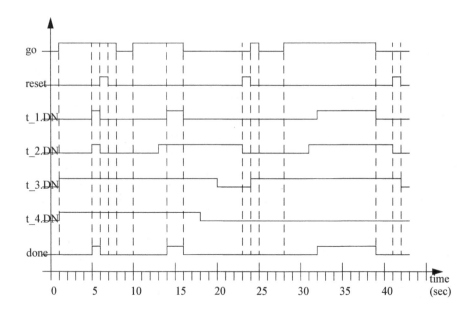

Figure 8.12   *A Timing Diagram for Figure 8.1*

Consider the short ladder logic program in Figure 8.1 for control of a heating oven. The system is started with a *Start* button that seals in the *Auto* mode. This can be stopped if the *Stop* button is pushed. (Remember: Stop buttons are normally closed.) When the *Auto* goes on initially the TON timer is used to sound the horn for the first 10 seconds to warn that the oven will start, and after that the horn stops and the heating coils start. When the oven is turned off the fan continues to blow for 300s or 5 minutes after.

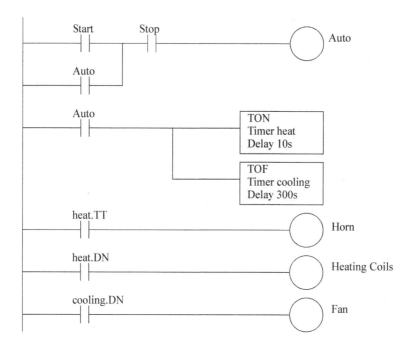

Note: For the remainder of the text I will use the shortened notation for timers shown above. This will save space and reduce confusion.

Figure 8.13   *A Timer Example*

A program is shown in Figure 8.1 that will flash a light once every second. When the PLC starts, the second timer will be off and the *t_on.DN* bit will be off, therefore the normally closed input to the first timer will be on. *t_off* will start timing until it reaches 0.5s, when it is done the second timer will start timing, until it reaches 0.5s. At that point *t_on.DN* will become true, and the input to the first time will become false. *t_off* is then set back to zero, and then *t_on* is set back to zero. And, the process starts again from the beginning. In this example the first timer is used to drive the second timer. This type of arrangement is normally called cascading, and can use more that two timers.

Figure 8.14   Another Timer Example

8.3 COUNTERS

There are two basic counter types: count-up and count-down. When the input to a count-up counter goes true the accumulator value will increase by 1 (no matter how long the input is true.) If the accumulator value reaches the preset value the counter *DN* bit will be set. A count-down counter will decrease the accumulator value until the preset value is reached.

An Allen Bradley count-up (CTU) instruction is shown in Figure 8.1. The instruction requires memory in the PLC to store values and status, in this case is *example*. The preset value is 4 and the value in the accumulator is 2. If the input *A* were to go from false to true the value in the accumulator would increase to 3. If *A* were to go off, then on again the accumulator value would increase to 4, and the *DN* bit would go on. The count can continue above the preset value. If input *B* becomes true the value in the counter accumulator will become zero.

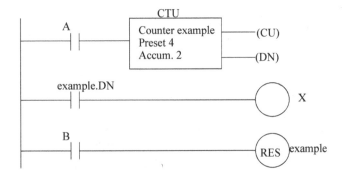

Note: The notations for older Allen-Bradley equipment are very similar to the newer notations. The examples below show the older (PLC-5 and micrologix notations compared to the new RS-Logix (5000) notations. In the older PLCs the counter is given a unique name, in the RSLogix 5000 processors it is given a name (in this case 'c') and the type 'COUNTER'.

Older	Newer
C5:0/DN	c.DN
C5:0/CU	c.CU
C5:0.PRE	c.PRE
C5:0.ACC	c.ACC
C5:0/CD	c.CD

Figure 8.15   An Allen Bradley Counter

Count-down counters are very similar to count-up counters. And, they can actually both be used on the same counter memory location. Consider the example in Figure 8.1, the example input *cnt_up* drives the count-up instruction for counter *example*. Input *cnt_down* drives the count-down instruction for the same counter location. The preset value for a counter is stored in memory location *example* so both the count-up and count-down instruction must have the same preset. Input *reset* will reset the counter.

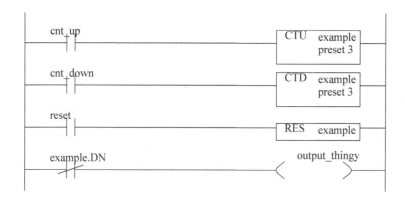

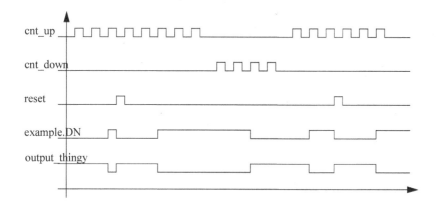

Figure 8.16   **A Counter Example**

The timing diagram in Figure 8.1 illustrates the operation of the counter. If we assume that the value in the accumulator starts at *0*, then the positive edges on the *cnt_up* input will cause it to count up to 3 where it turns the counter *example* done bit on. It is then reset by input *reset* and the accumulator value goes to zero. Input *cnt_up* then pulses again and causes the accumulator value to increase again, until it reaches a maximum of 5. Input *cnt_down* then causes the accumulator value to decrease down below 3, and the counter turns off again. Input *cnt_up* then causes it to increase, but input *reset* resets the accumulator back to zero again, and the pulses continue until 3 is reached near the end.

The program in Figure 8.1 is used to remove 5 out of every 10 parts from a conveyor with a pneumatic cylinder. When the part is detected both counters will increase their values by 1. When the sixth part arrives the first counter will then be done, thereby allowing the pneumatic cylinder to actuate for any part after the fifth. The second counter will continue until the eleventh part is detected and then both of the counters will be reset.

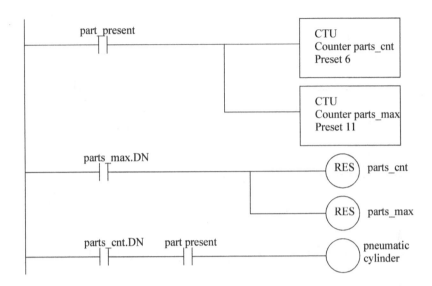

Figure 8.17   *A Counter Example*

8.4 MASTER CONTROL RELAYS (MCRs)

In an electrical control system a Master Control Relay (MCR) is used to shut down a section of an electrical system, as shown earlier in the electrical wiring chapter. This concept has been implemented in ladder logic also. A section of ladder logic can be put between two lines containing MCR's. When the first MCR coil is active, all of the intermediate ladder logic is executed up to the second line with an MCR coil. When the first MCR coil in inactive, the ladder logic is still examined, but all of the outputs are forced off.

Consider the example in Figure 8.1. If *A* is true, then the ladder logic after will be executed as normal. If *A* is false the following ladder logic will be examined, but all of the outputs will be forced off. The second MCR function appears on a line by itself and marks the end of the MCR block. After the second MCR the program execution returns to normal. While *A* is true, *X* will equal *B*, and *Y* can be turned on by *C*, and off by *D*. But, if *A* becomes false *X* will be forced off, and *Y* will be left in its last state. Using MCR blocks to remove sections of programs will not increase the speed of program execution significantly because the logic is still examined.

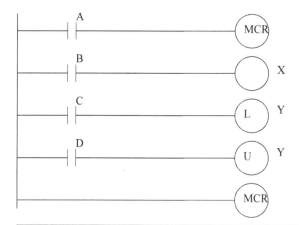

Note: If a normal input is used inside an MCR block it will be forced off. If the output is also used in other MCR blocks the last one will be forced off. The MCR is designed to fully stop an entire section of ladder logic, and is best used this way in ladder logic designs.

Figure 8.18   **MCR Instructions**

If the MCR block contained another function, such as a TON timer, turning off the MCR block would force the timer off. As a general rule normal outputs should be outside MCR blocks, unless they must be forced off when the MCR block is off.

8.5 INTERNAL BITS

Simple programs can use inputs to set outputs. More complex programs also use internal memory locations that are not inputs or outputs. These Boolean memory locations are sometimes referred to as 'internal relays' or 'control relays'. Knowledgeable programmers will often refer to these as 'bit memory'. In the newer Allen Bradley PLCs these can be defined as variables with the type 'BOOL'. The programmer is free to use these memory locations however they see fit.

	bit number	memory location	bit number	memory location
NOTE: In the older Allen Bradley PLCs these addresses begin with 'B3' by default. The first bit in memory is 'B3:0/0', where the first zero represents the first 16 bit word, and the second zero represents the first bit in the word. The sequence of bits is shown to the right.	0	B3:0/0	18	B3:1/2
	1	B3:0/1	19	B3:1/3
	2	B3:0/2	20	B3:1/4
	3	B3:0/3	21	B3:1/5
	4	B3:0/4	22	B3:1/6
	5	B3:0/5	23	B3:1/7
	6	B3:0/6	24	B3:1/8
	7	B3:0/7	25	B3:1/9
	8	B3:0/8	26	B3:1/10
	9	B3:0/9	27	B3:1/11
	10	B3:0/10	28	B3:1/12
	11	B3:0/11	29	B3:1/13
	12	B3:0/12	30	B3:1/14
	13	B3:0/13	31	B3:1/15
	14	B3:0/14	32	B3:2/0
	15	B3:0/15	33	B3:2/1
	16	B3:1/0	34	B3:2/2
	17	B3:1/1	etc...	etc...

An example of bit memory usage is shown in Figure 8.1. The first ladder logic rung will turn on the internal memory bit 'A_pushed' (e.g., B3:0/0) when input 'hand_A' is activated, and input 'clear' is off. (Notice that the Boolean memory is being used as both an input and output.) The second line of ladder logic similar. In this case when both inputs have been activated, the output 'press on' is active.

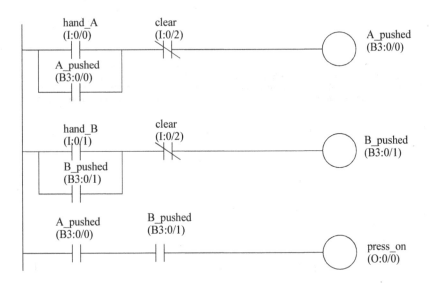

Figure 8.19   An example using bit memory (older notations are in parentheses)

148

Bit memory was presented briefly here because it is important for design techniques in the following chapters, but it will be presented in greater depth after that.

8.6 DESIGN CASES

The following design cases are presented to help emphasize the principles presented in this chapter. I suggest that you try to develop the ladder logic before looking at the provided solutions.

8.6.1 Basic Counters And Timers

Problem: Develop the ladder logic that will turn on an output light, 15 seconds after switch *A* has been turned on.

Solution:

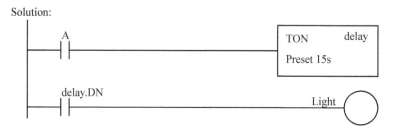

Figure 8.20   *A Simple Timer Example*

Problem: Develop the ladder logic that will turn on a light, after switch *A* has been closed 10 times. Push button *B* will reset the counters.

Solution:

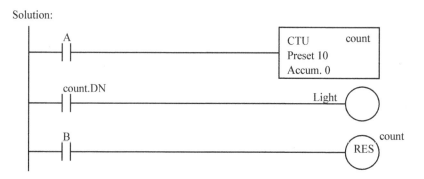

Figure 8.21   *A Simple Counter Example*

8.6.2 More Timers And Counters

Problem: Develop a program that will latch on an output *B* 20 seconds after input A has been turned on.

After *A* is pushed, there will be a 10 second delay until *A* can have any effect again. After *A* has been pushed 3 times, *B* will be turned off.

Solution:

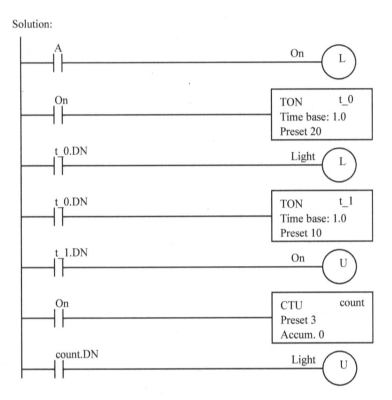

Figure 8.22   *A More Complex Timer Counter Example*

8.6.3 Deadman Switch

Problem: A motor will be controlled by two switches. The *Go* switch will start the motor and the *Stop* switch will stop it. If the *Stop* switch was used to stop the motor, the *Go* switch must be thrown twice to start the *motor*. When the *motor* is active a *light* should be turned on. The *Stop* switch will be wired as normally closed.

Solution:

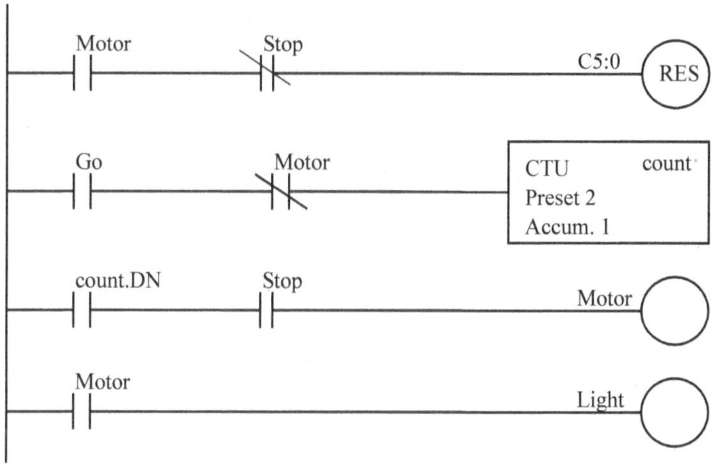

Consider:
What will happen if stop is pushed and the motor is not running?

Figure 8.23   **A Motor Starter Example**

8.6.4 Conveyor

Problem: A conveyor is run by switching on or off a motor. We are positioning parts on the conveyor with an optical detector. When the optical sensor goes on, we want to wait 1.5 seconds, and then stop the conveyor. After a delay of 2 seconds the conveyor will start again. We need to use a start and stop button - a light should be on when the system is active.

Solution:

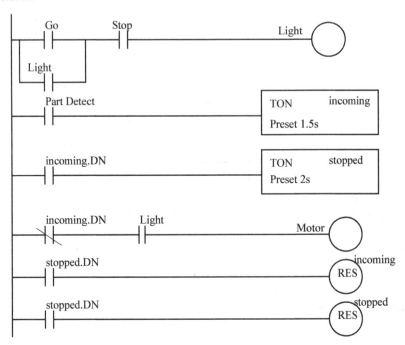

Consider: What is assumed about part arrival and departure?

Figure 8.24   **A Conveyor Controller Example**

8.6.5 Accept/Reject Sorting

Problem: For the conveyor in the last case we will add a sorting system. Gages have been attached that indicate good or bad. If the part is good, it continues on. If the part is bad, we do not want to delay for 2 seconds, but instead actuate a pneumatic cylinder.

Solution:

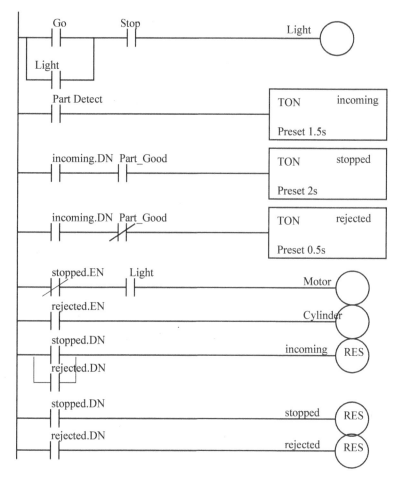

Figure 8.25   *A Conveyor Sorting Example*

8.6.6 Shear Press

Problem: The basic requirements are,

1. A toggle start switch (TS1) and a limit switch on a safety gate (LS1) must both be on before a solenoid (SOL1) can be energized to extend a stamping cylinder to the top of a part.
2. While the stamping solenoid is energized, it must remain energized until a limit switch (LS2) is activated. This second limit switch indicates the end of a stroke. At this point the solenoid should be de-energized, thus retracting the cylinder.
3. When the cylinder is fully retracted a limit switch (LS3) is activated. The cycle may not begin again until this limit switch is active.
4. A cycle counter should also be included to allow counts of parts produced. When this value exceeds 5000 the machine should shut down and a light lit up.
5. A safety check should be included. If the cylinder solenoid has been on for more than 5 seconds, it suggests that the cylinder is jammed or the machine has a fault. If this is the case, the

machine should be shut down and a maintenance light turned on.

Solution:

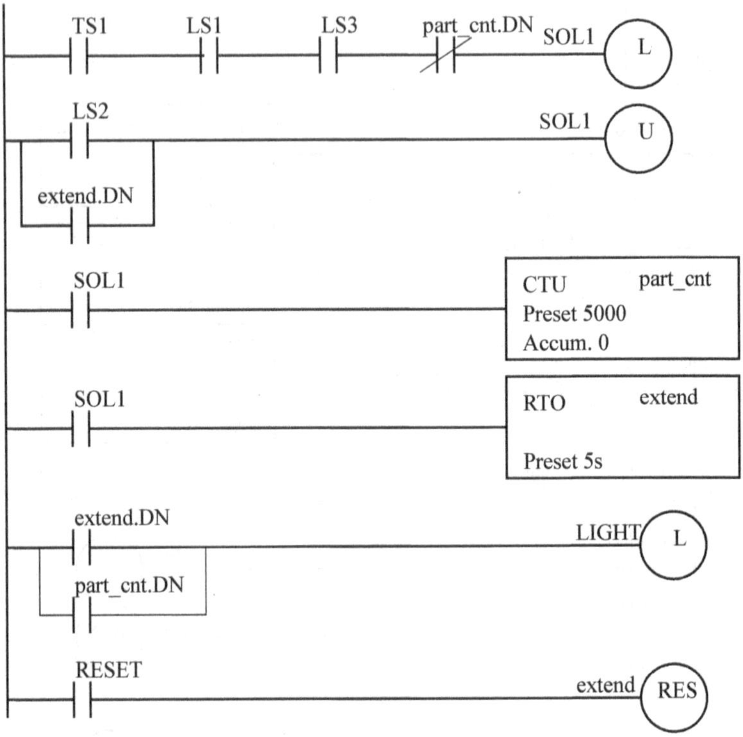

- what do we need to do when the machine is reset?

Figure 8.26   *A Shear Press Controller Example*

8.7 SUMMARY

- Latch and unlatch instructions will hold outputs on, even when the power is turned off.
- Timers can delay turning on or off. Retentive timers will keep values, even when inactive. Resets are needed for retentive timers.
- Counters can count up or down.
- When timers and counters reach a preset limit the *DN* bit is set.
- MCRs can force off a section of ladder logic.

8.8 PRACTICE PROBLEMS

(Note: Problem solutions are available at http://sites.google.com/site/automatedmanufacturingsystems/)

1. What does edge triggered mean? What is the difference between positive and negative edge triggered?

2. Are reset instructions necessary for all timers and counters?

3. What are the numerical limits for typical timers and counters?

4. If a counter goes below the bottom limit which counter bit will turn on?

5. a) Write ladder logic for a motor starter that has a start and stop button that uses latches. b) Write the same ladder logic without latches.

6. Use a timing diagram to explain how an on delay and off delay timer are different.

7. For the retentive off timer below, draw out the status bits.

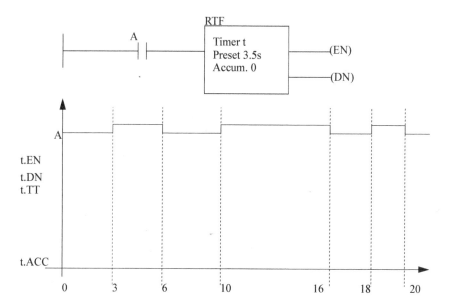

8. Complete the timing diagrams for the two timers below.

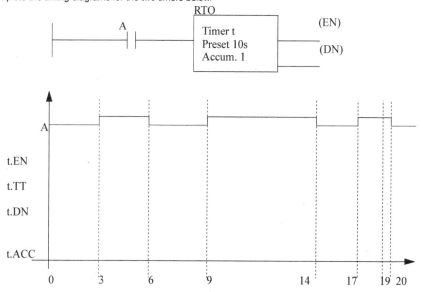

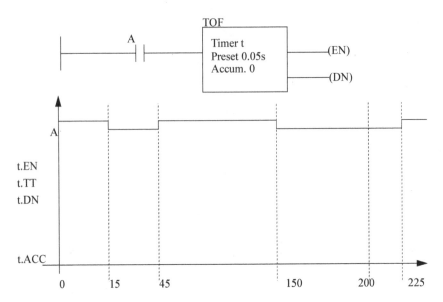

9. Given the following timing diagram, draw the done bits for all four fundamental timer types. Assume all start with an accumulated value of zero, and have a preset of 1.5 seconds.

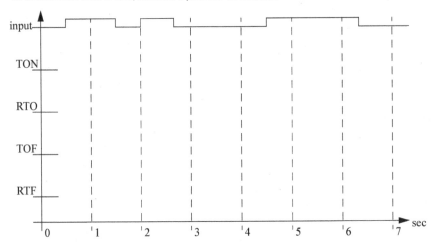

10. Design ladder logic that allows an RTO to behave like a TON.

11. Design ladder logic that uses a timer and counter to measure a time of 50.0 days.

12. Develop the ladder logic that will turn on an output (light), 15 seconds after switch (A) has been turned on.

13. Develop the ladder logic that will turn on a output (light), after a switch (A) has been closed 10 times. Push button (B) will reset the counters.

14. Develop a program that will latch on an output (B), 20 seconds after input (A) has been turned on. The timer will continue to cycle up to 20 seconds, and reset itself, until A has been turned off. After the third time the timer has timed to 20 seconds, B will be unlatched.

15. A motor will be connected to a PLC and controlled by two switches. The GO switch will start the motor, and the STOP switch will stop it. If the motor is going, and the GO switch is thrown, this will also stop the motor. If the

STOP switch was used to stop the motor, the GO switch must be thrown twice to start the motor. When the motor is running, a light should be turned on (a small lamp will be provided).

16. In dangerous processes it is common to use two palm buttons that require a operator to use both hands to start a process (this keeps hands out of presses, etc.). To develop this there are two inputs that must be turned on within 0.25s of each other before a machine cycle may begin.

17. Design a conveyor control system that follows the design guidelines below.
    - The conveyor has an optical sensor S1 that detects boxes entering a workcell
    - There is also an optical sensor S2 that detects boxes leaving the workcell
    - The boxes enter the workcell on a conveyor controlled by output C1
    - The boxes exit the workcell on a conveyor controlled by output C2
    - The controller must keep a running count of boxes using the entry and exit sensors
    - If there are more than five boxes in the workcell the entry conveyor will stop
    - If there are no boxes in the workcell the exit conveyor will be turned off
    - If the entry conveyor has been stopped for more than 30 seconds the count will be reset to zero, assuming that the boxes in the workcell were scrapped.

18. Write a ladder logic program that does what is described below.
    - When button A is pushed, a light will flash for 5 seconds.
    - The flashing light will be on for 0.25 sec and off for 0.75 sec.
    - If button A has been pushed 5 times the light will not flash until the system is reset.
    - The system can be reset by pressing button B

19. Write a program that will turn on a flashing light for the first 15 seconds after a PLC is turned on. The light should flash for half a second on and half a second off.

20. A buffer can hold up to 10 parts. Parts enter the buffer on a conveyor controller by output *conveyor*. As parts arrive they trigger an input sensor *enter*. When a part is removed from the buffer they trigger the *exit* sensor. Write a program to stop the conveyor when the buffer is full, and restart it when there are fewer than 10 parts in the buffer. As normal the system should also include a start and stop button.

21. What is wrong with the following ladder logic? What will happen if it is used?

22. We are using a pneumatic cylinder in a process. The cylinder can become stuck, and we need to detect this. Proximity sensors are added to both endpoints of the cylinder's travel to indicate when it has reached the end of motion. If the cylinder takes more than 2 seconds to complete a motion this will indicate a problem. When this occurs the machine should be shut down and a light turned on. Develop ladder logic that will cycle the cylinder in and out repeatedly, and watch for failure.

8.9 ASSIGNMENT PROBLEMS

1. Draw the timer and counter done bits for the ladder logic below. Assume that the accumulators of all the timers

and counters are reset to begin with.

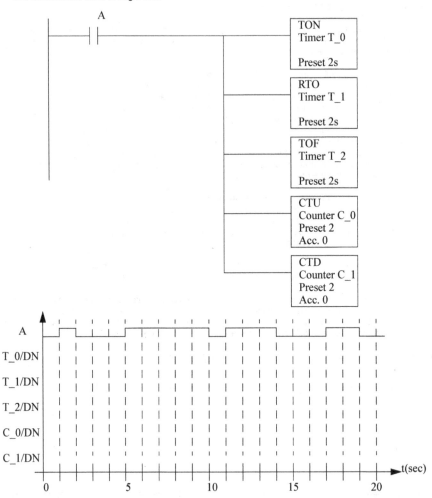

2. Write a ladder logic program that will count the number of parts in a buffer. As parts arrive they activate input *A*. As parts leave they will activate input *B*. If the number of parts is less than 8 then a conveyor motor, output *C*, will be turned on.

3. Explain what would happen in the following program when A is on or off.

4. Write a simple program that will use one timer to flash a light. The light should be on for 1.0 seconds and off for 0.5 seconds. Do not include start or stop buttons.

5. We are developing a safety system (using a PLC-5) for a large industrial press. The press is activated by turning on the compressor power relay (R, connected to O:013/05). After R has been on for 30 seconds the press can be activated to move (P connected to O:013/06). The delay is needed for pressure to build up. After the press has been activated (with P) the system must be shut down (R and P off), and then the cycle may begin again. For safety, there is a sensor that detects when a worker is inside the press (S, connected to I:011/02), which must be off before the press can be activated. There is also a button that must be pushed 5 times (B, connected to I:011/01) before the press cycle can begin. If at any time the worker enters the press (and S becomes active) the press will be shut down (P and R turned off). Develop the ladder logic. State all assumptions, and show all work.

6. Write a program that only uses one timer. When an input A is turned on a light will be on for 10 seconds. After that it will be off for two seconds, and then again on for 5 seconds. After that the light will not turn on again until the input A is turned off.

7. A new printing station will add a logo to parts as they travel along an assembly line. When a part arrives a 'part' sensor will detect it. After this the 'clamp' output is turned on for 10 seconds to hold the part during the operation. For the first 2 seconds the part is being held a 'spray' output will be turned on to apply the thermoset ink. For the last 8 seconds a 'heat' output will be turned on to cure the ink. After this the part is released and allowed to continue along the line. Write the ladder logic for this process.

8. Write a ladder logic program. that will turn on an output Q five seconds after an input A is turned on. If input B is on the delay will be eight seconds. YOU MAY ONLY USE ONE TIMER.

9. Use the timing diagram below to design ladder logic. The sequence should start when input X turns on. X may only be on momentarily, but the sequence should execute anyway. Note that output A is normally on.

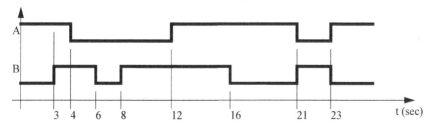

159

8.1 PRACTICE PROBLEM SOLUTIONS

1. edge triggered means the event when a logic signal goes from false to true (positive edge) or from true to false (negative edge).

2. no, but they are essential for retentive timers, and very important for counters.

3. Timers on PLC-5s and Micrologix are 16 bit, so they are limited to a range of -32768 to +32767. ControlLogix timers are 32 bit and have a range of -2,147,483,648 to 2,147,483,647.

4. the *un* underflow bit. This may result in a fault in some PLCs.

5.

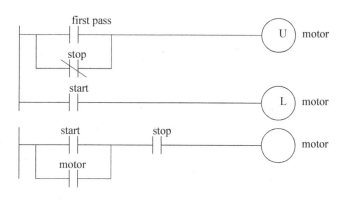

6.

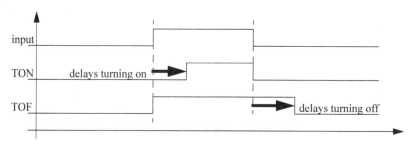

7.

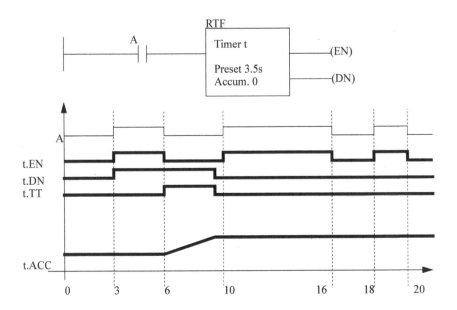

8.

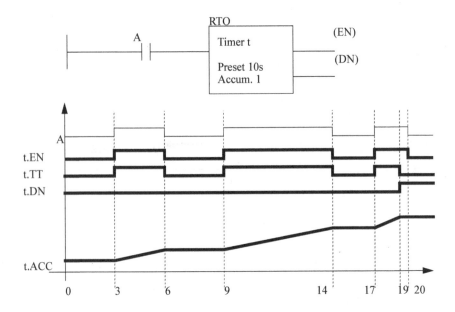

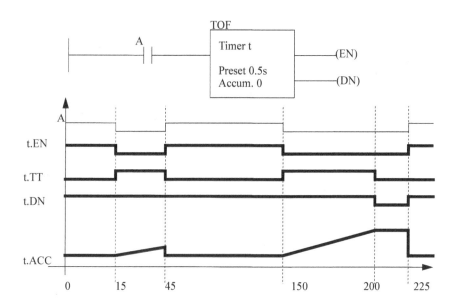

9.

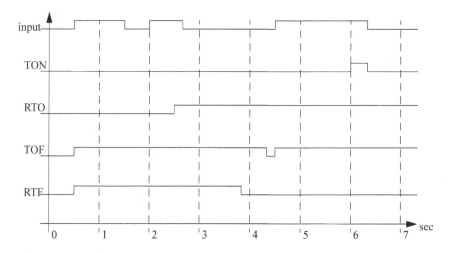

10.

11.

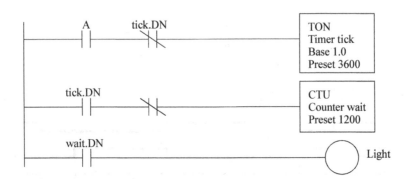

12.

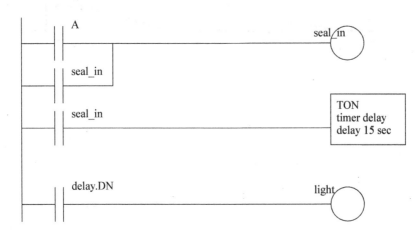

13.

164

14.

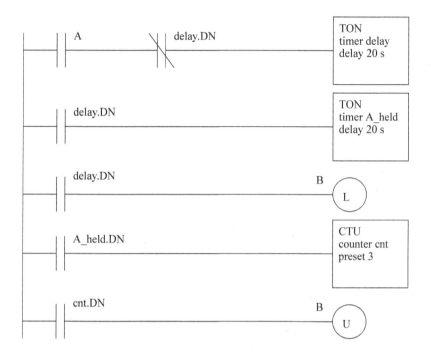

15.

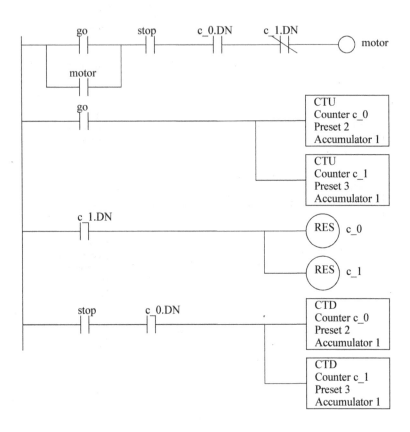

16.

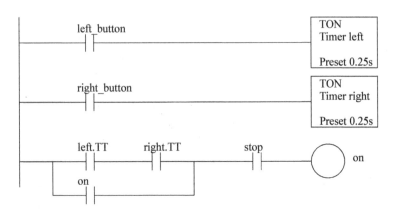

17.

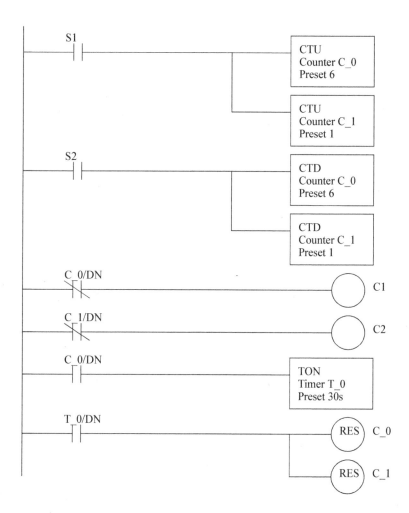

18.

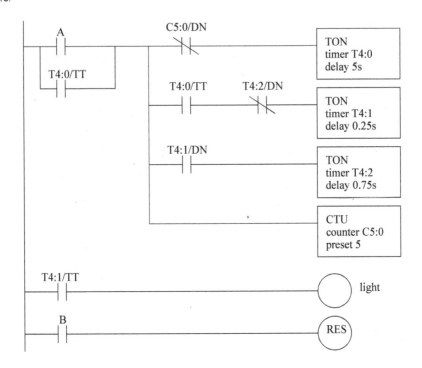

19.

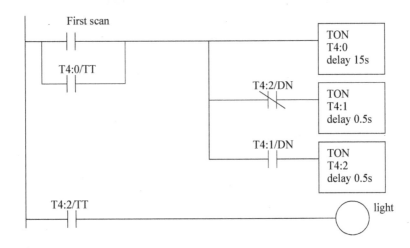

20.

```
         start            stop
    ├─────┤ ├──────────────┤/├─────────────────────────( )  active
    │
    │     active
    └─────┤ ├──────┘

          enter                                    ┌──────────────────┐
    ├──────┤ ├──────────────────────────────────── │ CTU              │
    │                                               │ counter C5:0     │
    │                                               │ preset 10        │
    │                                               └──────────────────┘
    │     exit                                      ┌──────────────────┐
    ├──────┤ ├──────────────────────────────────── │ CTD              │
    │                                               │ counter C5:0     │
    │                                               │ preset 10        │
    │                                               └──────────────────┘
    │     active          C5:0/DN
    └──────┤ ├──────────────┤/├─────────────────────────( )  active
```

21. The normal output 'Y' is repeated twice. In this example the value of 'Y' would always match 'B', and the earlier rung with 'A' would have no effect on 'Y'.

9STRUCTURED LOGIC DESIGN

Topics:

- Timing diagrams
- Design examples
- Designing ladder logic with process sequence bits and timing diagrams

Objectives:

- Know examples of applications to industrial problems.
- Know how to design time base control programs.

Traditionally ladder logic programs have been written by thinking about the process and then beginning to write the program. This always leads to programs that require debugging. And, the final program is always the subject of some doubt. Structured design techniques, such as Boolean algebra, lead to programs that are predictable and reliable. The structured design techniques in this and the following chapters are provided to make ladder logic design routine and predictable for simple sequential systems.

Note: Structured design is very important in engineering, but many engineers will write software without taking the time or effort to design it. This often comes from previous experience with programming where a program was written, and then debugged. This approach is not acceptable for mission critical systems such as industrial controls. The time required for a poorly designed program is 10% on design, 30% on writing, 40% debugging and testing, 10% documentation. The time required for a high quality program design is 30% design, 10% writing software, 10% debugging and testing, 10% documentation. Yes, a well designed program requires less time! Most beginners perceive the writing and debugging as more challenging and productive, and so they will rush through the design stage. If you are spending time debugging ladder logic programs you are doing something wrong. Structured design also allows others to verify and modify your programs.

Axiom: Spend as much time on the design of the program as possible. Resist the temptation to implement an incomplete design.

Most control systems are sequential in nature. Sequential systems are often described with words such as mode and behavior. During normal operation these systems will have multiple steps or states of operation. In each operational state the system will behave differently. Typical states include start-up, shut-down, and normal operation. Consider a set of traffic lights - each light pattern constitutes a state. Lights may be green or yellow in one direction and red in the other. The lights change in a predictable sequence. Sometimes traffic lights are equipped with special features such as cross walk buttons that alter the behavior of the lights to give pedestrians time to cross busy roads.

Sequential systems are complex and difficult to design. In the previous chapter timing charts and process sequence bits were discussed as basic design techniques. But, more complex systems require more mature techniques, such as those shown in Figure 9.1. For simpler controllers we can use limited design techniques such as process sequence bits and flow charts. More complex processes, such as traffic lights, will have many states of operation and controllers can be designed using state diagrams. If the control problem involves multiple states of operation, such as one controller for two independent traffic lights, then Petri net or SFC based designs are preferred.

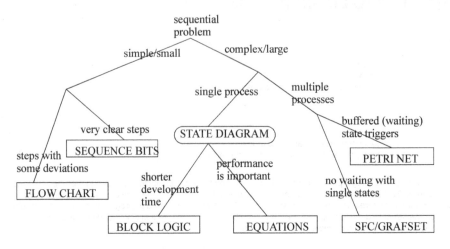

Figure 9.1   *Sequential Design Techniques*

9.1 PROCESS SEQUENCE BITS

A typical machine will use a sequence of repetitive steps that can be clearly identified. Ladder logic can be written that follows this sequence. The steps for this design method are;

1. Understand the process.
2. Write the steps of operation in sequence and give each step a number.
3. For each step assign a bit.
4. Write the ladder logic to turn the bits on/off as the process moves through its states.
5. Write the ladder logic to perform machine functions for each step.
6. If the process is repetitive, have the last step go back to the first.

Consider the example of a flag raising controller in Figure 9.1 and Figure 9.1. The problem begins with a written description of the process. This is then turned into a set of numbered steps. Each of the numbered steps is then converted to ladder logic.

Description:

A flag raiser that will go up when an up button is pushed, and down when a down button is pushed, both push buttons are momentary. There are limit switches at the top and bottom to stop the flag pole. When turned on at first the flag should be lowered until it is at the bottom of the pole.

Steps:

1. The flag is moving down the pole waiting for the bottom limit switch.
2. The flag is idle at the bottom of the pole waiting for the up button.
3. The flag moves up, waiting for the top limit switch.
4. The flag is idle at the top of the pole waiting for the down button.

Ladder Logic:

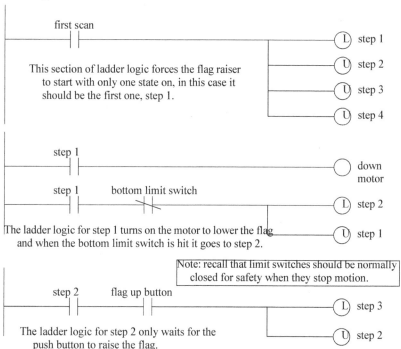

Figure 9.2   *A Process Sequence Bit Design Example*

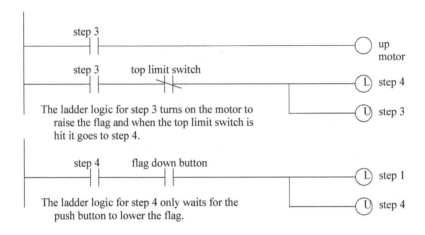

The ladder logic for step 3 turns on the motor to raise the flag and when the top limit switch is hit it goes to step 4.

The ladder logic for step 4 only waits for the push button to lower the flag.

Figure 9.3   *A Process Sequence Bit Design Example (continued)*

The previous method uses latched bits, but the use of latches is sometimes discouraged. A more common method of implementation, without latches, is shown in Figure 9.1.

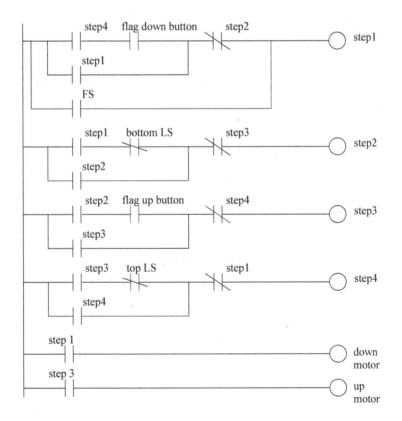

Figure 9.4   ***Process Sequence Bits Without Latches***

Similar methods are explored in further detail in the book Cascading Logic (Kirckof, 2003).

9.2 TIMING DIAGRAMS

Timing diagrams can be valuable when designing ladder logic for processes that are only dependant on time. The timing diagram is drawn with clear start and stop times. Ladder logic is constructed with timers that are used to turn outputs on and off at appropriate times. The basic method is;

1. Understand the process.
2. Identify the outputs that are time dependant.
3. Draw a timing diagram for the outputs.
4. Assign a timer for each time when an output turns on or off.
5. Write the ladder logic to examine the timer values and turn outputs on or off.

Consider the handicap door opener design in Figure 9.1 that begins with a verbal description. The verbal description is converted to a timing diagram, with t=0 being when the door open button is pushed. On the timing diagram the critical times are 2s, 10s, 14s. The ladder logic is constructed in a careful order. The first item is the latch to seal-in the open button, but shut off after the last door closes. *auto* is used to turn on the three timers for the critical times. The logic for opening the doors is then written to use the timers.

Description: A handicap door opener has a button that will open two doors. When the button is pushed (momentarily) the first door will start to open immediately, the second door will start to open 2 seconds later. The first door power will stay open for a total of 10 seconds, and the second door power will stay on for 14 seconds. Use a timing diagram to design the ladder logic.

Timing Diagram:

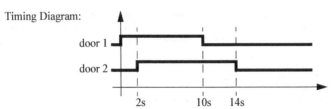

Ladder Logic:

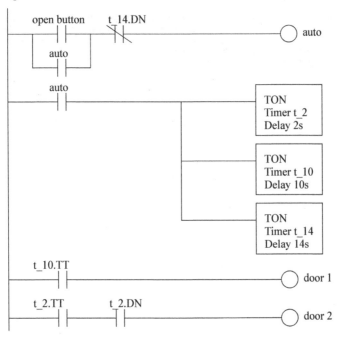

Figure 9.5   *Design With a Timing Diagram*

9.3 SUMMARY

- Timing diagrams can show how a system changes over time.
- Process sequence bits can be used to design a process that changes over time.
- Timing diagrams can be used for systems with a time driven performance.

9.4 PRACTICE PROBLEMS

(Note: Problem solutions are available at http://sites.google.com/site/automatedmanufacturingsystems/)

1. Write ladder logic that will give the following timing diagram for *B* after input *A* is pushed. After *A* is pushed any changes in the state of *A* will be ignored.

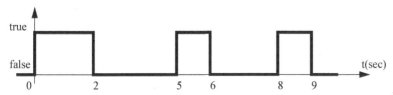

2. Design ladder logic for the timing diagram below. When an input *A* becomes active the sequence should start.

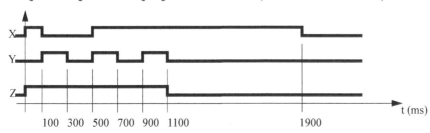

3. A wrapping process is to be controlled with a PLC. The general sequence of operations is described below. Develop the ladder logic using process sequence bits.
   1. The folder is idle until a part arrives.
   2. When a part arrives it triggers the *part* sensor and the part is held in place by actuating the *hold* actuator.
   3. The first wrap is done by turning on output *paper* for 1 second.
   4. The paper is then folded by turning on the *crease* output for 0.5 seconds.
   5. An adhesive is applied by turning on output *tape* for 0.75 seconds.
   6. The part is release by turning off output *hold*.
   7. The process pauses until the *part* sensors goes off, and then the machine returns to idle.

4. Draw a timing diagram for the following ladder logic.

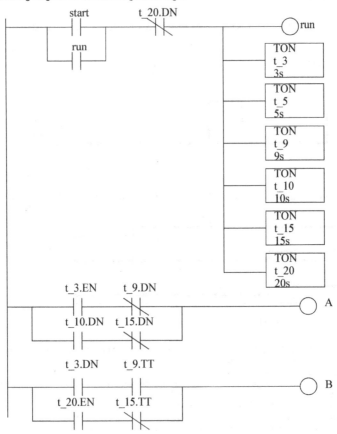

9.5 ASSIGNMENT PROBLEMS

1. Convert the following timing diagram to ladder logic. It should begin when input 'A' becomes true.

2. Use the timing diagram below to design ladder logic. The sequence should start when input X turns on. X may

only be on momentarily, but the sequence should continue to execute until it ends at 26 seconds.

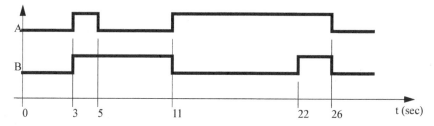

3. Use the timing diagram below to design ladder logic. The sequence should start when input X turns on. X may only be on momentarily, but the sequence should execute anyway.

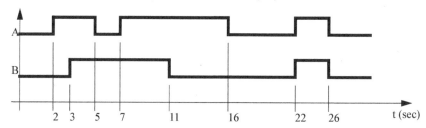

4. Write a program that will execute the following steps. When in steps b) or d), output C will be true. Output X will be true when in step c).
   a) Start in an idle state. If input G becomes true go to b)
   b) Wait until P becomes true before going to step c).
   c) Wait for 3 seconds then go to step d).
   d) Wait for P to become false, and then go to step b).

5. Write a program that will execute the following steps. When in steps b) or d), output C will be true. Output X will be true when in step c).
   a) Start in an idle state. If input G becomes true go to b)
   b) Wait until P becomes true before going to step c). If input S becomes true then go to step a).
   c) Wait for 3 seconds then go to step d).
   d) Wait for P to become false, and then go to step b).

6. A PLC is to control an amusement park water ride. The ride will fill a tank of water and splash a tour group. 10 seconds later a water jet will be ejected at another point. Develop ladder logic for the process that follows the steps listed below.
   1. The process starts in 'idle'.
   2. The 'cart_detect' opens the 'filling' valve.
   3. After a delay of 30 seconds from the start of the filling of the tank the tank 'outlet' valve opens. When the tank Is 'full' the 'filling' valve closes.
   4. When the tank is empty the 'outlet' valve is closed.
   5. After a 10 second delay, from the tank outlet valve opening, a water 'jet' is opened.
   6. After '2' seconds the water 'jet' is closed and the process returns to the 'idle state.

7. Write a ladder logic program to extend and retract a cylinder after a start button is pushed. There are limit switches at the ends of travel. If the cylinder is extending if more than 5 seconds the machine should shut down and turn on a fault light. If it is retracting for more than 3 seconds it should also shut down and turn on the fault light. It can be reset with a reset button.

8. Design a program with sequence bits for a hydraulic press that will advance when two palm buttons are pushed. Top and bottom limit switches are used to reverse the advance and stop after a retract. At any time the hands removed from the palm button will stop an advance and retract the press. Include start and stop buttons to put the press in and out of an active mode.

9. A machine has been built for filling barrels. Use process sequence bits to design ladder logic for the sequential process as described below.
   1. The process begins in an idle state.
   2. If the 'fluid_pressure' and 'barrel_present' inputs are on, the system will open a flow valve for 2 seconds with output 'flow'.
   3. The 'flow' valve will then be turned off for 10 seconds.
   4. The 'flow' valve will then be turned on until the 'full' sensor indicates the barrel is full.
   5. The system will wait until the 'barrel_present' sensor goes off before going to the idle state.

10. Design ladder logic for an oven using process sequence bits. (Note: the solution will only be graded if the process sequence bit method is used.) The operations are as listed below.
   1. The oven begins in an IDLE state.
   2. An operator presses a start button and an ALARM output is turned on for 1 minute.
   3. The ALARM output is turned off and the HEAT is turned on for 3 minutes to allow the temperature to rise to the acceptable range.
   4. The CONVEYOR output is turned on.
   5. If the STOP input is activated (turned off) the HEAT will be turned off, but the CONVEYOR output will be kept on for two minutes. After this the oven returns to IDLE.

11. We are developing a safety system (using a PLC-5) for a large industrial press. The press is activated by turning on the compressor power relay (R, connected to O:013/05). After R has been on for 30 seconds the press can be activated to move (P connected to O:013/06). The delay is needed for pressure to build up. After the press has been activated (with P for 1.0 seconds) the system must be shut down (R and P off), and then the cycle may begin again. For safety, there is a sensor that detects when a worker is inside the press (S, connected to I:011/02), which must be off before the press can be activated. There is also a button that must be pushed 5 times (B, connected to I:011/01) before the press cycle can begin. If at any time the worker enters the press (and S becomes active) the press will be shut down (P and R turned off). Develop the process sequence and sequence bits, and then ladder logic for the states. State all assumptions, and show all work.

12. A machine is being designed to wrap boxes of chocolate. The boxes arrive at the machine on a conveyor belt. The list below shows the process steps in sequence.
   1. The box arrives and is detected by an optical sensor (P), after this the conveyor is stopped (C) and the box is clamped in place (H).
   2. A wrapping mechanism (W) is turned on for 2 seconds.
   3. A sticker cylinder (S) is turned on for 1 second to put consumer labelling on the box.
   4. The clamp (H) is turned off and the conveyor (C) is turned on.
   5. After the box leaves the system returns to an idle state.
   Develop ladder logic programs for the system using the following methods. Don't forget to include regular start and stop inputs.
      i) a timing diagram
      ii) process sequence bits

9.1 PRACTICE PROBLEM SOLUTIONS

1.

2.

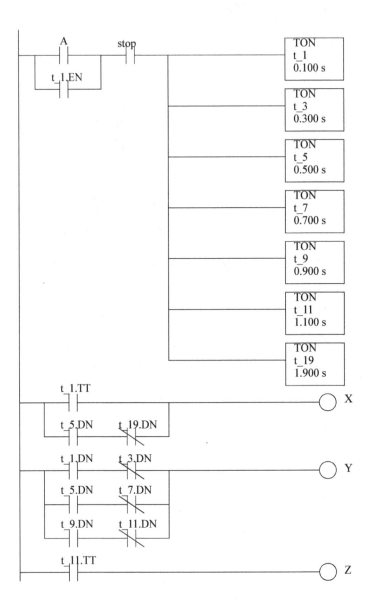

3.

(for both solutions

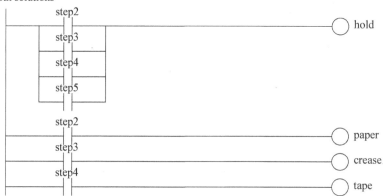

4.

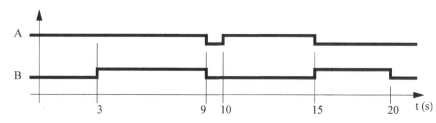

(without latches

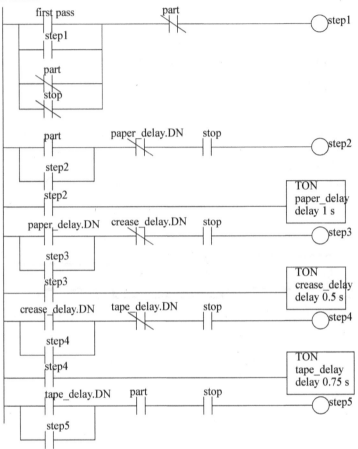

(with latches

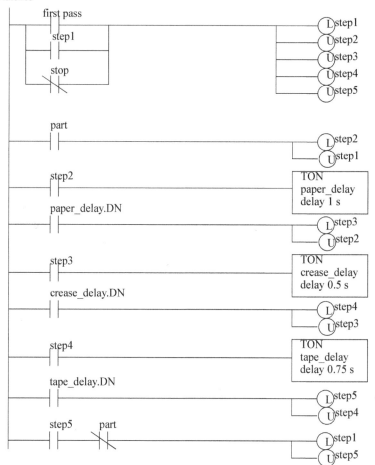

10 FLOWCHART BASED DESIGN

Topics:

- Describing process control using flowcharts
- Conversion of flowcharts to ladder logic

Objectives:

- Ba able to describe a process with a flowchart.
- Be able to convert a flowchart to ladder logic.

A flowchart is ideal for a process that has sequential process steps. The steps will be executed in a simple order that may change as the result of some simple decisions. The symbols used for flowcharts are shown in Figure 10.1. These blocks are connected using arrows to indicate the sequence of the steps. The different blocks imply different types of program actions. Programs always need a *start* block, but PLC programs rarely stop so the *stop* block is rarely used. Other important blocks include *operations* and *decisions*. The other functions may be used but are not necessary for most PLC applications.

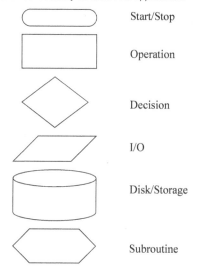

Start/Stop

Operation

Decision

I/O

Disk/Storage

Subroutine

Figure 10.1   *Flowchart Symbols*

A flowchart is shown in Figure 10.1 for a control system for a large water tank. When a start button is pushed the tank will start to fill, and the flow out will be stopped. When full, or the stop button is pushed the outlet will open up, and the flow in will be stopped. In the flowchart the general flow of execution starts at the top. The first operation is to open the outlet valve and close the inlet valve. Next, a single decision block is used to wait for a button to be pushed. when the button is pushed the *yes* branch is followed and the inlet valve is opened, and the outlet valve is closed. Then the flow chart goes into a loop that uses two decision blocks to wait until the tank is full, or the stop button is pushed. If either case occurs the inlet valve is closed and the outlet valve is opened. The system then goes back to wait for the start button to be pushed again. When the controller is on the program should always be running, so only a start block is needed. Many beginners will neglect to put in checks for stop buttons.

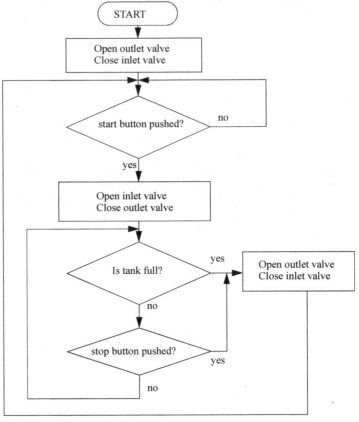

Figure 10.2   **A Flowchart for a Tank Filler**

The general method for constructing flowcharts is:

        1. Understand the process.
        2. Determine the major actions, these are drawn as blocks.
        3. Determine the sequences of operations, these are drawn with arrows.
        4. When the sequence may change use decision blocks for branching.

Once a flowchart has been created ladder logic can be written. There are two basic techniques that can be used, the first presented uses blocks of ladder logic code. The second uses normal ladder logic.

10.1 BLOCK LOGIC

The first step is to name each block in the flowchart, as shown in Figure 10.1. Each of the numbered steps will then be converted to ladder logic

STEP 1: Add labels to each block in the flowchart

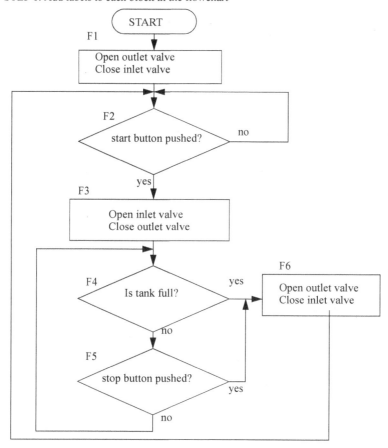

Figure 10.3   **Labeling Blocks in the Flowchart**

Each block in the flowchart will be converted to a block of ladder logic. To do this we will use the MCR (Master Control Relay) instruction (it will be discussed in more detail later.) The instruction is shown in Figure 10.1, and will appear as a matched pair of outputs labelled *MCR*. If the first MCR line is true then the ladder logic on the following lines will be scanned as normal to the second MCR. If the first line is false the lines to the next MCR block will all be forced off. If a normal output is used inside an MCR block, it may be forced off. Therefore latches will be used in this method.

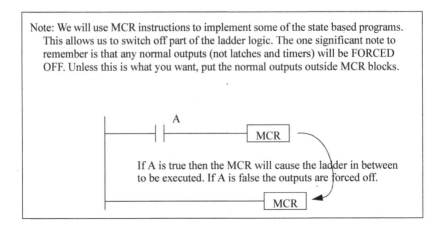

Note: We will use MCR instructions to implement some of the state based programs. This allows us to switch off part of the ladder logic. The one significant note to remember is that any normal outputs (not latches and timers) will be FORCED OFF. Unless this is what you want, put the normal outputs outside MCR blocks.

If A is true then the MCR will cause the ladder in between to be executed. If A is false the outputs are forced off.

Figure 10.4   ***The MCR Function***

The first part of the ladder logic required will reset the logic to an initial condition, as shown in Figure 10.1. The line will only be true for the first scan of the PLC, and at that time it will turn on the flowchart block *F1* which is the *reset all values off* operation. All other operations will be turned off.

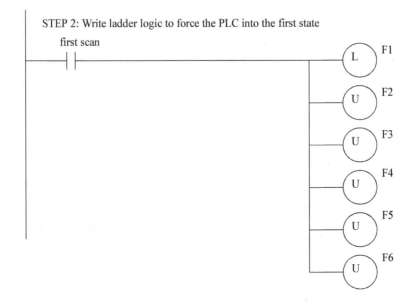

Figure 10.5   ***Initial Reset of States***

The ladder logic for the first state is shown in Figure 10.1. When *F1* is true the logic between the MCR lines will be scanned, if *F1* is false the logic will be ignored. This logic turns on the outlet valve and turns off the inlet valve. It then turns off operation *F1*, and turns on the next operation *F2*.

STEP 3: Write ladder logic for each function in the flowchart

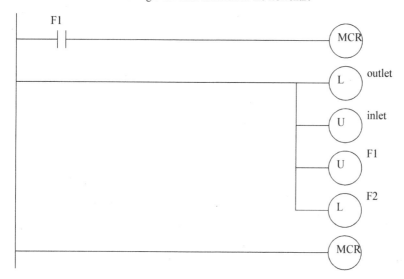

Figure 10.6   ***Ladder Logic for the Operation F1***

The ladder logic for operation $F2$ is simple, and when the start button is pushed, it will turn off $F2$ and turn on $F3$. The ladder logic for operation $F3$ opens the inlet valve and moves to operation $F4$.

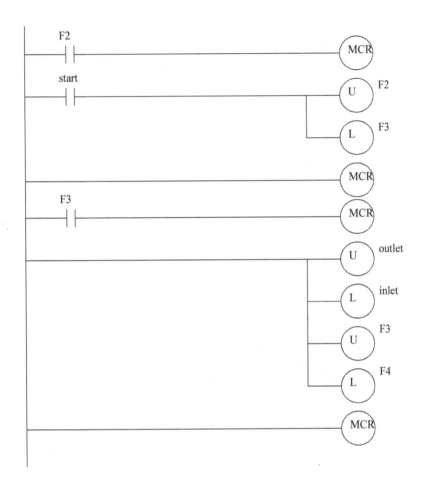

Figure 10.7   *Ladder Logic for Flowchart Operations F2 and F3*

The ladder logic for operation *F4* turns off *F4*, and if the tank is full it turns on *F6*, otherwise *F5* is turned on. The ladder logic for operation *F5* is very similar.

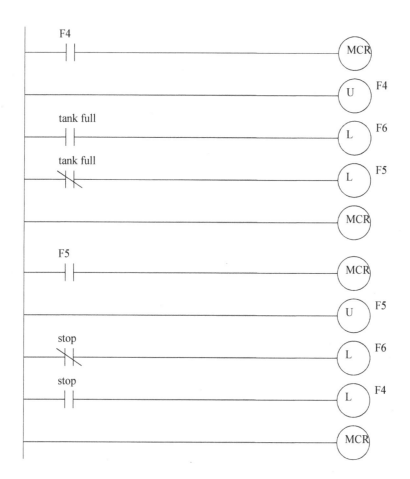

Figure 10.8   *Ladder Logic for Operations F4 and F5*

The ladder logic for operation *F6* turns the outlet valve on and turns off the inlet valve. It then ends operation *F6* and returns to operation *F2*.

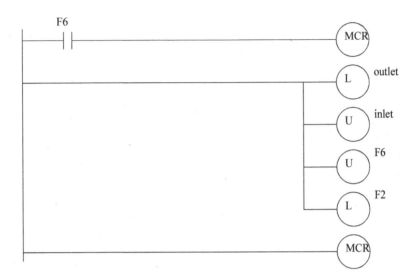

Figure 10.9   **Ladder Logic for Operation F6**

10.2 SEQUENCE BITS

In general there is a preference for methods that do not use MCR statements or latches. The flowchart used in the previous example can be implemented without these instructions using the following method. The first step to this process is shown in Figure 10.1. As before each of the blocks in the flowchart are labelled, but now the connecting arrows (transitions) in the diagram must also be labelled. These transitions indicate when another function block will be activated.

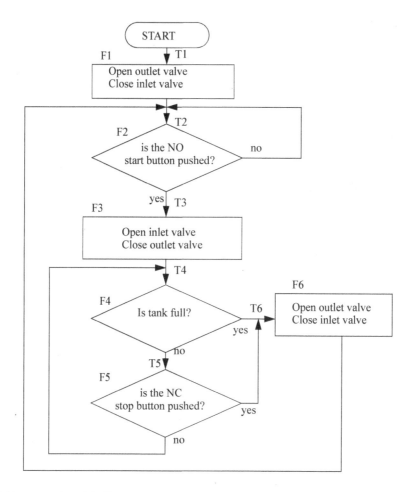

Figure 10.10   *Label the Flowchart Blocks and Arrows*

The first section of ladder logic is shown in Figure 10.1. This indicates when the transitions between functions should occur. All of the logic for the transitions should be kept together, and appear before the state logic that follows in Figure 10.1.

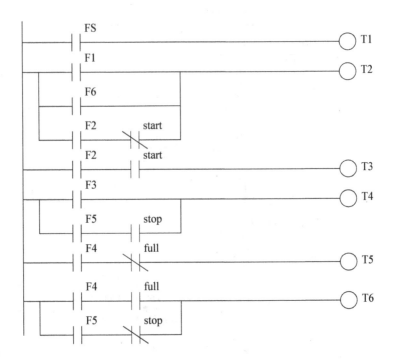

Figure 10.11   **The Transition Logic**

The logic shown in Figure 10.1 will keep a function on, or switch to the next function. Consider the first ladder rung for *F1*, it will be turned on by transition *T1* and once function *F1* is on it will keep itself on, unless *T2* occurs shutting it off. If *T2* has occurred the next line of ladder logic will turn on *F2*. The function logic is followed by output logic that relates output values to the active functions.

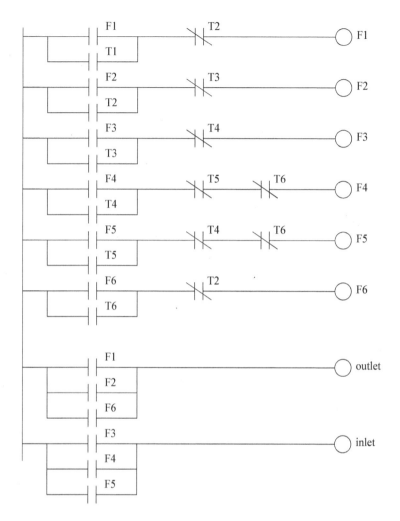

Figure 10.12   *The Function Logic and Outputs*

10.3 SUMMARY

- Flowcharts are suited to processes with a single flow of execution.
- Flowcharts are suited to processes with clear sequences of operation.

10.4 PRACTICE PROBLEMS

(Note: Problem solutions are available at http://sites.google.com/site/automatedmanufacturingsystems/)

1. Convert the following flow chart to ladder logic.

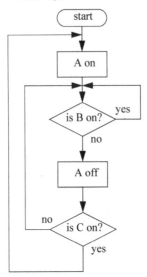

2. Draw a flow chart for cutting the grass, then develop ladder logic for three of the actions/decisions.

3. Design a garage door controller using a flowchart. The behavior of the garage door controller is as follows,
   - there is a single button in the garage, and a single button remote control.
   - when the button is pushed the door will move up or down.
   - if the button is pushed once while moving, the door will stop, a second push will start motion again in the opposite direction.
   - there are top/bottom limit switches to stop the motion of the door.
   - there is a light beam across the bottom of the door. If the beam is cut while the door is closing the door will stop and reverse.
   - there is a garage light that will be on for 5 minutes after the door opens or closes.

10.5 ASSIGNMENT PROBLEMS

1. Develop ladder logic for the flowchart below.

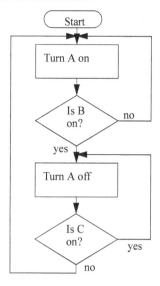

2. Use a flow chart to design a parking gate controller.

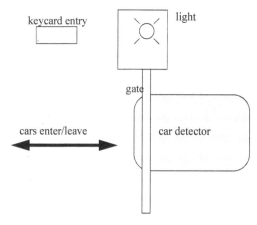

- the gate will be raised by one output and lowered by another. If the gate gets stuck an over current detector will make a PLC input true. If this is the case the gate should reverse and the light should be turned on indefinitely.
- if a valid keycard is entered a PLC input will be true. The gate is to rise and stay open for 10 seconds.
- when a car is over the car detector a PLC input will go true. The gate is to open while this detector is active. If it is active for more that 30 seconds the light should also turn on until the gate closes.

3. A welding station is controlled by a PLC. On the outside is a safety cage that must be closed while the cell is active. A belt moves the parts into the welding station and back out. An inductive proximity sensor detects when a part is in place for welding, and the belt is stopped. To weld, an actuator is turned on for 3 seconds. As normal the cell has start and stop push buttons.
        a) Draw a flow chart
        b) Implement the chart in ladder logic

Inputs

DOOR OPEN (NC)
START (NO)
STOP (NC)
PART PRESENT

Outputs

CONVEYOR ON
WELD

4. Convert the following flowchart to ladder logic.

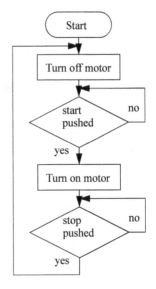

5. A machine is being designed to wrap boxes of chocolate. The boxes arrive at the machine on a conveyor belt. The list below shows the process steps in sequence.

    1. The box arrives and is detected by an optical sensor (P), after this the conveyor is stopped (C) and the box is clamped in place (H).
    2. A wrapping mechanism (W) is turned on for 2 seconds.
    3. A sticker cylinder (S) is turned on for 1 second to put consumer labelling on the box.
    4. The clamp (H) is turned off and the conveyor (C) is turned on.
    5. After the box leaves the system returns to an idle state.

Develop ladder logic for the system using a flowchart. Don't forget to include regular start and stop inputs.

10.1 PRACTICE PROBLEM SOLUTIONS

1.

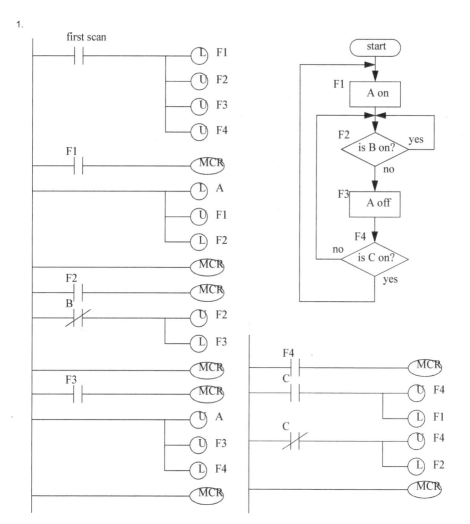

2.

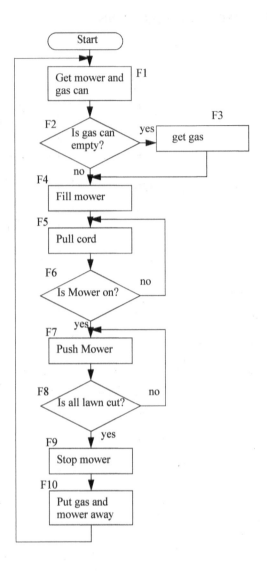

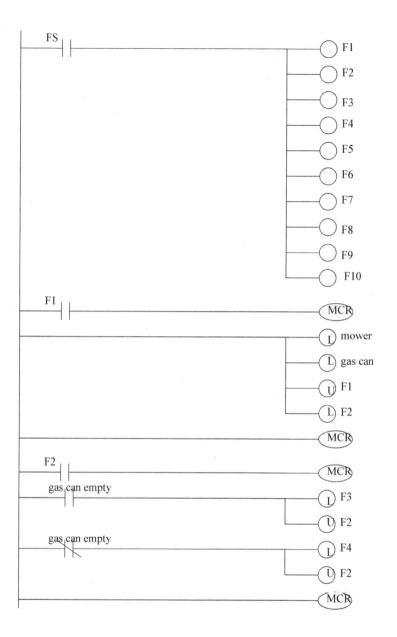

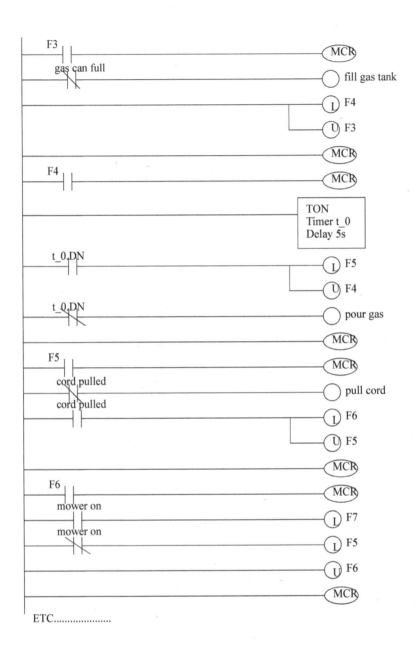

ETC.....................

3.

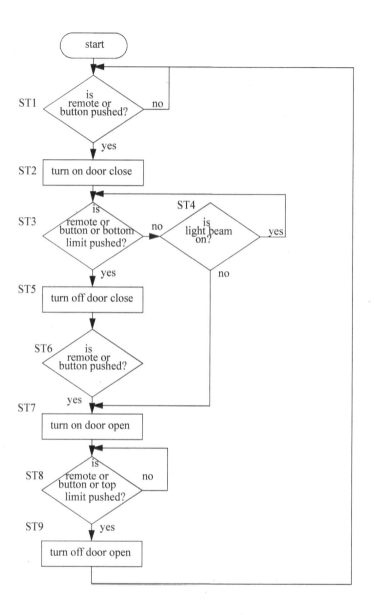

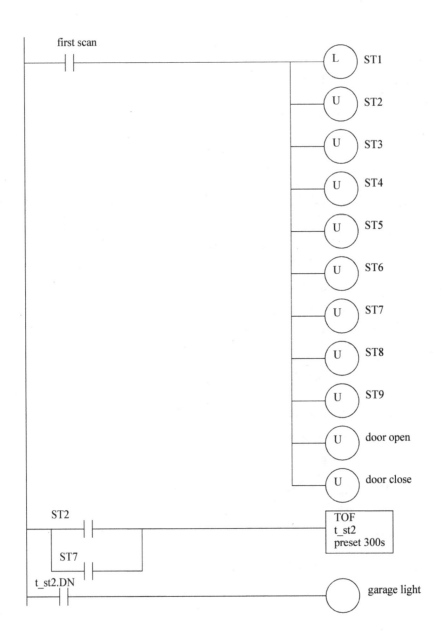

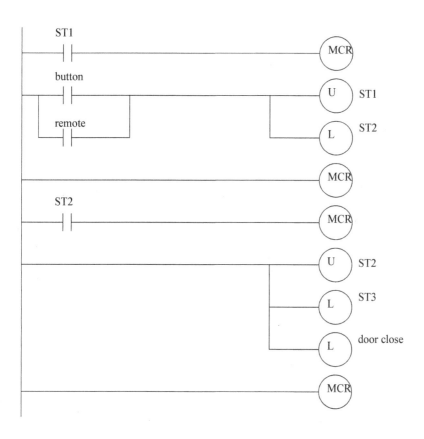

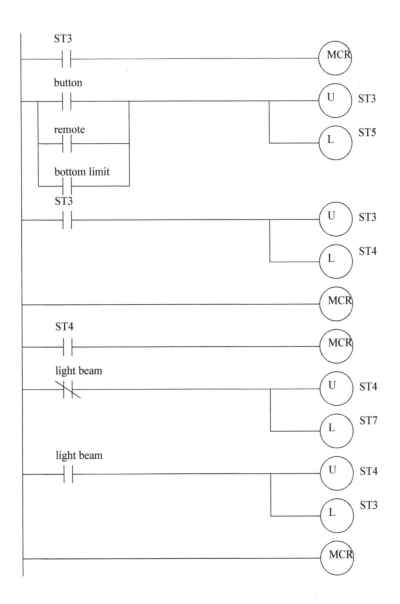

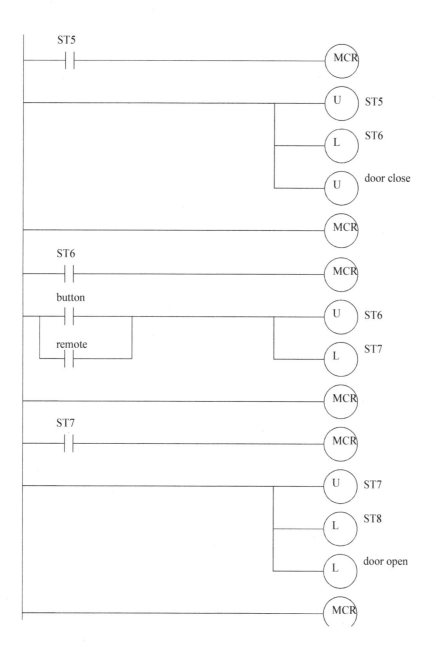

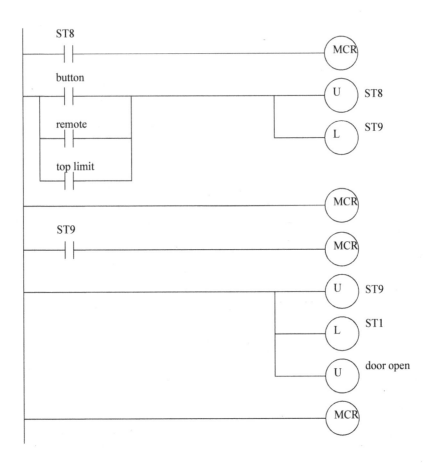

11STATE BASED DESIGN

Topics:

- Describing process control using state diagrams
- Conversion of state diagrams to ladder logic
- MCR blocks

Objectives:

- Be able to construct state diagrams for a process.
- Be able to convert a state diagram to ladder logic directly.
- Be able to convert state diagrams to ladder logic using equations.

A system state is a mode of operation. Consider a bank machine that will go through very carefully selected states. The general sequence of states might be idle, scan card, get secret number, select transaction type, ask for amount of cash, count cash, deliver cash/return card, then idle.

A State based system can be described with system states, and the transitions between those states. A state diagram is shown in Figure 11.1. The diagram has two states, *State 1* and *State 2*. If the system is in state 1 and *A* happens the system will then go into state 2, otherwise it will remain in State 1. Likewise if the system is in state 2, and *B* happens the system will return to state 1. As shown in the figure this state diagram could be used for an automatic light controller. When the power is turned on the system will go into the lights off state. If motion is detected or an on push button is pushed the system will go to the lights on state. If the system is in the lights on state and 1 hour has passed, or an off push button is pushed then the system will go to the lights off state. The else statements are omitted on the second diagram, but they are implied.

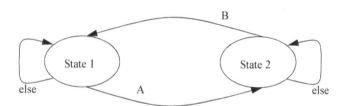

This diagram could describe the operation of energy efficient lights in a room operated by two push buttons. State 1 might be lights off and state 2 might be lights on. The arrows between the states are called transitions and will be followed when the conditions are true. In this case if we were in state 1 and A occurred we would move to state 2. The *else* loop indicate that a state will stay active if a transition are is not followed. These are so obvious they are often omitted from state diagrams.

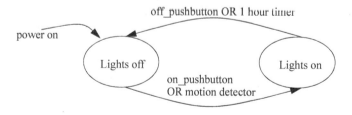

Figure 11.1   ***A State Diagram***

The most essential part of creating state diagrams is identifying states. Some key questions to ask are,

1. Consider the system,
   What does the system do normally?
   Does the system behavior change?
   Can something change how the system behaves?
   Is there a sequence to actions?
2. List *modes* of operation where the system is doing one identifiable activity that will start and stop. Keep in mind that some activities may just be to wait.

Consider the design of a coffee vending machine. The first step requires the identification of vending machine states as shown in Figure 11.1. The main state is the idle state. There is an inserting coins state where the total can be displayed. When enough coins have been inserted the user may select their drink of choice. After this the make coffee state will be active while coffee is being brewed. If an error is detected the service needed state will be activated.

STATES

idle - the machine has no coins and is doing nothing
inserting coins - coins have been entered and the total is displayed
user choose - enough money has been entered and the user is making coffee selection
make coffee - the selected type is being made
service needed - the machine is out of coffee, cups, or another error has occurred

Notes:
1. These states can be subjective, and different designers might pick others.
2. The states are highly specific to the machine.
3. The previous/next states are not part of the states.
4. There is a clean difference between states.

Figure 11.2   *Definition of Vending Machine States*

The states are then drawn in a state diagram as shown in Figure 11.1. Transitions are added as needed between the states. Here we can see that when powered up the machine will start in an idle state. The transitions here are based on the inputs and sensors in the vending machine. The state diagram is quite subjective, and complex diagrams will differ from design to design. These diagrams also expose the controller behavior. Consider that if the machine needs maintenance, and it is unplugged and plugged back in, the service needed statement would not be reentered until the next customer paid for but did not receive their coffee. In a commercial design we would want to fix this oversight.

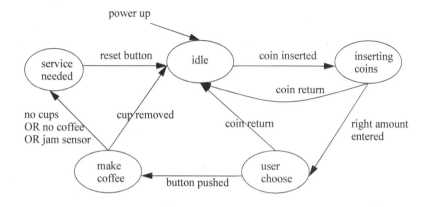

power up

service needed — reset button → idle — coin inserted → inserting coins

no cups OR no coffee OR jam sensor

cup removed

coin return

coin return

right amount entered

make coffee ← button pushed — user choose

Figure 11.3   *State Diagram for a Coffee Machine*

11.1 LADDER LOGIC BY DESIGN

11.1.1 State Diagram Example

Consider the traffic lights in Figure 11.1. The normal sequences for traffic lights are a green light in one direction for a long period of time, typically 10 or more seconds. This is followed by a brief yellow light, typically 4 seconds. This is then followed by a similar light pattern in the other direction. It is understood that a green or yellow light in one direction implies a red light in the other direction. Pedestrian buttons are provided so that when pedestrians are present a cross walk light can be turned on and the duration of the green light increased.

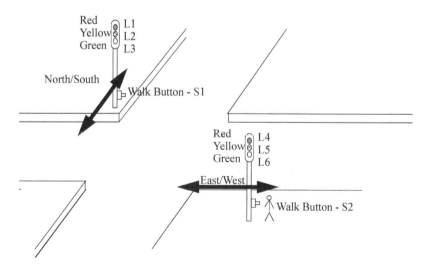

Red   L1
Yellow L2
Green L3

North/South

Walk Button - S1

Red   L4
Yellow L5
Green L6

East/West

Walk Button - S2

Figure 11.4   *Traffic Lights*

The first step for developing a controller is to define the inputs and outputs of the system as shown in Figure 11.1. First we will describe the system variables. These will vary as the system moves from state to state. Please note that some of these together can define a state (alone they are not the states). The inputs are used when defining the transitions. The outputs can be used to define the system state.

We have eight items that are ON or OFF

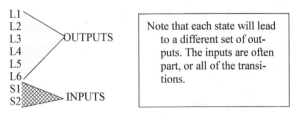

Note that each state will lead to a different set of outputs. The inputs are often part, or all of the transitions.

A simple diagram can be drawn to show sequences for the lights

Figure 11.5   *Inputs and Outputs for Traffic Light Controller*

Previously state diagrams were used to define the system, it is possible to use a state table as shown in Figure 11.1. Here the light sequences are listed in order. Each state is given a name to ease interpretation, but the corresponding output pattern is also given. The system state is defined as the bit pattern of the 6 lights. Note that there are only 4 patterns, but 6 binary bits could give as many as 64.

Step 1: Define the System States and put them (roughly) in sequence

System State

| L1 L2 L3 L4 L5 L6 |

A binary number
0 = light off
1 = light on

State Table

State Description	#	L1	L2	L3	L4	L5	L6
Green East/West	1	1	0	0	0	0	1
Yellow East/West	2	1	0	0	0	1	0
Green North/South	3	0	0	1	1	0	0
Yellow North/South	4	0	1	0	1	0	0

Here the four states determine how the 6 outputs are switched on/off.

Figure 11.6   *System State Table for Traffic Lights*

Transitions can be added to the state table to clarify the operation, as shown in Figure 11.1. Here the transition from Green E/W to Yellow E/W is S1. What this means is that a cross walk button must be pushed to end the green light. This is not normal, normally the lights would use a delay. The transition from Yellow E/W to Green N/S is caused by a 4 second delay (this is normal.) The next transition is also abnormal, requiring that the cross walk button be pushed to end the Green N/S state. The last state has a 4 second delay before returning to the first state in the table. In this state table the sequence will always be the same, but the times will vary for the green lights.

Step 2: Define State Transition Triggers, and add them to the list of states

Description	#	L1	L2	L3	L4	L5	L6	transition
Green East/West	1	1	0	0	0	0	1	S1
Yellow East/West	2	1	0	0	0	1	0	delay 4sec
Green North/South	3	0	0	1	1	0	0	S2
Yellow North/South	4	0	1	0	1	0	0	

delay 4 sec

Figure 11.7   *State Table with Transitions*

A state diagram for the system is shown in Figure 11.1. This diagram is equivalent to the state table in Figure 11.1, but it can be valuable for doing visual inspection.

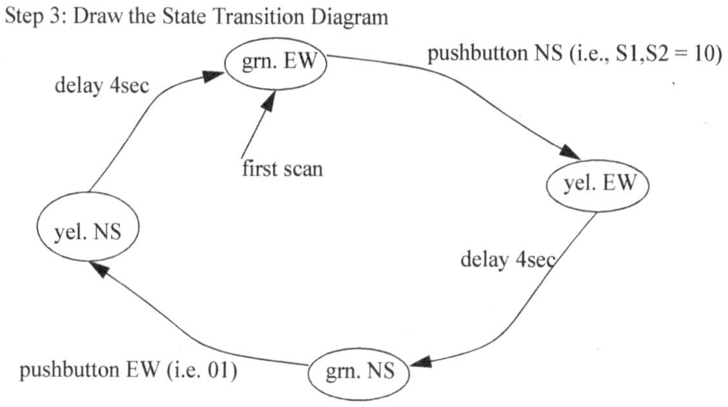

Step 3: Draw the State Transition Diagram

Figure 11.8   *A Traffic Light State Diagram*

11.1.2 Conversion to Ladder Logic

11.1.2.1 - Block Logic Conversion

State diagrams can be converted directly to ladder logic using block logic. This technique will produce larger programs, but it is a simple method to understand, and easy to debug. The previous traffic light example is to be implemented in ladder logic. The inputs and outputs are defined in Figure 11.1, assuming it will be implemented on an Allen Bradley Micrologix. *first scan* is the address of the first scan in the PLC. The locations state_1 to state_4 are internal memory locations that will be used to track which states are on. The behave like outputs, but are not available for connection outside the PLC. The input and output values are determined by the PLC layout.

STATES	OUTPUTS	INPUTS
state_1 - green E/W	L1 - red N/S	S1 - cross
state_2 - yellow E/W	L2 - yellow N/S	S2 - cross
state_3 - green N/S	L3 - green N/S	S:FS - first scan
state_4 - yellow N/S	L4 - red E/W	
	L5 - yellow E/W	
	L6 - green E/W	

Figure 11.9 *Inputs and Outputs for Traffic Light Controller*

The initial ladder logic block shown in Figure 11.1 will initialize the states of the PLC, so that only state 1 is on. The first scan indicator *first scan* will execute the MCR block when the PLC is first turned on, and the latches will turn on the value for state_1 and turn off the others.

RESET THE STATES

Figure 11.10 *Ladder Logic to Initialize Traffic Light Controller*

Note: We will use MCR instructions to implement some of the state based programs. This allows us to switch off part of the ladder logic. The one significant note to remember is that any normal outputs (not latches and timers) will be FORCED OFF. Unless this is what you want, put the normal outputs outside MCR blocks.

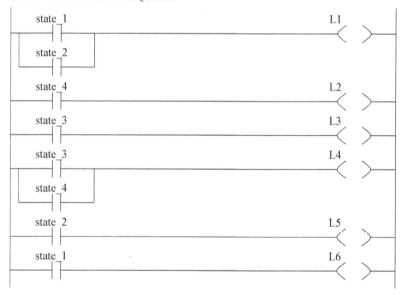

A

──┤ ├──────────────────[MCR]─

If A is true then the MCR will cause the ladder in between to be executed. If A is false the outputs are forced off.

──────────────────────────[MCR]◄

The next section of ladder logic only deals with outputs. For example the output *O/1* is the N/S red light, which will be on for states 1 and 2, or *B3/1* and *B3/2* respectively. Putting normal outputs outside the MCR blocks is important. If they were inside the blocks they could only be on when the MCR block was active, otherwise they would be forced off. Note: Many beginners will make the careless mistake of repeating outputs in this section of the program.

TURN ON LIGHTS AS REQUIRED

```
     state_1                                        L1
──┬───┤ ├────────────────────────────────────────( )──
  │  state_2
  └───┤ ├──
     state_4                                        L2
──────┤ ├────────────────────────────────────────( )──
     state_3                                        L3
──────┤ ├────────────────────────────────────────( )──
     state_3                                        L4
──┬───┤ ├────────────────────────────────────────( )──
  │  state_4
  └───┤ ├──
     state_2                                        L5
──────┤ ├────────────────────────────────────────( )──
     state_1                                        L6
──────┤ ├────────────────────────────────────────( )──
```

Figure 11.11　　　*General Output Control Logic*

The first state is implemented in Figure 11.1. If state_1 is active this will be active. The transition is S1 which will end state_1 and start state_2.

FIRST STATE WAIT FOR TRANSITIONS

Figure 11.12   *Ladder Logic for First State*

The second state is more complex because it involves a time delay, as shown in Figure 11.1. When the state is active the TON timer will be timing. When the timer is done state 2 will be unlatched, and state 3 will be latched on. The timer is nonretentive, so if state_2 if off the MCR block will force all of the outputs off, including the timer, causing it to reset.

SECOND STATE WAIT FOR TRANSITIONS

Figure 11.13   *Ladder Logic for Second State*

The third and fourth states are shown in Figure 11.1 and Figure 11.1. Their layout is very similar to that of the first two states.

THIRD STATE WAIT FOR TRANSITIONS

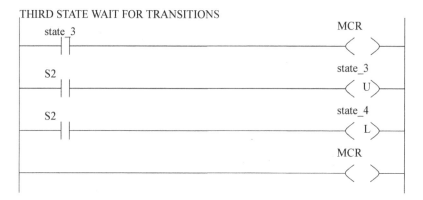

Figure 11.14   **Ladder Logic for State Three**

FOURTH STATE WAIT FOR TRANSITIONS

Figure 11.15   **Ladder Logic for State Four**

The previous example only had one path through the state tables, so there was never a choice between states. The state diagram in Figure 11.1 could potentially have problems if two transitions occur simultaneously. For example if state *STB* is active and A and C occur simultaneously, the system could go to either *STA* or *STC* (or both in a poorly written program.) To resolve this problem we should choose one of the two transitions as having a higher priority, meaning that it should be chosen over the other transition. This decision will normally be clear, but if not an arbitrary decision is still needed.

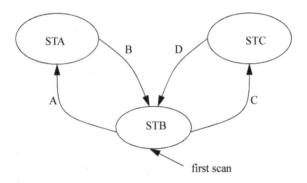

first scan

Figure 11.16   *A State Diagram with Priority Problems*

The state diagram in Figure 11.1 is implemented with ladder logic in Figure 11.1 and Figure 11.1. The implementation is the same as described before, but for state *STB* additional ladder logic is added to disable transition *A* if transition *C* is active, therefore giving priority to *C*.

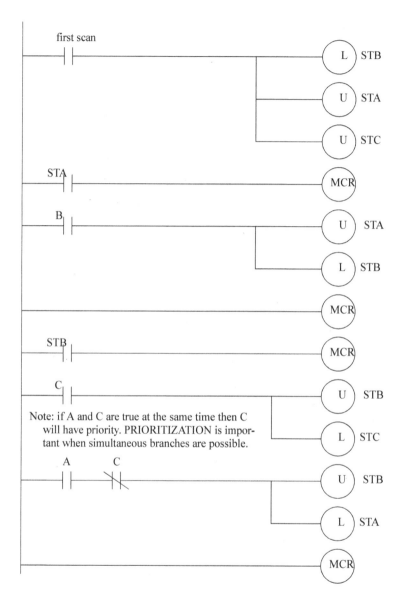

first scan

Note: if A and C are true at the same time then C will have priority. PRIORITIZATION is important when simultaneous branches are possible.

Figure 11.17   *State Diagram for Prioritization Problem*

221

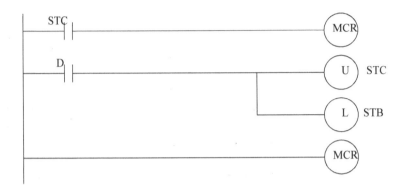

The Block Logic technique described does not require any special knowledge and the programs can be written directly from the state diagram. The final programs can be easily modified, and finding problems is easier. But, these programs are much larger and less efficient.

11.1.2.2 - State Equations

State diagrams can be converted to Boolean equations and then to Ladder Logic. The first technique that will be described is state equations. These equations contain three main parts, as shown below in Figure 11.1. To describe them simply - a state will be on if it is already on, or if it has been turned on by a transition from another state, but it will be turned off if there was a transition to another state. An equation is required for each state in the state diagram.

Informally,

State X = (State X + just arrived from another state) and has not left for another state

Formally,

$$STATE_i = \left(STATE_i + \sum_{j=1}^{n} (T_{j,i} \bullet STATE_j) \right) \bullet \prod_{k=1}^{m} \overline{(T_{i,k} \bullet STATE_i)}$$

where, $STATE_i$ = A variable that will reflect if state i is on

   n = the number of transitions to state i

   m = the number of transitions out of state i

   $T_{j,i}$ = The logical condition of a transition from state j to i

   $T_{i,k}$ = The logical condition of a transition out of state i to k

Figure 11.19   ***State Equations***

The state equation method can be applied to the traffic light example in Figure 11.1. The first step in the process is to define variable names (or PLC memory locations) to keep track of which states are on or off. Next, the state diagram is examined, one state at a time. The first equation if for ST1, or *state 1 - green NS*. The start of the equation can be read as ST1 will be on if it is on, or if ST4 is on, and it has been on for 4s, or if it is the first scan of the PLC. The end of the equation can be read as ST1 will be turned off if it is on, but S1 has been pushed and S2 is off. As discussed before, the first half of the equation will turn the state on, but the second half will turn it off. The first scan is also used to turn on ST1 when the PLC starts. It is put outside the terms to force ST1 on, even if the exit conditions are true.

Defined state variables:

$$ST1 \;=\; \text{state 1 - green NS}$$

$$ST2 \;=\; \text{state 2 - yellow NS}$$

$$ST3 \;=\; \text{state 3 - green EW}$$

$$ST4 \;=\; \text{state 4 - yellow EW}$$

The state entrance and exit condition equations:

$$ST1 \;=\; (ST1 + ST4 \cdot TON_2(ST4, 4s)) \cdot \overline{\overline{ST1} \cdot S1 \cdot \overline{S2}} + FS$$

$$ST2 \;=\; (ST2 + ST1 \cdot S1 \cdot \overline{S2}) \cdot \overline{\overline{ST2} \cdot TON_1(ST2, 4s)}$$

$$ST3 \;=\; (ST3 + ST2 \cdot TON_1(ST2, 4s)) \cdot \overline{\overline{ST3} \cdot \overline{S1} \cdot S2}$$

$$ST4 \;=\; (ST4 + ST3 \cdot \overline{S1} \cdot S2) \cdot \overline{\overline{ST4} \cdot TON_2(ST4, 4s)}$$

Note: Timers are represented in these equations in the form *TONi(A, delay)*. *TON* indicates that it is an on-delay timer, *A* is the input to the timer, and *delay* is the timer delay value. The subscript *i* is used to differentiate timers.

Figure 11.20   *State Equations for the Traffic Light Example*

The equations in Figure 11.1 cannot be implemented in ladder logic because of the NOT over the last terms. The equations are simplified in Figure 11.1 so that all NOT operators are only over a single variable.

Now, simplify these for implementation in ladder logic.

$$ST1 \;=\; (ST1 + ST4 \cdot TON_2(ST4, 4)) \cdot (\overline{\overline{ST1}} + \overline{S1} + S2) + FS$$

$$ST2 \;=\; (ST2 + ST1 \cdot S1 \cdot \overline{S2}) \cdot (\overline{ST2} + \overline{TON_1(ST2, 4)})$$

$$ST3 \;=\; (ST3 + ST2 \cdot TON_1(ST2, 4)) \cdot (\overline{ST3} + S1 + \overline{S2})$$

$$ST4 \;=\; (ST4 + ST3 \cdot \overline{S1} \cdot S2) \cdot (\overline{ST4} + \overline{TON_2(ST4, 4)})$$

Figure 11.21   *Simplified Boolean Equations*

These equations are then converted to the ladder logic shown in Figure 11.1 and Figure 11.1. At the top of the program the two timers are defined. (Note: it is tempting to combine the timers, but it is better to keep them separate.) Next, the Boolean state equations are implemented in ladder logic. After this we use the states to turn specific lights on.

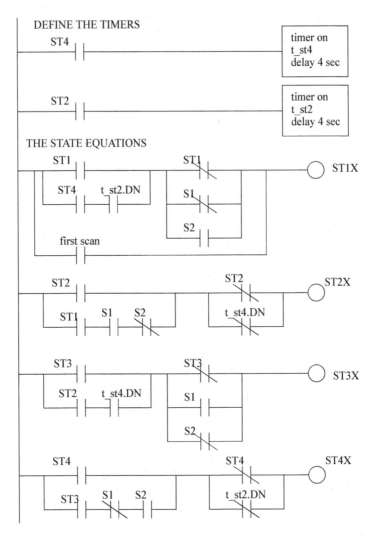

Figure 11.22   *Ladder Logic for the State Equations*

OUTPUT LOGIC FOR THE LIGHTS

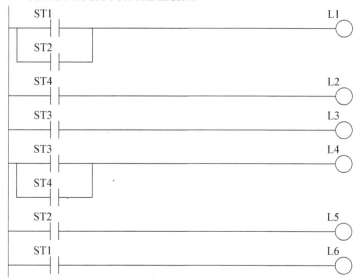

Figure 11.23   *Ladder Logic for the State Equations*

This method will provide the most compact code of all techniques, but there are potential problems. Consider the example in Figure 11.1. If push button *S1* has been pushed the line for ST1 should turn off, and the line for ST2 should turn on. But, the line for ST2 depends upon the value for *ST1* that has just been turned off. This will cause a problem if the value of ST1 goes off immediately after the line of ladder logic has been scanned. In effect the PLC will get *lost* and none of the states will be on. This problem arises because the equations are normally calculated in parallel, and then all values are updated simultaneously. To overcome this problem the ladder logic could be modified to the form shown in Figure 11.1. Here some temporary variables are used to hold the new state values. After all the equations are solved the states are updated to their new values.

THE STATE EQUATIONS

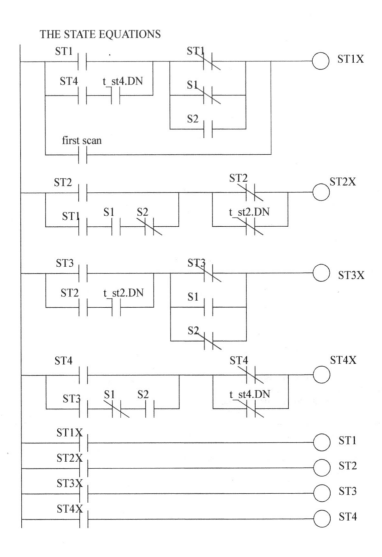

Figure 11.24   *Delayed State Updating*

When multiple transitions out of a state exist we must take care to add priorities. Each of the alternate transitions out of a state should be give a priority, from highest to lowest. The state equations can then be written to suppress transitions of lower priority when one or more occur simultaneously. The state diagram in Figure 11.1 has two transitions *A* and *C* that could occur simultaneously. The equations have been written to give *A* a higher priority. When *A* occurs, it will block *C* in the equation for *STC*. These equations have been converted to ladder logic in Figure 11.1.

226

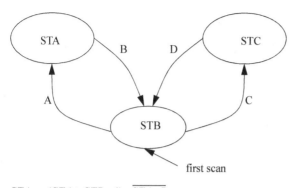

first scan

$$STA = (STA + STB \cdot A) \cdot \overline{STA \cdot B}$$

$$STB = (STB + STA \cdot B + STC \cdot D) \cdot \overline{STB \cdot A} \cdot \overline{STB \cdot C} + FS$$

$$STC = (STC + STB \cdot C \cdot \overline{A}) \cdot \overline{STC \cdot D}$$

Figure 11.25   *State Equations with Prioritization*

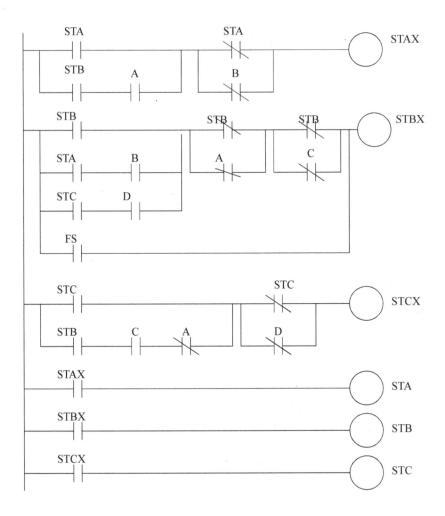

Figure 11.26   *Ladder Logic with Prioritization*

11.1.2.3 - State-Transition Equations

A state diagram may be converted to equations by writing an equation for each state and each transition. A sample set of equations is seen in Figure 11.1 for the traffic light example of Figure 11.1. Each state and transition needs to be assigned a unique variable name. (Note: It is a good idea to note these on the diagram) These are then used to write the equations for the diagram. The transition equations are written by looking at the each state, and then determining which transitions will end that state. For example, if ST1 is true, and crosswalk button *S1* is pushed, and S2 is not, then transition *T1* will be true. The state equations are similar to the state equations in the previous State Equation method, except they now only refer to the transitions. Recall, the basic form of these equations is that the state will be on if it is already on, or it has been turned on by a transition. The state will be turned off if an exiting transition occurs. In this example the first scan was given it's own transition, but it could have also been put into the equation for T4.

defined state and transition variables:

$ST1$ = state 1 - green NS

$ST2$ = state 2 - yellow NS

$ST3$ = state 3 - green EW

$ST4$ = state 4 - yellow EW

T1 = transition from ST1 to ST2

T2 = transition from ST2 to ST3

T3 = transition from ST3 to ST4

T4 = transition from ST4 to ST1

T5 = transition to ST1 for first scan

state and transition equations:

$$T4 = ST4 \cdot TON_2(ST4, 4)$$

$$T1 = ST1 \cdot S1 \cdot \overline{S2}$$

$$T2 = ST2 \cdot TON_1(ST2, 4)$$

$$T3 = ST3 \cdot \overline{S1} \cdot S2$$

$$T5 = FS$$

$$ST1 = (ST1 + T4 + T5) \cdot \overline{T1}$$

$$ST2 = (ST2 + T1) \cdot \overline{T2}$$

$$ST3 = (ST3 + T2) \cdot \overline{T3}$$

$$ST4 = (ST4 + T3) \cdot \overline{T4}$$

Figure 11.27   *State-Transition Equations*

These equations can be converted directly to the ladder logic in Figure 11.1, Figure 11.1 and Figure 11.1. It is very important that the transition equations all occur before the state equations. By updating the transition equations first and then updating the state equations the problem of state variable values changing is negated - recall this problem was discussed in the State Equations section.

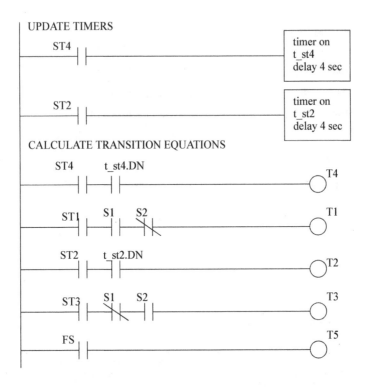

UPDATE TIMERS

ST4

timer on
t_st4
delay 4 sec

ST2

timer on
t_st2
delay 4 sec

CALCULATE TRANSITION EQUATIONS

ST4   t_st4.DN   T4

ST1   S1   S2   T1

ST2   t_st2.DN   T2

ST3   S1   S2   T3

FS   T5

Figure 11.28   *Ladder Logic for the State-Transition Equations*

230

CALCULATE STATE EQUATIONS

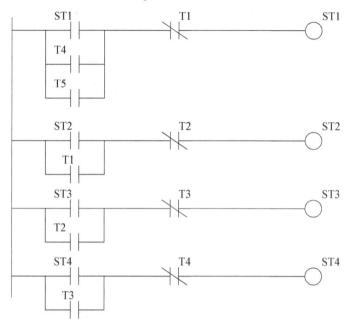

Figure 11.29   *Ladder Logic for the State-Transition Equations*

UPDATE OUTPUTS

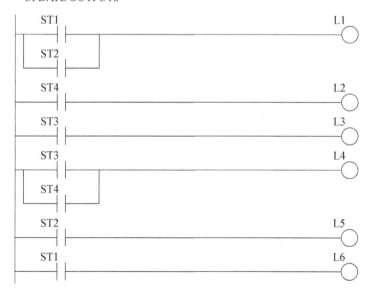

Figure 11.30   *Ladder Logic for the State-Transition Equations*

231

The problem of prioritization also occurs with the State-Transition equations. Equations were written for the State Diagram in Figure 11.1. The problem will occur if transitions *A* and *C* occur simultaneously. In the example transition *T2* is given a higher priority, and if it is true, then the transition *T3* will be suppressed when calculating *STC*. In this example the transitions have been considered in the state update equations, but they can also be used in the transition equations.

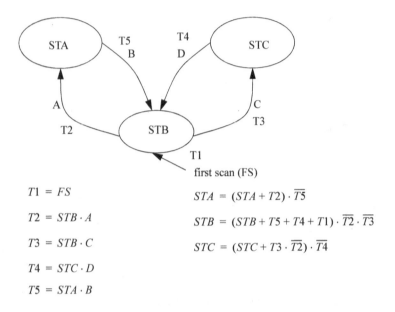

$$T1 = FS$$
$$T2 = STB \cdot A$$
$$T3 = STB \cdot C$$
$$T4 = STC \cdot D$$
$$T5 = STA \cdot B$$

$$STA = (STA + T2) \cdot \overline{T5}$$
$$STB = (STB + T5 + T4 + T1) \cdot \overline{T2} \cdot \overline{T3}$$
$$STC = (STC + T3 \cdot \overline{T2}) \cdot \overline{T4}$$

Figure 11.31   *Prioritization for State Transition Equations*

11.2 SUMMARY

- State diagrams are suited to processes with a single flow of execution.
- State diagrams are suited to problems that has clearly defines modes of execution.
- Controller diagrams can be converted to ladder logic using MCR blocks
- State diagrams can also be converted to ladder logic using equations
- The sequence of operations is important when converting state diagrams to ladder logic.

11.3 PRACTICE PROBLEMS

(Note: Problem solutions are available at http://sites.google.com/site/automatedmanufacturingsystems/)

1. Draw a state diagram for a microwave oven.

2. Convert the following state diagram to equations.

Inputs	Outputs
A	P
B	Q
C	R
D	
E	
F	

state	P	Q	R
S0	0	1	1
S1	1	0	1
S2	1	1	0

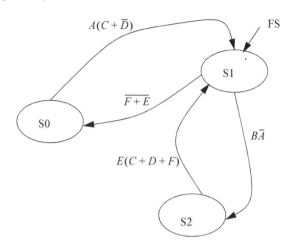

3. Implement the following state diagram with equations.

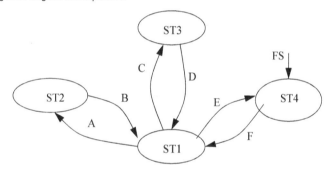

4. Given the following state diagram, use equations to implement ladder logic.

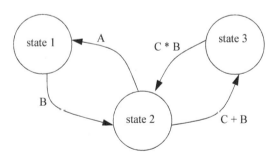

5. Convert the following state diagram to logic using equations.

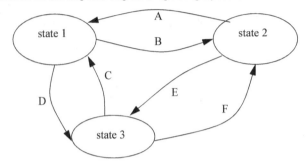

6. You have been asked to program a PLC that is controlling a handicapped access door opener. The client has provided the electrical wiring diagram below to show how the PLC inputs and outputs have been wired. Button A is located inside and button B is located outside. When either button is pushed the motor will be turned on to open the door. The motor is to be kept on for a total of 15 seconds to allow the person to enter. After the motor is turned off the door will fall closed. In the event that somebody gets caught in the door the thermal relay will go off, and the motor should be turned off. After 20,000 cycles the door should stop working and the light should go on to indicate that maintenance is required.

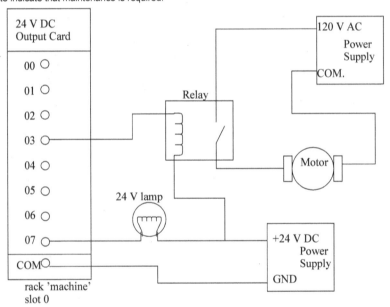

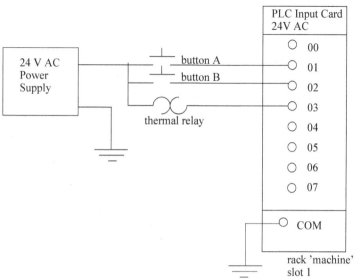

a) Develop a state diagram for the control of the door.

b) Convert the state diagram to ladder logic. (list the input and the output addresses first)

c) Convert the state diagram to delayed update equations.

7. Design a garage door controller using a) block logic, and b) state-transition equations. The behavior of the garage door controller is as follows,

- there is a single button in the garage, and a single button remote control.
- when the button is pushed the door will move up or down.
- if the button is pushed once while moving, the door will stop, a second push will start motion again in the opposite direction.
- there are top/bottom limit switches to stop the motion of the door.
- there is a light beam across the bottom of the door. If the beam is cut while the door is closing the door will stop and reverse.
- there is a garage light that will be on for 5 minutes after the door opens or closes.

8. Convert the following ladder logic to delayed update equations and then draw the state diagram for the system.

Is something missing from the system?

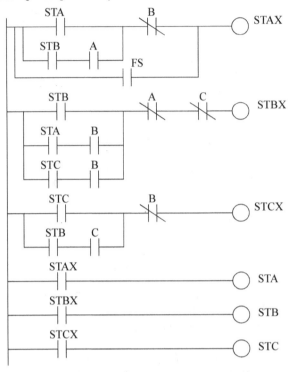

9. A program is to perform the following actions for a self-service security check. The device will allow bags to be inserted to the test chamber through an entrance door. If the bag passes the check it can be removed through an exit door, otherwise an alarm is sounded. Create a state diagram using the steps below.
   1. The machine starts in an 'idle' state. The 'open_entry' output is activated to open the input door. The 'open_exit' output is deactivated to close the output door.
   2. When a bag is inserted the 'bag_detected' input goes high. The 'open_entry' output should be deactivated to close the door.
   3. When the 'entry_door_closed' and 'exit_door_closed' inputs are active then a 'test' output will be set high to start a scan of the bags.
   4. When the scan of the bags is complete a 'scan_done' input is set. The 'test' output should be turned off.
   5. The scan results in two real values 'nitrates' and 'mass'. The calculation below is performed. If the 'risk' is below 0.3, or above 23.5, then the machine enters an alarm state (step 8), otherwise it continues to step 6.

$$risk = 4^{nitrates} + sqrt(mass)nitrates$$

   6. The 'open_exit' output is activated to open the exit door. The machine waits until the 'bag_detected' input goes low.
   7. The 'open_exit' output is deactivated to close the door. The machine waits until the 'exit_door_closed' input is high before returning to the 'idle state.
   8. In the alarm state an operator input 'key' must be active to open the exit door. After this input is released the door will close and return to the 'idle' state.

11.4 ASSIGNMENT PROBLEMS

1. Describe the difference between the block logic, delayed update, and transition equation methods for converting state diagrams to ladder logic.

2. Write the ladder logic for the state diagram below using the block logic method.

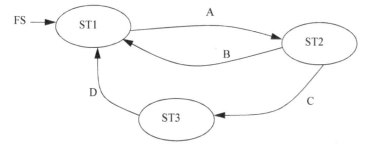

3. Convert the following state diagram to ladder logic using the block logic method. Give the stop button higher priority.

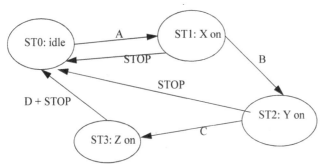

4. Convert the following state diagram to ladder logic using the delayed update method.

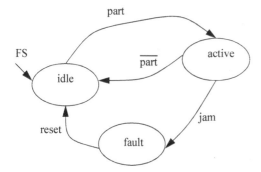

5. Use equations to develop ladder logic for the state diagram below using the delayed update method. Be sure to

deal with the priority problems.

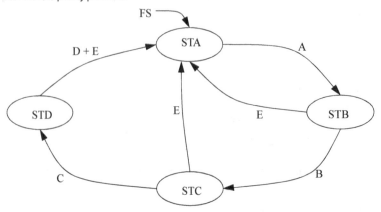

6. Implement the State-Transition equations in the figure below with ladder logic.

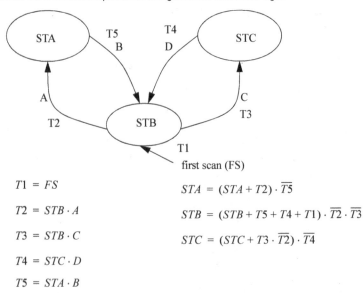

$T1 = FS$

$T2 = STB \cdot A$

$T3 = STB \cdot C$

$T4 = STC \cdot D$

$T5 = STA \cdot B$

$STA = (STA + T2) \cdot \overline{T5}$

$STB = (STB + T5 + T4 + T1) \cdot \overline{T2} \cdot \overline{T3}$

$STC = (STC + T3 \cdot \overline{T2}) \cdot \overline{T4}$

7. Write ladder logic to implement the state diagram below using state transition equations.

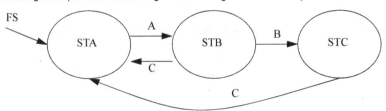

8. Convert the following state diagram to ladder logic using a) an equation based method, b) a method that is not

238

based on equations.

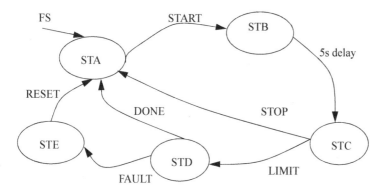

9. The state diagram below is for a simple elevator controller. a) Develop a ladder logic program that implements it with state transition equations. b) Develop the ladder logic using the block logic technique. c) Develop the ladder logic using the delayed update method.

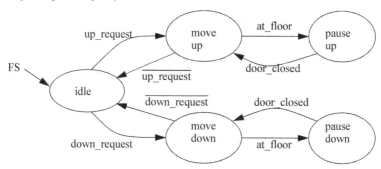

10. Write ladder logic for the state diagram below a) using an equation based method. b) without using an equation based method.

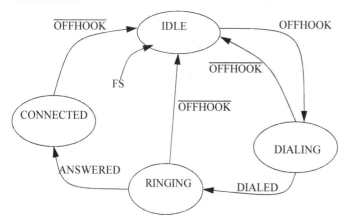

11. For the state diagram for the traffic light example, add a 15 second green light timer and speed up signal for an emergency vehicle. A strobe light mounted on fire trucks will cause the lights to change so that the truck doesn't need to stop. Modify the state diagram to include this option. Implement the new state diagram with ladder logic.

12. Design a program with a state diagram for a hydraulic press that will advance when two palm buttons are pushed. Top and bottom limit switches are used to reverse the advance and stop after a retract. At any time the hands removed from the palm button will stop an advance and retract the press. Include start and stop buttons to put the press in and out of an active mode.

13. In dangerous processes it is common to use two palm buttons that require a operator to use both hands to start a process (this keeps hands out of presses, etc.). To develop this there are two inputs (P1 and P2) that must both be turned on within 0.25s of each other before a machine cycle may begin.

Develop ladder logic with a state diagram to control a process that has a start (START) and stop (STOP) button for the power. After the power is on the palm buttons (P1 and P2) may be used as described above to start a cycle. The cycle will consist of turning on an output (MOVE) for 2 seconds. After the press has been cycled 1000 times the press power should turn off and an output (LIGHT) should go on.

14. Use a state diagram to design a parking gate controller.

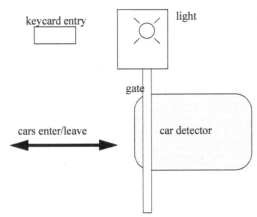

- the gate will be raised by one output and lowered by another. If the gate gets stuck an over current detector will make a PLC input true. If this is the case the gate should reverse and the light should be turned on indefinitely.
- if a valid keycard is entered a PLC input will be true. The gate is to rise and stay open for 10 seconds.
- when a car is over the car detector a PLC input will go true. The gate is to open while this detector is active. If it is active for more that 30 seconds the light should also turn on until the gate closes.

15. This morning you received a call from Mr. Ian M. Daasprate at the Old Fashioned Widget Company. In the past when they built a new machine they would used punched paper cards for control, but their supplier of punched paper readers went out of business in 1972 and they have decided to try using PLCs this time. He explains that the machine will dip wooden parts in varnish for 2 seconds, and then apply heat for 5 minutes to dry the coat, after this they are manually removed from the machine, and a new part is put in. They are also considering a premium line of parts that would call for a dip time of 30 seconds, and a drying time of 10 minutes. He then refers you to the project manager, Ann Nooyed.

You call Ann and she explains how the machine should operate. There should be start and stop buttons. The start button will be pressed when the new part has been loaded, and is ready to be coated. A light should be mounted to indicate when the machine is in operation. The part is mounted on a wheel that is rotated by a motor. To dip the part, the motor is turned on until a switch is closed. To remove the part from the dipping bath the motor is turned on until a second switch is closed. If the motor to rotate the wheel is on for more that 10 seconds before hitting a switch, the machine should be turned off, and a fault light turned on. The fault condition will be cleared by manually setting the machine back to its initial state, and hitting the start button twice. If the part has been dipped and dried properly, then a done light should be lit. To select a premium product you will use an input switch that needs to be pushed before the start button is pushed. She closes by saying she will be going on vacation and you need to have it done before she returns.

You hang up the phone and, after a bit of thought, decide to use the following outputs and inputs,

INPUTS
- I/1 - start push button
- I/2 - stop button
- I/3 - premium part push button
- I/4 - switch - part is in bath on wheel
- I/5 - switch - part is out of bath on wheel

OUTPUTS
- O/1 - start button
- O/2 - in operation
- O/3 - fault light
- O/4 - part done light
- O/5 - motor on
- O/6 - heater power supply

a) Draw a state diagram for the process.
b) List the variables needed to indicate when each state is on, and list any timers and counters used.
c) Write a Boolean expression for each transition in the state diagram.
d) Do a simple wiring diagram for the PLC.
e) Write the ladder logic for the state that involves moving the part into the dipping bath.

16. Design ladder logic with a state diagram for the following process description.

a) A toggle start switch (TS1) and a limit switch on a safety gate (LS1) must both be on before a solenoid (SOL1) can be energized to extend a stamping cylinder to the top of a part. Should a part detect sensor (PS1) also be considered? Explain your answer.

b) While the stamping solenoid is energized, it must remain energized until a limit switch (LS2) is activated. This second limit switch indicates the end of a stroke. At this point the solenoid should be de-energized, thus retracting the cylinder.

c) When the cylinder is fully retracted a limit switch (LS3) is activated. The cycle may not begin again until this limit switch is active. This is one way to ensure that a new part is present, is there another?

d) A cycle counter should also be included to allow counts of parts produced. When this value exceeds some variable amount (from 1 to 5000) the machine should shut down, and a job done light lit up.

e) A safety check should be included. If the cylinder solenoid has been on for more than 5 seconds, it suggests that the cylinder is jammed, or the machine has a fault. If this is the case the machine should be shut down, and a maintenance light turned on.

f) Implement the ladder diagram on a PLC in the laboratory.

g) Fully document the ladder logic and prepare a short report - This should be of use to another engineer that will be maintaining the system.

17. a) Write ladder logic to implement the state diagram below using the state transition equation method.

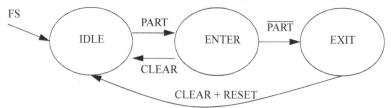

b) Write ladder logic to implement the state diagram below using the delayed update equation method.

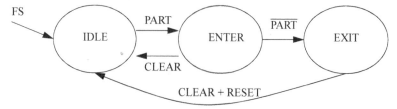

11.1 PRACTICE PROBLEM SOLUTIONS

1.

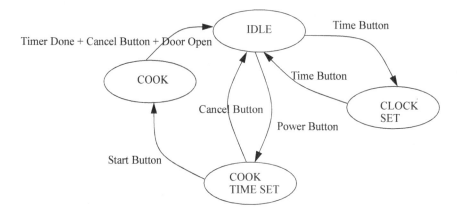

2.

$$T1 = FS$$

$$T2 = S1(B\bar{A})$$

$$T3 = S2(E(C + D + F))$$

$$T4 = S1(\overline{F + E})$$

$$T5 = S0(A(C + \bar{D}))$$

$$S1 = (S1 + T1 + T3 + T5)\overline{T2}\,\overline{T4}$$

$$S2 = (S2 + T2)\overline{T3}$$

$$S0 = (S0 + T4\overline{T2})\overline{T5}$$

$$P = S1 + S2$$

$$Q = S0 + S2$$

$$R = S0 + S1$$

3.

$$T1 = ST1 \bullet A$$
$$T2 = ST2 \bullet B$$
$$T3 = ST1 \bullet C$$
$$T4 = ST3 \bullet D$$
$$T5 = ST1 \bullet E$$
$$T6 = ST4 \bullet F$$

$$ST1 = (ST1 + T2 + T4 + T6) \cdot \overline{T1} \cdot \overline{T3} \cdot \overline{T5}$$
$$ST2 = (ST2 + T1 \cdot \overline{T3} \cdot \overline{T5}) \cdot \overline{T2}$$
$$ST3 = (ST3 + T3 \cdot \overline{T5}) \cdot \overline{T4}$$
$$ST4 = (ST4 + T5 + FS) \cdot \overline{T6}$$

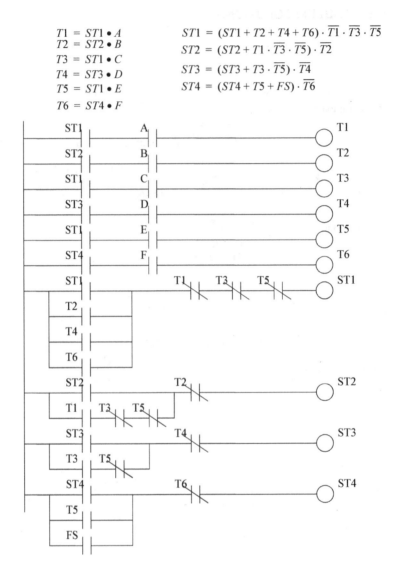

244

4.

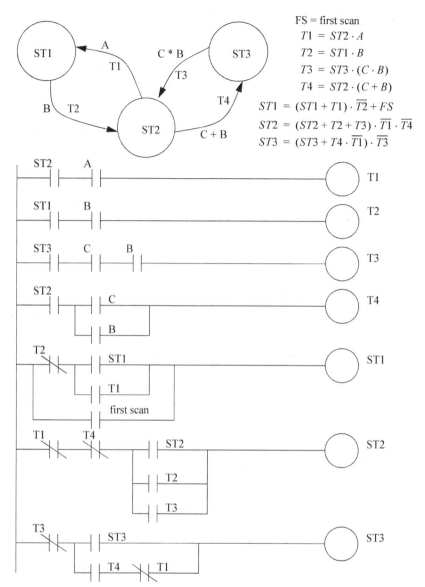

FS = first scan
$T1 = ST2 \cdot A$
$T2 = ST1 \cdot B$
$T3 = ST3 \cdot (C \cdot B)$
$T4 = ST2 \cdot (C + B)$
$ST1 = (ST1 + T1) \cdot \overline{T2} + FS$
$ST2 = (ST2 + T2 + T3) \cdot \overline{T1} \cdot \overline{T4}$
$ST3 = (ST3 + T4 \cdot \overline{T1}) \cdot \overline{T3}$

5.

$$TA = ST2 \cdot A$$
$$TB = ST1 \cdot B$$
$$TC = ST3 \cdot C$$
$$TD = ST1 \cdot D \cdot \bar{B}$$
$$TE = ST2 \cdot E \cdot \bar{A}$$
$$TF = ST3 \cdot F \cdot \bar{C}$$

$$ST1 = (ST1 + TA + TC) \cdot \overline{TB} \cdot \overline{TD}$$
$$ST2 = (ST2 + TB + TF) \cdot \overline{TA} \cdot \overline{TE}$$
$$ST3 = (ST3 + TD + TE) \cdot \overline{TC} \cdot \overline{TF}$$

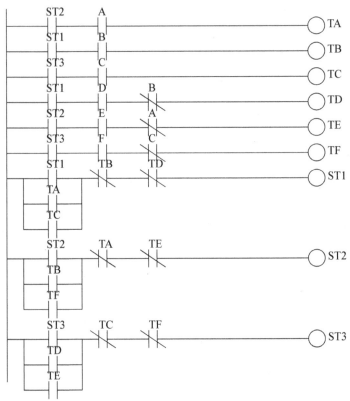

6.

a)

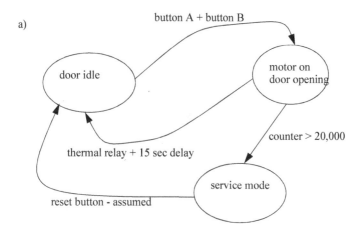

button A + button B

door idle

motor on
door opening

counter > 20,000

thermal relay + 15 sec delay

service mode

reset button - assumed

b)

Legend
button A   Machine:0.I.Data.1
button B   Machine:0.I.Data.2
motor   Machine:1.O.Data.3
thermal relay   Machine:0.I.Data.3
reset button   Machine:0.I.Data.4 - assumed
state 1
state 2
state 3
lamp   Machine:1.O.Data.7

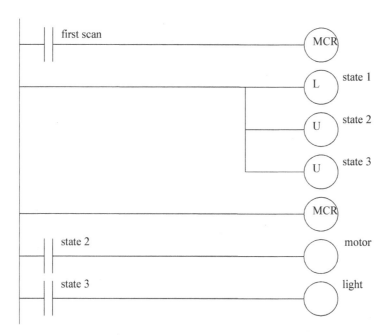

first scan   MCR

                                              L   state 1

                                              U   state 2

                                              U   state 3

                                              MCR

state 2   motor

state 3   light

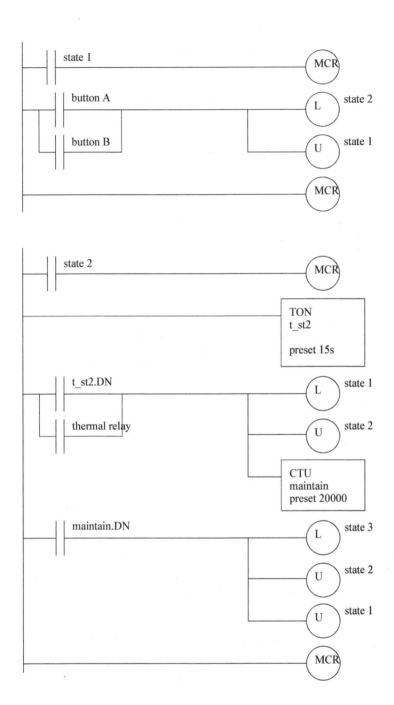

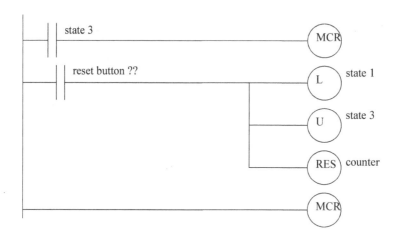

c) $S0 = (S0 + S1(delay(15) + thermal))\overline{S0(buttonA + buttonB)}$

   $S1 = (S1 + S0(buttonA + buttonB))\overline{S1(delay(15) + thermal)}\overline{S3(counter)}$

   $S3 = (S3 + S2(counter))\overline{S3(reset)}$

   $motor = S1$

   $light = S3$

7.

a) block logic method

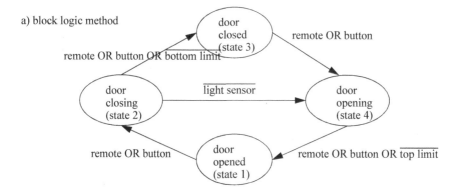

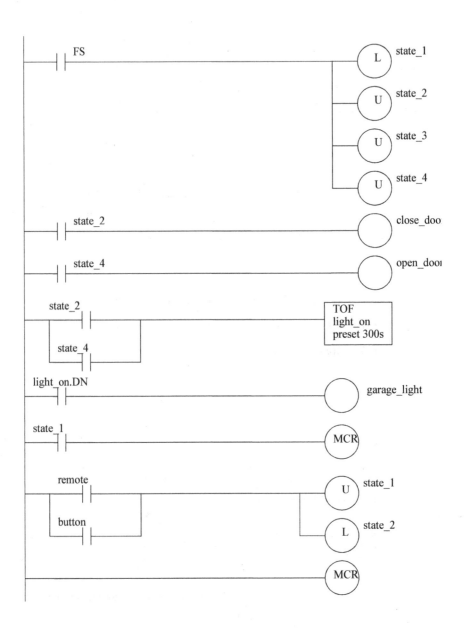

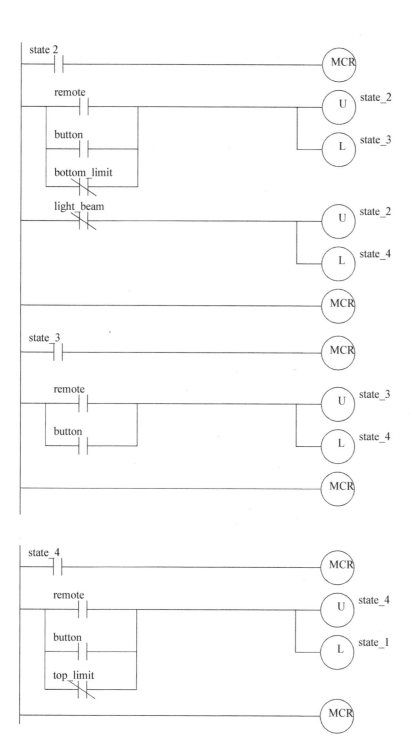

b) state-transition equations

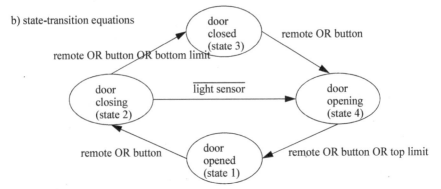

using the previous state diagram.

ST1 = state 1   T1 = state 1 to state 2
ST2 = state 2   T2 = state 2 to state 3
ST3 = state 3   T3 = state 2 to state 4
ST4 = state 4   T4 = state 3 to state 4
FS = first scan   T5 = state 4 to state 1

$$ST1 = (ST1 + T5) \cdot \overline{T1}$$
$$ST2 = (ST2 + T1) \cdot \overline{T2} \cdot \overline{T3}$$
$$ST3 = (ST3 + T2) \cdot \overline{T4}$$
$$ST4 = (ST4 + T3 + T4) \cdot \overline{T5}$$

$$T1 = ST1 \cdot (remote + button)$$
$$T2 = ST2 \cdot (remote + button + bottomlimit)$$
$$T3 = ST2 \cdot (remote + button)$$
$$T4 = ST3 \cdot (\overline{lighbeam})$$
$$T5 = ST4 \cdot (remote + button + toplimit) + FS$$

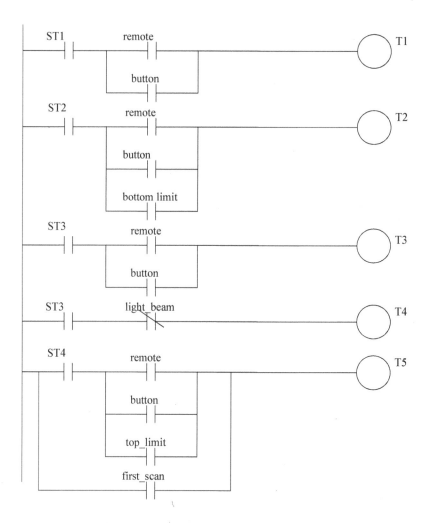

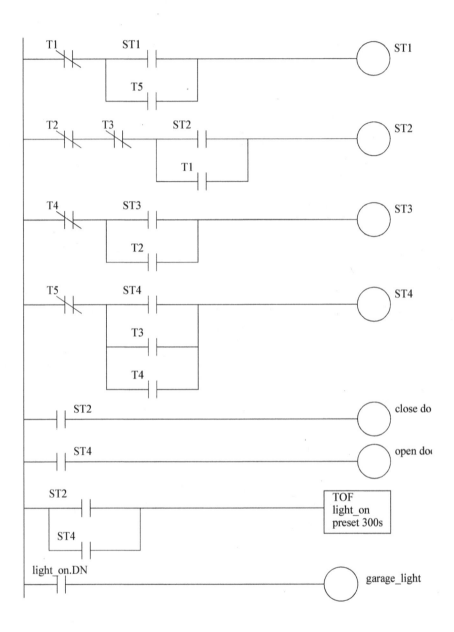

8.

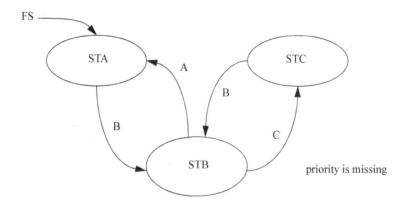

priority is missing

9.

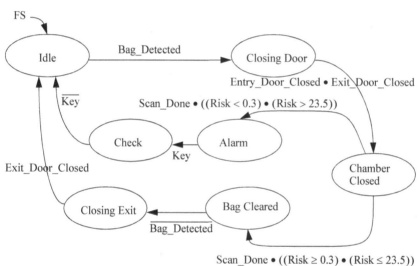

12NUMBERS AND DATA

Topics:

- Number bases; binary, octal, decimal, hexadecimal
- Binary calculations; 2s compliments, addition, subtraction and Boolean operations
- Encoded values; BCD and ASCII
- Error detection; parity, gray code and checksums

Objectives:

- To be familiar with binary, octal and hexadecimal numbering systems. .
- To be able to convert between different numbering systems.
- To understand 2s compliment negative numbers.
- To be able to convert ASCII and BCD values.
- To be aware of basic error detection techniques.

Base 10 (decimal) numbers developed naturally because the original developers (probably) had ten fingers, or 10 digits. Now consider logical systems that only have wires that can be on or off. When counting with a wire the only digits are 0 and 1, giving a base 2 numbering system. Numbering systems for computers are often based on base 2 numbers, but base 4, 8, 16 and 32 are commonly used. A list of numbering systems is give in Figure 12.1. An example of counting in these different numbering systems is shown in Figure 12.1.

Base	Name	Data Unit
2	Binary	Bit
8	Octal	Nibble
10	Decimal	Digit
16	Hexadecimal	Byte

Figure 12.1   Numbering Systems

decimal	binary	octal	hexadecimal
0	0	0	0
1	1	1	1
2	10	2	2
3	11	3	3
4	100	4	4
5	101	5	5
6	110	6	6
7	111	7	7
8	1000	10	8
9	1001	11	9
10	1010	12	a
11	1011	13	b
12	1100	14	c
13	1101	15	d
14	1110	16	e
15	1111	17	f
16	10000	20	10
17	10001	21	11
18	10010	22	12
19	10011	23	13
20	10100	24	14

Note: As with all numbering systems most significant digits are at left, least significant digits are at right.

Figure 12.2   *Numbers in Decimal, Binary, Octal and Hexadecimal*

The effect of changing the base of a number does not change the actual value, only how it is written. The basic rules of mathematics still apply, but many beginners will feel disoriented. This chapter will cover basic topics that are needed to use more complex programming instructions later in the book. These will include the basic number systems, conversion between different number bases, and some data oriented topics.

12.1 NUMERICAL VALUES

12.1.1 Binary

Binary numbers are the most fundamental numbering system in all computers. A single binary digit (a bit) corresponds to the condition of a single wire. If the voltage on the wire is true the bit value is *1*. If the voltage is off the bit value is *0*. If two or more wires are used then each new wire adds another significant digit. Each binary number will have an equivalent digital value. Figure 12.1 shows how to convert a binary number to a decimal equivalent. Consider the digits, starting at the right. The least significant digit is *1*, and is in the 0th position. To convert this to a decimal equivalent the number base (2) is raised to the position of the digit, and multiplied by the digit. In this case the least significant digit is a trivial conversion. Consider the most significant digit, with a value of *1* in the 6th position. This is converted by the number base to the exponent 6 and multiplying by the digit value of 1. This method can also be used for converting the other number system to decimal.

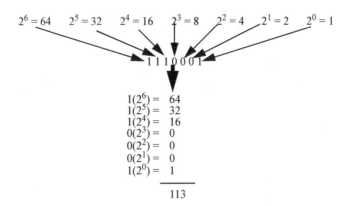

Figure 12.3   *Conversion of a Binary Number to a Decimal Number*

Decimal numbers can be converted to binary numbers using division, as shown in Figure 12.1. This technique begins by dividing the decimal number by the base of the new number. The fraction after the decimal gives the least significant digit of the new number when it is multiplied by the number base. The whole part of the number is now divided again. This process continues until the whole number is zero. This method will also work for conversion to other number bases.

start with decimal number 932

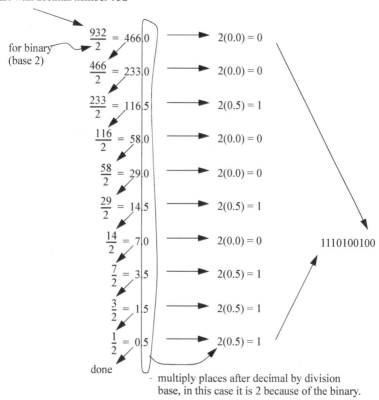

for binary
(base 2)

$$\frac{932}{2} = 466|0 \longrightarrow 2(0.0) = 0$$

$$\frac{466}{2} = 233|0 \longrightarrow 2(0.0) = 0$$

$$\frac{233}{2} = 116|5 \longrightarrow 2(0.5) = 1$$

$$\frac{116}{2} = 58|0 \longrightarrow 2(0.0) = 0$$

$$\frac{58}{2} = 29|0 \longrightarrow 2(0.0) = 0$$

$$\frac{29}{2} = 14|5 \longrightarrow 2(0.5) = 1$$

$$\frac{14}{2} = 7|0 \longrightarrow 2(0.0) = 0 \qquad 1110100100$$

$$\frac{7}{2} = 3|5 \longrightarrow 2(0.5) = 1$$

$$\frac{3}{2} = 1|5 \longrightarrow 2(0.5) = 1$$

$$\frac{1}{2} = 0|5 \longrightarrow 2(0.5) = 1$$

done

· multiply places after decimal by division
base, in this case it is 2 because of the binary.

* This method works for other number bases also, the divisor and multipliers
should be changed to the new number bases.

Figure 12.4   **Conversion from Decimal to Binary**

Most scientific calculators will convert between number bases. But, it is important to understand the conversions between number bases. And, when used frequently enough the conversions can be done in your head.

Binary numbers come in three basic forms - a bit, a byte and a word. A bit is a single binary digit, a byte is eight binary digits, and a word is 16 digits. Words and bytes are shown in Figure 12.1. Notice that on both numbers the least significant digit is on the right hand side of the numbers. And, in the word there are two bytes, and the right hand one is the least significant byte.

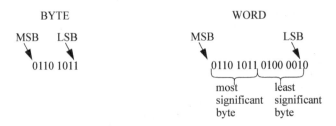

Figure 12.5 **Bytes and Words**

Binary numbers can also represent fractions, as shown in Figure 12.1. The conversion to and from binary is identical to the previous techniques, except that for values to the right of the decimal the equivalents are fractions.

binary: 101.011

$$1(2^2) = 4 \quad 0(2^1) = 0 \quad 1(2^0) = 1 \quad 0(2^{-1}) = 0 \quad 1(2^{-2}) = \frac{1}{4} \quad 1(2^{-3}) = \frac{1}{8}$$

$$= 4 + 0 + 1 + 0 + \frac{1}{4} + \frac{1}{8} = 5.375 \quad \text{decimal}$$

Figure 12.6   *A Binary Decimal Number*

12.1.1.1 - Boolean Operations

In the next chapter you will learn that entire blocks of inputs and outputs can be used as a single binary number (typically a word). Each bit of the number would correspond to an output or input as shown in Figure 12.1.

There are three motors M_1, M_2 and M_3 represented with three bits in a binary number. When any bit is on the corresponding motor is on.

100 = Motor 1 is the only one on
111 = All three motors are on
in total there are 2^n or 2^3 possible combinations of motors on.

Figure 12.7   *Motor Outputs Represented with a Binary Number*

We can then manipulate the inputs or outputs using Boolean operations. Boolean algebra has been discussed before for variables with single values, but it is the same for multiple bits. Common operations that use multiple bits in numbers are shown in Figure 12.1. These operations compare only one bit at a time in the number, except the shift instructions that move all the bits one place left or right.

Name	Example	Result	
AND	0010 * 1010	0010	
OR	0010 + 1010	1010	
NOT	$\overline{0010}$	1101	
EOR	0010 eor 1010	1000	
NAND	$\overline{0010 * 1010}$	1101	
shift left	111000	110001	(other results are possible)
shift right	111000	011100	(other results are possible)
etc.			

Figure 12.8   *Boolean Operations on Binary Numbers*

12.1.1.2 - Binary Mathematics

Negative numbers are a particular problem with binary numbers. As a result there are three common numbering systems used as shown in Figure 12.1. Unsigned binary numbers are common, but they can only be used for positive values. Both signed and 2s compliment numbers allow positive and negative values, but the maximum positive values is reduced by half. 2s compliment numbers are very popular because the hardware and software to add and subtract is simpler and faster. All three types of numbers will be found in PLCs.

Type	Description	Range for Byte
unsigned	binary numbers can only have positive values.	0 to 255
signed	the most significant bit (MSB) of the binary number is used to indicate positive/negative.	-127 to 127
2s compliment	negative numbers are represented by complimenting the binary number and then adding 1.	-128 to 127

Figure 12.9   *Binary (Integer) Number Types*

Examples of signed binary numbers are shown in Figure 12.1. These numbers use the most significant bit to indicate when a number is negative.

decimal	binary byte	
2	00000010	
1	00000001	
0	00000000	
-0	10000000	Note: there are two zeros
-1	10000001	
-2	10000010	

Figure 12.10   *Signed Binary Numbers*

An example of 2s compliment numbers are shown in Figure 12.1. Basically, if the number is positive, it will be a regular binary number. If the number is to be negative, we start the positive number, compliment it (reverse all

261

the bits), then add 1. Basically when these numbers are negative, then the most significant bit is set. To convert from a negative 2s compliment number, subtract 1, and then invert the number.

decimal	binary byte	METHOD FOR MAKING A NEGATIVE NUMBER
2	00000010	1. write the binary number for the positive
1	00000001	
0	00000000	for -30 we write 30 = 00011110
-1	11111111	2. Invert (compliment) the number
-2	11111110	

2. Invert (compliment) the number

   00011110 becomes 11100001

3. Add 1

   11100001 + 00000001 = 11100010

Figure 12.11   *2s Compliment Numbers*

Using 2s compliments for negative numbers eliminates the redundant zeros of signed binaries, and makes the hardware and software easier to implement. As a result most of the integer operations in a PLC will do addition and subtraction using 2s compliment numbers. When adding 2s compliment numbers, we don't need to pay special attention to negative values. And, if we want to subtract one number from another, we apply the twos compliment to the value to be subtracted, and then apply it to the other value.

Figure 12.1 shows the addition of numbers using 2s compliment numbers. The three operations result in zero, positive and negative values. Notice that in all three operation the top number is positive, while the bottom operation is negative (this is easy to see because the MSB of the numbers is set). All three of the additions are using bytes, this is important for considering the results of the calculations. In the left and right hand calculations the additions result in a 9th bit - when dealing with 8 bit numbers we call this bit the carry C. If the calculation started with a positive and negative value, and ended up with a carry bit, there is no problem, and the carry bit should be ignored. If doing the calculation on a calculator you will see the carry bit, but when using a PLC you must look elsewhere to find it.

$$00000001 = 1$$
$$+ \quad 11111111 = -1$$
$$C+00000000 = 0$$

$$00000001 = 1$$
$$+ \quad 11111110 = -2$$
$$11111111 = -1$$

$$00000010 = 2$$
$$+ \quad 11111111 = -1$$
$$C+00000001 = 1$$

ignore the carry bits

Note: Normally the carry bit is ignored during the operation, but some additional logic is required to make sure that the number has not *overflowed* and moved outside of the range of the numbers. Here the 2s compliment byte can have values from -128 to 127.

Figure 12.12   *Adding 2s Compliment Numbers*

The integers have limited value ranges, for example a 16 bit word ranges from -32,768 to 32,767 whereas a 32 bit word ranges from -2,147,483,648 to 2,147,483,647. In some cases calculations will give results outside this range, and the Overflow O bit will be set. (Note: an overflow condition is a major error, and the PLC will probably halt when this happens.) For an addition operation the Overflow bit will be set when the sign of both numbers is the same, but the sign of the result is opposite. When the signs of the numbers are opposite an overflow cannot occur. This can be seen in Figure 12.1 where the numbers two of the three calculations are outside the range. When this happens the result goes from positive to negative, or the other way.

$$01111111 = 127$$
$$+ \quad 00000011 = 3$$
$$\overline{}$$
$$10000010 = -126$$
$$C = 0$$
$$O = 1 \text{ (error)}$$

$$10000001 = -127$$
$$+ \quad 11111111 = -1$$
$$\overline{}$$
$$10000000 = -128$$
$$C = 1$$
$$O = 0 \text{ (no error)}$$

$$10000001 = -127$$
$$+ \quad 11111110 = -2$$
$$\overline{}$$
$$01111111 = 127$$
$$C = 1$$
$$O = 1 \text{ (error)}$$

> Note: If an overflow bit is set this indicates that a calculation is outside and acceptable range. When this error occurs the PLC will halt. Do not ignore the limitations of the numbers.

Figure 12.13   **Carry and Overflow Bits**

These bits also apply to multiplication and division operations. In addition the PLC will also have bits to indicate when the result of an operation is zero Z and negative N.

12.1.2 Other Base Number Systems

Other number bases are typically converted to and from binary for storage and mathematical operations. Hexadecimal numbers are popular for representing binary values because they are quite compact compared to binary. (Note: large binary numbers with a long string of 1s and 0s are next to impossible to read.) Octal numbers are also popular for inputs and outputs because they work in counts of eight; inputs and outputs are in counts of eight.

An example of conversion to, and from, hexadecimal is shown in Figure 12.1 and Figure 12.1. Note that both of these conversions are identical to the methods used for binary numbers, and the same techniques extend to octal numbers also.

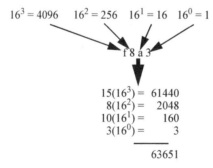

$$16^3 = 4096 \quad 16^2 = 256 \quad 16^1 = 16 \quad 16^0 = 1$$

$$f\,8\,a\,3$$

$$15(16^3) = 61440$$
$$8(16^2) = 2048$$
$$10(16^1) = 160$$
$$3(16^0) = 3$$
$$\overline{63651}$$

Figure 12.14   **Conversion of a Hexadecimal Number to a Decimal Number**

$$\frac{5724}{16} = 357.75 \longrightarrow 16(0.75) = 12 \text{ 'c'}$$

$$\frac{357}{16} = 22.3125 \longrightarrow 16(0.3125) = 5$$

$$\frac{22}{16} = 1.375 \longrightarrow 16(0.375) = 6$$

$$\frac{1}{16} = 0.0625 \longrightarrow 16(0.0625) = 1$$

1 6 5 c

Figure 12.15   *Conversion from Decimal to Hexadecimal*

12.1.3 BCD (Binary Coded Decimal)

Binary Coded Decimal (BCD) numbers use four binary bits (a nibble) for each digit. (Note: this is not a base number system, but it only represents decimal digits.) This means that one byte can hold two digits from *00* to *99*, whereas in binary it could hold from 0 to 255. A separate bit must be assigned for negative numbers. This method is very popular when numbers are to be output or input to the computer. An example of a BCD number is shown in Figure 12.1. In the example there are four digits, therefore 16 bits are required. Note that the most significant digit and bits are both on the left hand side. The BCD number is the binary equivalent of each digit.

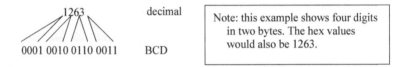

1263   decimal

0001 0010 0110 0011   BCD

Note: this example shows four digits in two bytes. The hex values would also be 1263.

Figure 12.16   *A BCD Encoded Number*

Most PLCs store BCD numbers in words, allowing values between *0000* and *9999*. They also provide functions to convert to and from BCD. It is also possible to calculations with BCD numbers, but this is uncommon, and when necessary most PLCs have functions to do the calculations. But, when doing calculations you should probably avoid BCD and use integer mathematics instead. Try to be aware when your numbers are BCD values and convert them to *integer* or binary value before doing any calculations.

12.2 DATA CHARACTERIZATION

12.2.1 ASCII (American Standard Code for Information Interchange)

When dealing with non-numerical values or data we can use plain text characters and strings. Each character is given a unique identifier and we can use these to store and interpret data. The ASCII (American Standard Code for Information Interchange) is a very common character encryption system is shown in Figure 12.1 and Figure 12.1. The table includes the basic written characters, as well as some special characters, and some control codes. Each one is given a unique number. Consider the letter *A*, it is readily recognized by most computers worldwide when they see the number 65.

decimal	hexadecimal	binary	ASCII	decimal	hexadecimal	binary	ASCII
0	0	00000000	NUL	32	20	00100000	space
1	1	00000001	SOH	33	21	00100001	!
2	2	00000010	STX	34	22	00100010	"
3	3	00000011	ETX	35	23	00100011	#
4	4	00000100	EOT	36	24	00100100	$
5	5	00000101	ENQ	37	25	00100101	%
6	6	00000110	ACK	38	26	00100110	&
7	7	00000111	BEL	39	27	00100111	'
8	8	00001000	BS	40	28	00101000	(
9	9	00001001	HT	41	29	00101001	)
10	A	00001010	LF	42	2A	00101010	*
11	B	00001011	VT	43	2B	00101011	+
12	C	00001100	FF	44	2C	00101100	,
13	D	00001101	CR	45	2D	00101101	-
14	E	00001110	S0	46	2E	00101110	.
15	F	00001111	S1	47	2F	00101111	/
16	10	00010000	DLE	48	30	00110000	0
17	11	00010001	DC1	49	31	00110001	1
18	12	00010010	DC2	50	32	00110010	2
19	13	00010011	DC3	51	33	00110011	3
20	14	00010100	DC4	52	34	00110100	4
21	15	00010101	NAK	53	35	00110101	5
22	16	00010110	SYN	54	36	00110110	6
23	17	00010111	ETB	55	37	00110111	7
24	18	00011000	CAN	56	38	00111000	8
25	19	00011001	EM	57	39	00111001	9
26	1A	00011010	SUB	58	3A	00111010	:
27	1B	00011011	ESC	59	3B	00111011	;
28	1C	00011100	FS	60	3C	00111100	<
29	1D	00011101	GS	61	3D	00111101	=
30	1E	00011110	RS	62	3E	00111110	>
31	1F	00011111	US	63	3F	00111111	?

Figure 12.17　　　**ASCII Character Table**

decimal	hexadecimal	binary	ASCII	decimal	hexadecimal	binary	ASCII
64	40	01000000	@	96	60	01100000	`
65	41	01000001	A	97	61	01100001	a
66	42	01000010	B	98	62	01100010	b
67	43	01000011	C	99	63	01100011	c
68	44	01000100	D	100	64	01100100	d
69	45	01000101	E	101	65	01100101	e
70	46	01000110	F	102	66	01100110	f
71	47	01000111	G	103	67	01100111	g
72	48	01001000	H	104	68	01101000	h
73	49	01001001	I	105	69	01101001	i
74	4A	01001010	J	106	6A	01101010	j
75	4B	01001011	K	107	6B	01101011	k
76	4C	01001100	L	108	6C	01101100	l
77	4D	01001101	M	109	6D	01101101	m
78	4E	01001110	N	110	6E	01101110	n
79	4F	01001111	O	111	6F	01101111	o
80	50	01010000	P	112	70	01110000	p
81	51	01010001	Q	113	71	01110001	q
82	52	01010010	R	114	72	01110010	r
83	53	01010011	S	115	73	01110011	s
84	54	01010100	T	116	74	01110100	t
85	55	01010101	U	117	75	01110101	u
86	56	01010110	V	118	76	01110110	v
87	57	01010111	W	119	77	01110111	w
88	58	01011000	X	120	78	01111000	x
89	59	01011001	Y	121	79	01111001	y
90	5A	01011010	Z	122	7A	01111010	z
91	5B	01011011	[	123	7B	01111011	{
92	5C	01011100	yen	124	7C	01111100		
93	5D	01011101	]	125	7D	01111101	}
94	5E	01011110	^	126	7E	01111110	r arr.
95	5F	01011111	_	127	7F	01111111	l arr.

Figure 12.18   **ASCII Character Table**

This table has the codes from 0 to 127, but there are more extensive tables that contain special graphics symbols, international characters, etc. It is best to use the basic codes, as they are supported widely, and should suffice for all controls tasks.

An example of a string of characters encoded in ASCII is shown in Figure 12.1.

e.g. The sequence of numbers below will convert to

A W e e T e s t

A	65
space	32
W	87
e	101
e	101
space	32
T	84
e	101
s	115
t	116

Figure 12.19 ***A String of Characters Encoded in ASCII***

When the characters are organized into a string to be transmitted and *LF* and/or *CR* code are often put at the end to indicate the end of a line. When stored in a computer an ASCII value of zero is used to end the string.

12.2.2 Parity

Errors often occur when data is transmitted or stored. This is very important when transmitting data in noisy factories, over phone lines, etc. Parity bits can be added to data as a simple check of transmitted data for errors. If the data contains error it can be retransmitted, or ignored.

A parity bit is normally a 9th bit added onto an 8 bit byte. When the data is encoded the number of true bits are counted. The parity bit is then set to indicate if there are an even or odd number of true bits. When the byte is decoded the parity bit is checked to make sure it that there are an even or odd number of data bits true. If the parity bit is not satisfied, then the byte is judged to be in error. There are two types of parity, even or odd. These are both based upon an even or odd number of data bits being true. The odd parity bit is true if there are an odd number of bits on in a binary number. On the other hand the Even parity is set if there are an even number of true bits. This is illustrated in Figure 12.1.

	data bits	parity bit
Odd Parity	10101110 10111000	1 0
Even Parity	00101010 10111101	0 1

Figure 12.20 ***Parity Bits on a Byte***

Parity bits are normally suitable for single bytes, but are not reliable for data with a number of bits.

> Note: Control systems perform important tasks that can be dangerous in certain circumstances. If an error occurs there could be serious consequences. As a result error detection methods are very important for control system. When error detection occurs the system should either be *robust* enough to recover from the error, or the system should *fail-safe*. If you ignore these design concepts you will eventually cause an accident.

12.2.3 Checksums

Parity bits are suitable for a few bits of data, but checksums are better for larger data transmissions. These are simply an algebraic sum of all of the data transmitted. Before data is transmitted the numeric values of all of the bytes are added. This sum is then transmitted with the data. At the receiving end the data values are summed again, and the total is compared to the checksum. If they match the data is accepted as good. An example of this method is shown in Figure 12.1.

DATA
    124
    43
    255
    9
    27
    47

CHECKSUM
    505

Figure 12.21   **A Simplistic Checksum**

Checksums are very common in data transmission, but these are also hidden from the average user. If you plan to transmit data to or from a PLC you will need to consider parity and checksum values to verify the data. Small errors in data can have major consequences in received data. Consider an oven temperature transmitted as a binary integer (1023d = 0000 0100 0000 0000b). If a single bit were to be changed, and was not detected the temperature might become (0000 0110 0000 0000b = 1535d) This small change would dramatically change the process.

12.2.4 Gray Code

Parity bits and checksums are for checking data that may have any value. Gray code is used for checking data that must follow a binary sequence. This is common for devices such as angular encoders. The concept is that as the binary number counts up or down, only one bit changes at a time. Thus making it easier to detect erroneous bit changes. An example of a gray code sequence is shown in Figure 12.1. Notice that only one bit changes from one number to the next. If more than a single bit changes between numbers, then an error can be detected.

ASIDE: When the signal level in a wire rises or drops, it induces a magnetic pulse that excites a signal in other nearby lines. This phenomenon is known as *cross-talk*. This signal is often too small to be noticed, but several simultaneous changes, coupled with background noise could result in erroneous values.

---

decimal	gray code
0	0000
1	0001
2	0011
3	0010
4	0110
5	0111
6	0101
7	0100
8	1100
9	1101
10	1111
11	1110
12	1010
13	1011
14	1001
15	1000

Figure 12.22   **Gray Code for a Nibble**

12.3 SUMMARY

- Binary, octal, decimal and hexadecimal numbers were all discussed.
- 2s compliments allow negative binary numbers.
- BCD numbers encode digits in nibbles.
- ASCII values are numerical equivalents for common alphanumeric characters.
- Gray code, parity bits and checksums can be used for error detection.

12.4 PRACTICE PROBLEMS

(Note: Problem solutions are available at http://sites.google.com/site/automatedmanufacturingsystems/)

1. Why are binary, octal and hexadecimal used for computer applications?

2. Is a word is 3 nibbles?

3. What are the specific purpose for Gray code and parity?

4. Convert the following numbers to/from binary

   a) from base 10: 54,321   b) from base 2: 110000101101

5. Convert the BCD number below to a decimal number,

   0110 0010 0111 1001

6. Convert the following binary number to a BCD number,

0100 1011

7. Convert the following binary number to a Hexadecimal value,

0100 1011

8. Convert the following binary number to a octal,

0100 1011

9. Convert the decimal value below to a binary byte, and then determine the odd parity bit,

   97

10. Convert the following from binary to decimal, hexadecimal, BCD and octal.

a)	101101	c)	10000000001
b)	11011011	d)	0010110110101

11. Convert the following from decimal to binary, hexadecimal, BCD and octal.

a)	1	c)	20456
b)	17	d)	-10

12. Convert the following from hexadecimal to binary, decimal, BCD and octal.

a)	1	c)	ABC
b)	17	d)	-A

13. Convert the following from BCD to binary, decimal, hexadecimal and octal.

a)	1001	c)	0011 0110 0001
b)	1001 0011	d)	0000 0101 0111 0100

14. Convert the following from octal to binary, decimal, hexadecimal and BCD.

a)	7	c)	777
b)	17	d)	32634

15.

   a) Represent the decimal value thumb wheel input, 3532, as a Binary Coded Decimal (BCD) and a
   Hexadecimal Value (without using a calculator).
       i) BCD
       ii) Hexadecimal
   b) What is the corresponding decimal value of the BCD value, 1001111010011011?

16. Add/subtract/multiply/divide the following numbers.

a) binary 101101101 + 01010101111011   i) octal 123 - 777

b) hexadecimal 101 + ABC   j) 2s complement bytes 10111011 + 00000011

c) octal 123 + 777   k) 2s complement bytes 00111011 + 00000011

d) binary 110110111 - 0101111   l) binary 101101101 * 10101

e) hexadecimal ABC - 123   m) octal 123 * 777

f) octal 777 - 123   n) octal 777 / 123

g) binary 0101111 - 110110111   o) binary 101101101 / 10101

h) hexadecimal 123-ABC   p) hexadecimal ABC / 123

17. Do the following operations with 8 bit bytes, and indicate the condition of the overflow and carry bits.

a) 10111011 + 00000011   d) 110110111 - 01011111

b) 00111011 + 00000011   e) 01101011 + 01111011

c) 11011011 + 11011111   f) 10110110 - 11101110

18. Consider the three BCD numbers listed below.

    1001 0110 0101 0001
    0010 0100 0011 1000
    0100 0011 0101 0001

    a) Convert these numbers to their decimal values.
    b) Convert the decimal values to binary.
    c) Calculate a checksum for all three binary numbers.
    d) What would the even parity bits be for the binary words found in b).

19. Is the 2nd bit set in the hexadecimal value F49?

20. Explain where grey code occurs when creating Karnaugh maps.

21. Convert the decimal number 1000 to a binary number, and then to hexadecimal.

12.5 ASSIGNMENT PROBLEMS

1. Why are hexadecimal numbers useful when working with PLCs?

12.1 PRACTICE PROBLEM SOLUTIONS

1. base 2, 4, 8, and 16 numbers translate more naturally to the numbers stored in the computer.

2. no, it is four nibbles

3. Both of these are coding schemes designed to increase immunity to noise. A parity bit can be used to check for a changed bit in a byte. Gray code can be used to check for a value error in a stream of continuous values.

4. a) 1101 0100 0011 0001, b) 3117

5. 6279

6. 0111 0101

7. 4B

8. 113

9. 1100001 odd parity bit = 1

10.

binary	101101	11011011	10000000001	0010110110101
BCD	0100 0101	0010 0001 1001	0001 0000 0010 0101	0001 0100 0110 0001
decimal	45	219	1025	1461
hex	2D	DB	401	5B5
octal	55	333	2001	2665

11.

decimal	1	17	20456	-10
BCD	0001	0001 0111	0010 0000 0100 0101 0110	-0001 0000
binary	1	10001	0100 1111 1110 1000	1111 1111 1111 0110
hex	1	11	4FE8	FFF6
octal	1	21	47750	177766

12.

hex	1	17	ABC	-A
BCD	0001	0010 0011	0010 0111 0100 1000	-0001 0000
binary	1	10111	0000 1010 1011 1100	1111 1111 1111 0110
decimal	1	23	2748	-10
octal	1	27	5274	177766

13.

BCD	1001	1001 0011	0011 0110 0001	0000 0101 0111 0100
binary	1001	101 1101	1 0110 1001	10 0011 1110
decimal	9	93	361	0574
hex	9	5D	169	23E
octal	11	135	551	1076

14.

octal	7	17	777	32634
binary	111	1111	1 1111 1111	0011 0101 1001 1100
decimal	7	15	511	13724
hex	7	F	1FF	359C
BCD	0111	0001 0101	0101 0001 0001	0001 0011 0111 0010 0100

15. a) 3532 = 0011 0101 0011 0010 = DCC, b0 the number is not a valid BCD

16.

a) 0001 0110 1110 1000   i) -654

b) BBD   j) 0000 0001 0111 1010

c) 1122   k) 0000 0000 0011 1110

d) 0000 0001 1000 1000   l) 0001 1101 1111 0001

e) 999   m) 122655

f) 654   n) 6

g) 1111 1110 0111 1000   o) 0000 0000 0001 0001

h) -999   p) 9

17.

a) 10111011 + 00000011=1011 1110   d) 110110111 - 01011111=0101 1000+C+O

b) 00111011 + 00000011=0011 1110   e) 01101011 + 01111011=1110 0110

c) 11011011 + 11011111=1011 1010+C+O   f) 10110110 - 11101110=1100 1000

18. a) 9651, 2438, 4351, b) 0010 0101 1011 0011, 0000 1001 1000 0110, 0001 0000 1111 1111, c) 16440, d) 1, 0, 0

19. The binary value is 1111 0100 1001, so the second bit is 0

20. when selecting the sequence of bit changes for Karnaugh maps, only one bit is changed at a time. This is the same method used for grey code number sequences. By using the code the bits in the map are naturally grouped.

21.

$$1000_{10} = 1111101000_2 = 3e8_{16}$$

13PLC MEMORY

Topics:

- ControlLogix memory types; program and data
- Data types; output, input, status, bit, timer, counter, integer, floating point, etc.
- Memory addresses; words, bits, data files, expressions, literal values and indirect.

Objectives:

- To know the basic memory types available
- To be able to use addresses for locations in memory

Advanced ladder logic functions such as timers and counters allow controllers to perform calculations, make decisions and do other complex tasks. They are more complex than basic input contacts and output coils and they rely upon data stored in the memory of the PLC. The memory of the PLC is organized to hold different types of programs and data. This chapter will discuss these memory types. Functions that use them will be discussed in following chapters.

13.1 PROGRAM VS VARIABLE MEMORY

The memory in a PLC is divided into program and variable memory. The program memory contains the instructions to be executed and cannot be changed while the PLC is running. (Note: some PLCs allow on-line editing to make minor program changes while a program is running.) The variable memory is changed while the PLC is running. In ControlLogix the memory is defined using variable names (also called tags and aliases).

ASIDE: In older Allen Bradley PLCs the memory was often organized as files. There are two fundamental types of memory used in Allen-Bradley PLCs - Program and Data memory. Memory is organized into blocks of up to 1000 elements in an array called a file. The Program file holds programs, such as ladder logic. There are eight Data files defined by default, but additional data files can be added if they are needed.

Program Files

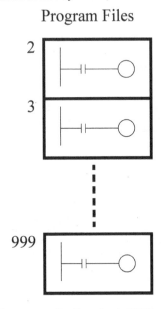

2	
3	
999	

Data Files

O0	Outputs
I1	Inputs
S2	Status
B3	Bits
T4	Timers
C5	Counters
R6	Control
N7	Integer
F8	Float

These are a collection of up to 1000 slots to store up to 1000 programs. The main program will be stored in program file 2. SFC programs must be in file 1, and file 0 is used for program and password information. All other program files from 3 to 999 can be used for *subroutines*.

This is where the variable data is stored that the PLC programs operate on. This is quite complicated, so a detailed explanation follows.

13.2 PROGRAMS

The PLC has a list of 'Main Tasks' that contain the main program(s) run each scan of the PLC. Additional programs can be created that are called as subroutines. Valid program types include Ladder Logic, Structured Text, Sequential Function Charts, and Function Block Diagrams.

Program files can also be created for 'Power-Up Handling' and 'Controller Faults'. The power-up programs are used to initialize the controller on the first scan. In previous chapters this was done in the main program using the 'S:FS' bit. Fault programs are used to respond to specific failures or issues that may lead to failure of the control system. Normally these programs are used to recover from minor failures, or shut down a system safely.

13.3 VARIABLES (TAGS)

Allen Bradley uses the terminology 'tags' to describe variables, status, and input/output (I/O) values for the controller. 'Controller Tags' include status values and I/O definitions. These are scoped, meaning that they can be global and used by all programs on the PLC. These can also be local, limiting their use to a program that owns it.

Variable tags can be an alias for another tags, or be given a data type. Some of the common tag types are listed below.

Type	Description
BOOL	Holds TRUE or FALSE values
CONTROL	General purpose memory for complex instructions
COUNTER	Counter memory
DINT	32 bit 2s compliment integer -2,147,483,648 to 2,147,483,647
INT	16 bit 2s compliment integer -32,768 to 32,767
MESSAGE	Used for communication with remote devices
PID	Used for PID control functions
REAL	32 bit floating point value +/-1.1754944e-38 to +/-3.4028237e38
SINT	8 bit 2s compliment integer -128 to 127
STRING	An ASCII string
TIMER	Timer memory

Figure 13.1   *Selected ControlLogic Data Types*

For older Allen Bradley PLCs data files are used for storing different information types, as shown below. These locations are numbered from 0 to 999. The letter in front of the number indicates the data type. For example, *F8:* is read as *floating point numbers* in *data file 8*. Numbers are not given for *O:* and *I:*, but they are implied to be *O0:* and *I1:*. The number that follows the *:* is the location number. Each file may contain from 0 to 999 locations that may store values. For the input *I:* and output *O:* files the locations are converted to physical locations on the PLC using rack and slot numbers. The addresses that can be used will depend upon the hardware configuration. The status *S2:* file is more complex and is discussed later. The other memory locations are simply slots to store data in. For example, *F8:35* would indicate the 36th value in the 8th data file which is floating point numbers.

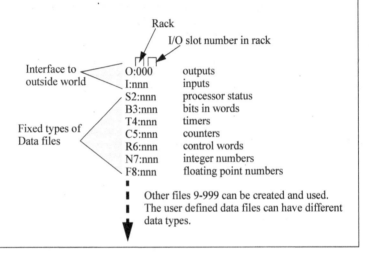

Rack

I/O slot number in rack

Interface to outside world

O:000   outputs
I:nnn   inputs
S2:nnn   processor status
B3:nnn   bits in words
T4:nnn   timers

Fixed types of Data files

C5:nnn   counters
R6:nnn   control words
N7:nnn   integer numbers
F8:nnn   floating point numbers

■   Other files 9-999 can be created and used.
■   The user defined data files can have different data types.

Data values do not always need to be stored in memory, they can be define literally. Figure 13.1 shows an example of two different data values. The first is an integer, the second is a real number. Hexadecimal numbers can be indicated by following the number with *H*, a leading zero is also needed when the first digit is *A, B, C, D, E* or *F*. A binary number is indicated by adding a *B* to the end of the number.

8 - an integer
8.5 - a floating point number
08FH - a hexadecimal value *8F*
01101101B - a binary number *01101101*

Figure 13.2   ***Literal Data Values***

Data types can be created in variable size 1D, 2D, or 3D arrays.

Sometimes we will want to refer to an array of values, as shown in Figure 13.1. This data type is indicated by beginning the number with a pound or hash sign '#'. The first example describes an array of floating point numbers staring in file *8* at location *5*. The second example is for an array of integers in file *7* starting at location *0*. The length of the array is determined elsewhere.

test[1, 4] - returns the value in the 2nd row and 5th column of array test

Expressions allow addresses and functions to be typed in and interpreted when the program is run. The example in Figure 13.1 will get a floating point number from 'test', perform a sine transformation, and then add 1.3. The text string is not interpreted until the PLC is running, and if there is an error, it may not occur until the program is running - so use this function cautiously.

expression - a text string that describes a complex operation.

"sin(test) + 1.3" - a simple calculation

Figure 13.4   **Expressions**

These data types and addressing modes will be discussed more as applicable functions are presented later in this chapter and book.

Figure 13.1 shows a simple example ladder logic with functions. The basic operation is such that while input *A* is true the functions will be performed. The first statement will move (MOV) the literal value of *130* into integer memory *X*. The next move function will copy the value from *X* to *Y*. The third statement will add integers value in *X* and *Y* and store the results in *Z*.

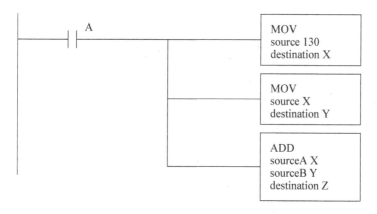

Figure 13.5   **An Example of Ladder Logic Functions**

13.3.1 Timer and Counter Memory

Previous chapters have discussed the basic operation of timers and counters. The ability to address their memory directly allows some powerful tools. The bits and words for timers are;

EN - timer enabled bit
TT - timer timing bit

279

DN - timer done bit
FS - timer first scan
LS - timer last scan
OV - timer value overflowed
ER - timer error
PRE - preset word
ACC - accumulated time word

Counter have the following bits and words.

CU - count up bit
CD - count down bit
DN - counter done bit
OV - overflow bit
UN - underflow bit
PRE - preset word
ACC - accumulated count word

As discussed before we can access timer and counter bits and words. Examples of these are shown in Figure 13.1. The bit values can only be read, and should not be changed. The presets and accumulators can be read and overwritten.

Words

timer.PRE - the preset value for timer T4:0
timer.ACC - the accumulated value for timer T4:0
counter.PRE - the preset value for counter C5:0
counter.ACC - the accumulated value for counter C5:0

Bits

timer.EN - indicates when the input to timer T4:0 is true
timer.TT - indicates when the timer T4:0 is counting
timer.DN - indicates when timer T4:0 has reached the maximum
counter.CU - indicates when the count up instruction is true for C5:0
counter.CD - indicates when the count down instruction is true for C5:0
counter.DN - indicates when the counter C5:0 has reached the preset
counter.OV - indicates when the counter C5:0 passes the maximum value (2,147,483,647)
counter.UN - indicates when the counter C5:0 passes the minimum value (-2,147,483,648)

Figure 13.6   *Examples of Timer and Counter Addresses*

Consider the simple ladder logic example in Figure 13.1. It shows the use of a timer timing *TT* bit to seal on the timer when a door input has gone true. While the timer is counting, the bit will stay true and keep the timer counting. When it reaches the 10 second delay the *TT* bit will turn off. The next line of ladder logic will turn on a light while the timer is counting for the first 10 seconds.

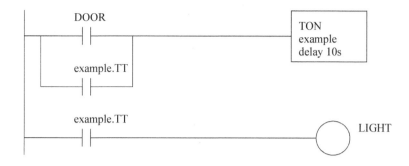

Figure 13.7 **Door Light Example**

13.3.2 PLC Status Bits

Status memory allows a program to check the PLC operation, and also make some changes. A selected list of status bits is shown in Figure 13.1 for Allen-Bradley ControlLogix PLCs. More complete lists are available in the manuals. The first six bits are commonly used and are given simple designations for use with simple ladder logic. More advanced instructions require the use of Get System Value (GSV) and Set System Value (SSV) functions. These functions can get/set different values depending upon the type of data object is being used. In the sample list given one data object is the 'WALLCLOCKTIME'. One of the attributes of the class is the DateTime that contains the current time. It is also possible to use the 'PROGRAM' object instance 'MainProgram' attribute 'Last-ScanTime' to determine how long the program took to run in the previous scan.

Immediately accessible status values

S:FS - First Scan Flag
S:N - The last calculation resulted in a negative value
S:Z - The last calculation resulted in a zero
S:V - The last calculation resulted in an overflow
S:C - The last calculation resulted in a carry
S:MINOR - A minor (non-critical/recoverable) error has occurred

Examples of SOME values available using the GSV and SSV functions

CONTROLLERDEVICE - information about the PLC
PROGRAM - information about the program running
            LastScanTime
            MaxScanTime
TASK
            EnableTimeout
            LastScanTime
            MaxScanTime
            Priority
            StartTime
            Watchdog
WALLCLOCKTIME - the current time
            DateTime
                        DINT[0] - year
                        DINT[1] - month 1=january
                        DINT[2] - day 1 to 31
                        DINT[3] - hour 0 to 24
                        DINT[4] - minute 0 to 59
                        DINT[5] - second 0 to 59
                        DINT[6] - microseconds 0 to 999,999

Figure 13.8   *Status Bits and Words for ControlLogix*

An example of getting and setting system status values is shown in Figure 13.1. The first line of ladder logic will get the current time from the class 'WALLCLOCKTIME'. In this case the class does not have an instance so it is blank. The attribute being recalled is the DateTime that will be written to the DINT array time[0..6]. For example 'time[3]' should give the current hour. In the second line the Watchdog time for the MainProgram is set to 200 ms. If the program MainProgram takes longer than 200ms to execute a fault will be generated.

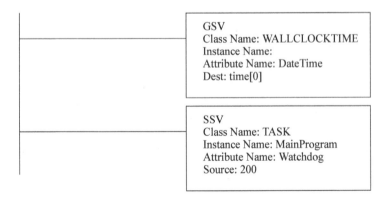

Figure 13.9 **Reading and Setting Status bits with GSV and SSV**

As always, additional classes and attributes for the status values can be found in the manuals for the processors and instructions being used.

A selected list of status bits is shown below for Allen-Bradley Micrologic and PLC-5 PLCs. More complete lists are available in the manuals. For example the first four bits $S2:0/x$ indicate the results of calculations, including carry, overflow, zero and negative/sign. The $S2:1/15$ will be true once when the PLC is turned on - this is the first scan bit. The time for the last scan will be stored in $S2:8$. The date and clock can be stored and read from locations $S2:18$ to $S2:23$.

S2:0/0 carry in math operation
S2:0/1 overflow in math operation
S2:0/2 zero in math operation
S2:0/3 sign in math operation
S2:1/15 first scan of program file
S2:8 the scan time (ms)
S2:18 year
S2:19 month
S2:20 day
S2:21 hour
S2:22 minute
S2:23 second
S2:28 watchdog setpoint
S2:29 fault routine file number
S2:30 STI (selectable timed interrupt) setpoint
S2:31 STI file number
S2:46-S2:54,S2:55-S2:56 PII (Programmable Input Interrupt) settings
S2:55 STI last scan time (ms)
S2:77 communication scan time (ms)

13.3.3 User Function Control Memory

Simple ladder logic functions can complete operations in a single scan of ladder logic. Other functions such as timers and counters will require multiple ladder logic scans to finish. While timers and counters have their own memory for control, a generic type of control memory is defined for other function. This memory contains the bits and words in Figure 13.1. Any given function will only use some of the values. The meaning of particular bits and words will be described later when discussing specific functions.

EN - enable bit
EU - enable unload
DN - done bit
EM - empty bit
ER - error bit
UL - unload bit
IN - inhibit bit
FD - found bit
LEN - length word
POS - position word

Figure 13.10 **Bits and Words for Control Memory**

13.4 SUMMARY

- Program are given unique names and can be for power-up, regular scans, and faults.
- Tags and aliases are used for naming variables and I/O.
- Files are like arrays and are indicated with [].
- Expressions allow equations to be typed in.
- Literal values for binary and hexadecimal values are followed by *B* and *H*.

13.5 PRACTICE PROBLEMS

(Note: Problem solutions are available at http://sites.google.com/site/automatedmanufacturingsystems/)

1. How are timer and counter memory similar?

2. What types of memory cannot be changed?

3. Develop Ladder Logic for a car door/seat belt safety system. When the car door is open, or the seatbelt is not done up, a buzzer will sound for 5 seconds if the key has been switched on. A cabin light will be switched on when the door is open and stay on for 10 seconds after it is closed, unless a key has started the ignition power.

4. Write ladder logic for the following problem description. When button *A* is pressed a value of 1001 will be stored in *X*. When button *B* is pressed a value of -345 will be stored in *Y*, when it is not pressed a value of 99 will be stored in *Y*. When button *C* is pressed *X* and *Y* will be added, and the result will be stored in *Z*.

5. Using the status memory locations, write a program that will flash a light for the first 15 seconds after it has been turned on. The light should flash once a second.

6. How many words are required for timer and counter memory?

7. A machine is being designed for a foreign parts supplier. As part of the contractual agreement the logic will run until February 26, 2008. However, after that date the machine will enable a 'contract_expired' value and no longer run. Write the ladder logic.

13.6 ASSIGNMENT PROBLEMS

1. Could timer 'T' and counter 'C' memory types be replaced with control 'R' memory types? Explain your answer.

13.1 PRACTICE PROBLEM SOLUTIONS

1. both are similar. The timer and counter memories both use double words for the accumulator and presets, and they use bits to track the status of the functions. These bits are somewhat different, but parallel in function.

2. Inputs cannot be changed by the program, and some of the status bits/words cannot be changed by the user.

3.

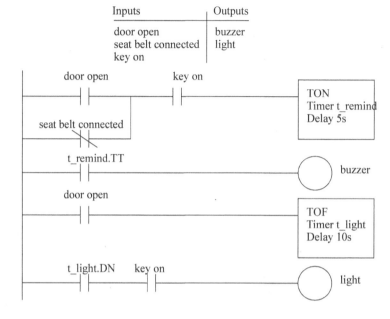

4.

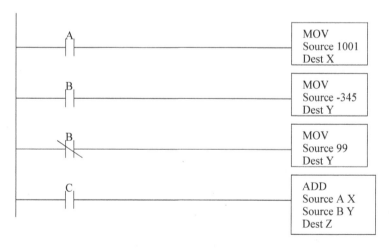

5.

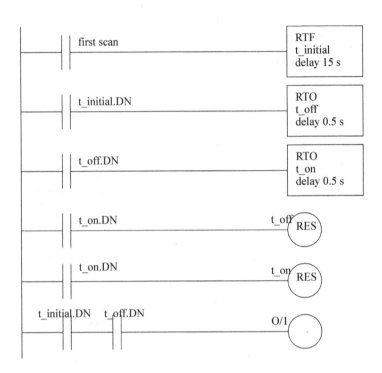

first scan	RTF t_initial delay 15 s
t_initial.DN	RTO t_off delay 0.5 s
t_off.DN	RTO t_on delay 0.5 s

6. three long words (3 * 32 bits) are used for a timer or a counter.

7.

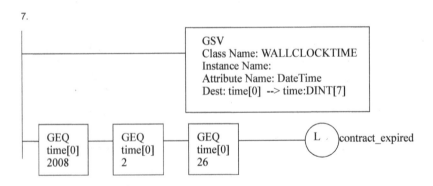

14LADDER LOGIC FUNCTIONS

Topics:

- Functions for data handling, mathematics, conversions, array operations, statistics, comparison and Boolean operations.
- Design examples

Objectives:

- To understand basic functions that allow calculations and comparisons
- To understand array functions using memory files

Ladder logic input contacts and output coils allow simple logical decisions. Functions extend basic ladder logic to allow other types of control. For example, the addition of timers and counters allowed event based control. A longer list of functions is shown in Figure 14.1. Combinatorial Logic and Event functions have already been covered. This chapter will discuss Data Handling and Numerical Logic. The next chapter will cover Lists and Program Control and some of the Input and Output functions. Remaining functions will be discussed in later chapters.

Combinatorial Logic
- relay contacts and coils
Events
- timer instructions
- counter instructions
Data Handling
- moves
- mathematics
- conversions
Numerical Logic
- boolean operations
- comparisons
Lists
- shift registers/stacks
- sequencers
Program Control
- branching/looping
- immediate inputs/outputs
- fault/interrupt detection
Input and Output
- PID
- communications
- high speed counters
- ASCII string functions

Figure 14.1   *Basic PLC Function Categories*

Most of the functions will use PLC memory locations to get values, store values and track function status. Most function will normally become active when the input is true. But, some functions, such as TOF timers, can remain active when the input is off. Other functions will only operate when the input goes from false to true, this is known as positive edge triggered. Consider a counter that only counts when the input goes from false to true, the length of time the input is true does not change the function behavior. A negative edge triggered function would be triggered when the input goes from true to false. Most functions are not edge triggered: unless stated assume functions are not edge triggered.

14.1 DATA HANDLING

14.1.1 Move Functions

There are two basic types of move functions;

MOV(value,destination) - moves a value to a memory location
MVM(value,mask,destination) - moves a value to a memory location, but with a mask to select
specific bits.

The simple MOV will take a value from one location in memory and place it in another memory location.
Examples of the basic MOV are given in Figure 14.1. When *A* is true the MOV function moves a floating point number from the source to the destination address. The data in the source address is left unchanged. When *B* is true
the floating point number in the source will be converted to an integer and stored in the destination address in integer memory. The floating point number will be rounded up or down to the nearest integer. When *C* is true the integer value of 123 will be placed in the integer file *test_int*.

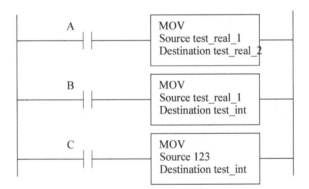

Figure 14.2   ***Examples of the MOV Function***

A more complex example of move functions is given in Figure 14.1. When *A* becomes true the first move
statement will move the value of 130 into *int_0*. And, the second move statement will move the value of -9385 from
int_1 to *int_2*. (Note: The number is shown as negative because we are using 2s compliment.) For the simple
MOVs the binary values are not needed, but for the MVM statement the binary values are essential. The statement
moves the binary bits from *int_3* to *int_5*, but only those bits that are also on in the mask *int_4*, other bits in the des-

tination will be left untouched. Notice that the first bit *int_5.0* is true in the destination address before and after, but it is not true in the mask. The MVM function is very useful for applications where individual binary bits are to be manipulated, but they are less useful when dealing with actual number values.

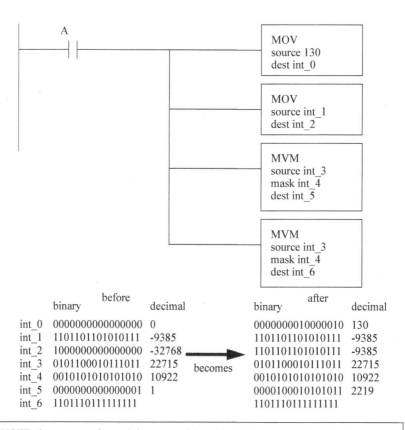

	before binary	decimal		after binary	decimal
int_0	0000000000000000	0		0000000010000010	130
int_1	1101101101010111	-9385		1101101101010111	-9385
int_2	1000000000000000	-32768	becomes	1101101101010111	-9385
int_3	0101100010111011	22715		0101100010111011	22715
int_4	0010101010101010	10922		0010101010101010	10922
int_5	0000000000000001	1		0000100010101011	2219
int_6	1101110111111111			1101110111111111	

NOTE: the concept of a mask is very useful, and it will be used in other functions. Masks allow instructions to change a couple of bits in a binary number without having to change the entire number. You might want to do this when you are using bits in a number to represent states, modes, status, etc.

Figure 14.3   ***Example of the MOV and MVM Statement with Binary Values***

14.1.2 Mathematical Functions

Mathematical functions will retrieve one or more values, perform an operation and store the result in memory. Figure 14.1 shows an *ADD* function that will retrieve values from *int_1* and *real_1*, convert them both to the type of the destination address, add the floating point numbers, and store the result in *real_2*. The function has two sources labelled *source A* and *source B*. In the case of ADD functions the sequence can change, but this is not true for other operations such as subtraction and division. A list of other simple arithmetic function follows. Some of the functions, such as the negative function are unary, so there is only one source.

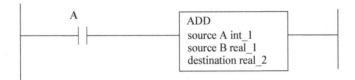

ADD(value,value,destination) - add two values
SUB(value,value,destination) - subtract
MUL(value,value,destination) - multiply
DIV(value,value,destination) - divide
NEG(value,destination) - reverse sign from positive/negative
CLR(value) - clear the memory location

NOTE: To save space the function types are shown in the shortened notation above. For example the function *ADD(value, value, destination)* requires two source values and will store it in a destination. It will use this notation in a few places to reduce the bulk of the function descriptions.

Figure 14.4   **Arithmetic Functions**

An application of the arithmetic function is shown in Figure 14.1. Most of the operations provide the results we would expect. The second ADD function retrieves a value from *int_3*, adds 1 and overwrites the source - this is normally known as an increment operation. The first DIV statement divides the integer 25 by 10, the result is rounded to the nearest integer, in this case 3, and the result is stored in *int_6*. The *NEG* instruction takes the new value of *-10*, not the original value of *0*, from *int_4* inverts the sign and stores it in *int_7*.

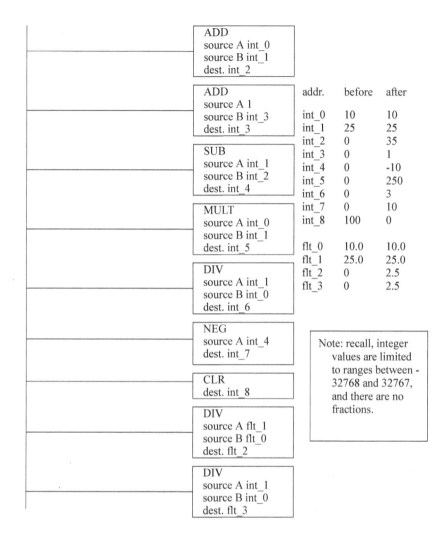

	ADD source A int_0 source B int_1 dest. int_2			

	ADD source A 1 source B int_3 dest. int_3	addr.	before	after
		int_0	10	10
		int_1	25	25
		int_2	0	35
	SUB source A int_1 source B int_2 dest. int_4	int_3	0	1
		int_4	0	-10
		int_5	0	250
		int_6	0	3
	MULT source A int_0 source B int_1 dest. int_5	int_7	0	10
		int_8	100	0
		flt_0	10.0	10.0
	DIV source A int_1 source B int_0 dest. int_6	flt_1	25.0	25.0
		flt_2	0	2.5
		flt_3	0	2.5

NEG
source A int_4
dest. int_7

Note: recall, integer
values are limited
to ranges between -
32768 and 32767,
and there are no
fractions.

CLR
dest. int_8

DIV
source A flt_1
source B flt_0
dest. flt_2

DIV
source A int_1
source B int_0
dest. flt_3

Figure 14.5   ***Arithmetic Function Example***

A list of more advanced functions are given in Figure 14.1. This list includes basic trigonometry functions, exponents, logarithms and a square root function. The last function *CPT* will accept an expression and perform a complex calculation.

ACS(value,destination) - inverse cosine
COS(value,destination) - cosine
ASN(value,destination) - inverse sine
SIN(value,destination) - sine
ATN(value,destination) - inverse tangent
TAN(value,destination) - tangent
XPY(value,value,destination) - X to the power of Y
LN(value,destination) - natural log
LOG(value,destination) - base 10 log
SQR(value,destination) - square root
CPT(destination,expression) - does a calculation

Figure 14.6   *Advanced Mathematical Functions*

Figure 14.1 shows an example where an equation has been converted to ladder logic. The first step in the conversion is to convert the variables in the equation to unused memory locations in the PLC. The equation can then be converted using the most nested calculations in the equation, such as the *LN* function. In this case the results of the *LN* function are stored in another memory location, to be recalled later. The other operations are implemented in a similar manner. (Note: This equation could have been implemented in other forms, using fewer memory locations.)

given

$$A = \sqrt{\ln B + e^C \text{acos}(D)}$$

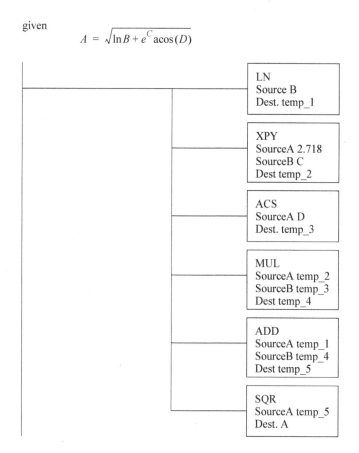

LN
Source B
Dest. temp_1

XPY
SourceA 2.718
SourceB C
Dest temp_2

ACS
SourceA D
Dest. temp_3

MUL
SourceA temp_2
SourceB temp_3
Dest temp_4

ADD
SourceA temp_1
SourceB temp_4
Dest temp_5

SQR
SourceA temp_5
Dest. A

Figure 14.7   ***An Equation in Ladder Logic***

The same equation in Figure 14.1 could have been implemented with a CPT function as shown in Figure 14.1. The equation uses the same memory locations chosen in Figure 14.1. The expression is typed directly into the PLC programming software.

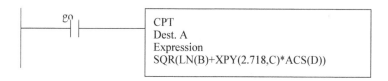

CPT
Dest. A
Expression
SQR(LN(B)+XPY(2.718,C)*ACS(D))

Figure 14.8   ***Calculations with a Compute Function***

Math functions can result in status flags such as overflow, carry, etc. care must be taken to avoid problems such as overflows. These problems are less common when using floating point numbers. Integers are more prone to these problems because they are limited to the range.

293

14.1.3 Conversions

Ladder logic conversion functions are listed in Figure 14.1. The example function will retrieve a BCD number from the *D* type (BCD) memory and convert it to a floating point number that will be stored in *F8:2*. The other function will convert from 2s compliment binary to BCD, and between radians and degrees.

TOD(value,destination) - convert from BCD to 2s compliment
FRD(value,destination) - convert from 2s compliment to BCD
DEG(value,destination) - convert from radians to degrees
RAD(value,destination) - convert from degrees to radians

Figure 14.9   *Conversion Functions*

Examples of the conversion functions are given in Figure 14.1. The functions load in a source value, do the conversion, and store the results. The TOD conversion to BCD could result in an overflow error.

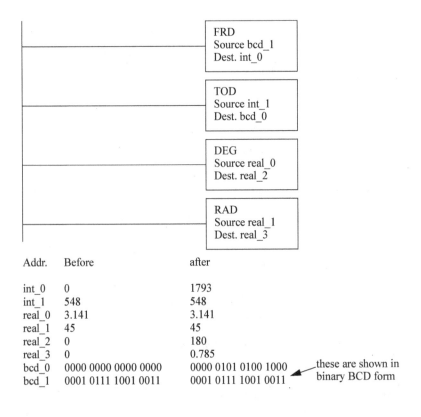

Addr.	Before	after	
int_0	0	1793	
int_1	548	548	
real_0	3.141	3.141	
real_1	45	45	
real_2	0	180	
real_3	0	0.785	
bcd_0	0000 0000 0000 0000	0000 0101 0100 1000	these are shown in
bcd_1	0001 0111 1001 0011	0001 0111 1001 0011	binary BCD form

Figure 14.10   *Conversion Example*

14.1.4 Array Data Functions

Arrays allow us to store multiple data values. In a PLC this will be a sequential series of numbers in integer, floating point, or other memory. For example, assume we are measuring and storing the weight of a bag of chips in floating point memory starting at *weight[0]*. We could read a weight value every 10 minutes, and once every hour find the average of the six weights. This section will focus on techniques that manipulate groups of data organized in arrays, also called blocks in the manuals.

14.1.4.1 - Statistics

Functions are available that allow statistical calculations. These functions are listed in Figure 14.1. When *A* becomes true the average (AVE) conversion will start at memory location *weight[0]* and average a total of *4* values. The control word *weight_control* is used to keep track of the progress of the operation, and to determine when the operation is complete. This operation, and the others, are edge triggered. The operation may require multiple scans to be completed. When the operation is done the average will be stored in *weight_avg* and the *weight_control.DN* bit will be turned on.

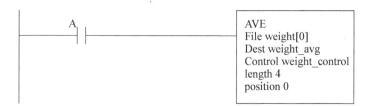

```
        A                          AVE
       ┤ ├                         File weight[0]
                                   Dest weight_avg
                                   Control weight_control
                                   length 4
                                   position 0
```

AVE(start value,destination,control,length) - average of values
STD(start value,destination,control,length) - standard deviation of values
SRT(start value,control,length) - sort a list of values

Figure 14.11   *Statistic Functions*

Examples of the statistical functions are given in Figure 14.1 for an array of data that starts at *weight[0]* and is 4 values long. When done the average will be stored in *weight_avg*, and the standard deviation will be stored in *weight_std*. The set of values will also be sorted in ascending order from weight[0] to weight[3]. Each of the function should have their own control memory to prevent overlap. It is not a good idea to activate the sort and the other calculations at the same time, as the sort may move values during the calculation, resulting in incorrect calculations.

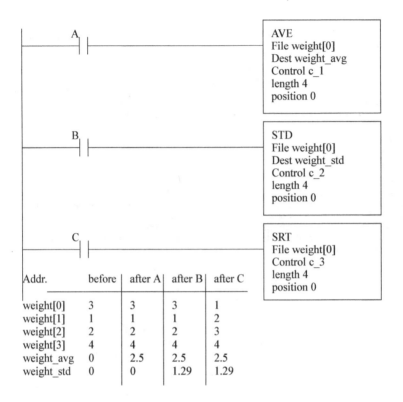

Addr.	before	after A	after B	after C
weight[0]	3	3	3	1
weight[1]	1	1	1	2
weight[2]	2	2	2	3
weight[3]	4	4	4	4
weight_avg	0	2.5	2.5	2.5
weight_std	0	0	1.29	1.29

Figure 14.12   ***Statistical Calculations***

ASIDE: These function will allow a real-time calculation of SPC data for control limits, etc. The only PLC function missing is a random function that would allow random sample times.

14.1.4.2 - Block Operations

A basic block function is shown in Figure 14.1. This COP (copy) function will copy an array of 10 values starting at *n[50]* to *n[40]*. The *FAL* function will perform mathematical operations using an expression string, and the *FSC* function will allow two arrays to be compared using an expression. The *FLL* function will fill a block of memory with a single value.

COP(start value,destination,length) - copies a block of values

FAL(control,length,mode,destination,expression) - will perform basic math operations to multiple values.

FSC(control,length,mode,expression) - will do a comparison to multiple values

FLL(value,destination,length) - copies a single value to a block of memory

Figure 14.13   **Block Operation Functions**

Figure 14.1 shows an example of the *FAL* function with different addressing modes. The first FAL function will do the following calculations $n[5]=n[0]+5$, $n[6]=n[1]+5$, $n[7]=n[2]+5$, $n[7]=n[3]+5$, $n[9]=n[4]+5$. The second FAL statement will be $n[5]=n[0]+5$, $n[6]=n[0]+5$, $n[7]=n[0]+5$, $n[7]=n[0]+5$, $n[9]=n[0]+5$. With a mode of 2 the instruction will do two of the calculations when there is a positive edge from B (i.e., a transition from false to true). The result of the last FAL statement will be $n[5]=n[0]+5$, $n[5]=n[1]+5$, $n[5]=n[2]+5$, $n[5]=n[3]+5$, $n[5]=n[4]+5$. The last operation would seem to be useless, but notice that the mode is *incremental*. This mode will do one calculation for each positive transition of C. The *all* mode will perform all five calculations in a single scan whenever there is a positive edge on the input. It is also possible to put in a number that will indicate the number of calculations per scan. The calculation time can be long for large arrays and trying to do all of the calculations in one scan may lead to a watchdog time-out fault.

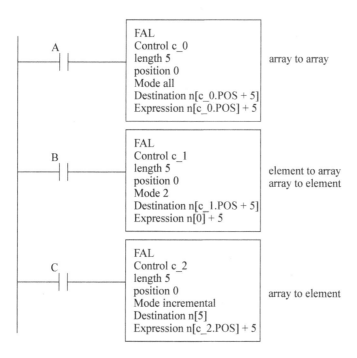

Figure 14.14   **File Algebra Example**

14.2 LOGICAL FUNCTIONS

14.2.1 Comparison of Values

Comparison functions are shown in Figure 14.1. Previous function blocks were outputs, these replace input contacts. The example shows an EQU (equal) function that compares two floating point numbers. If the numbers are equal, the output bit *light* is true, otherwise it is false. Other types of equality functions are also listed.

EQU(value,value) - equal
NEQ(value,value) - not equal
LES(value,value) - less than
LEQ(value,value) - less than or equal
GRT(value,value) - greater than
GEQ(value,value) - greater than or equal
CMP(expression) - compares two values for equality
MEQ(value,mask,threshold) - compare for equality using a mask
LIM(low limit,value,high limit) - check for a value between limits

Figure 14.15   **Comparison Functions**

The example in Figure 14.1 shows the six basic comparison functions. To the right of the figure are examples of the comparison operations.

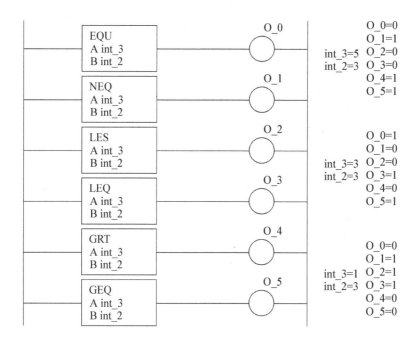

Figure 14.16   Comparison Function Examples

The ladder logic in Figure 14.1 is recreated in Figure 14.1 with the CMP function that allows text expressions.

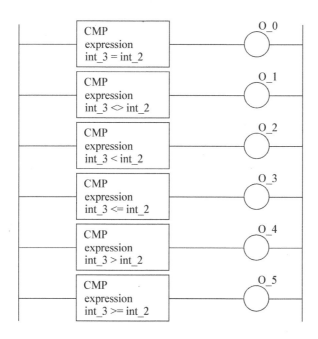

Figure 14.17 **Equivalent Statements Using CMP Statements**

Expressions can also be used to do more complex comparisons, as shown in Figure 14.1. The expression will determine if *B* is between *A* and *C*.

Figure 14.18 *A More Complex Comparison Expression*

The LIM and MEQ functions are shown in Figure 14.1. The first three functions will compare a test value to high and low limits. If the high limit is above the low limit and the test value is between or equal to one limit, then it will be true. If the low limit is above the high limit then the function is only true for test values outside the range. The masked equal will compare the bits of two numbers, but only those bits that are true in the mask.

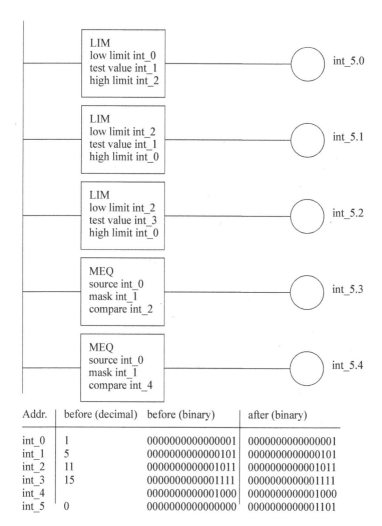

Addr.	before (decimal)	before (binary)	after (binary)
int_0	1	0000000000000001	0000000000000001
int_1	5	0000000000000101	0000000000000101
int_2	11	0000000000001011	0000000000001011
int_3	15	0000000000001111	0000000000001111
int_4		0000000000001000	0000000000001000
int_5	0	0000000000000000	0000000000001101

Figure 14.19　　　*Complex Comparison Functions*

Figure 14.1 shows a numberline that helps determine when the LIM function will be true.

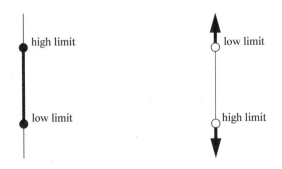

Figure 14.20   *A Number Line for the LIM Function*

File to file comparisons are also permitted using the FSC instruction shown in Figure 14.1. The instruction uses the control word c_0. It will interpret the expression 10 times, doing two comparisons per logic scan (the Mode is 2). The comparisons will be *f[10]<f[0], f[11]<f[0]* then *f[12]<f[0], f[13]<f[0]* then *f[14]<f[0], f[15]<f[0]* then *f[16]<f[0], f[17]<f[0]* then *f[18]<f[0], f[19]<f[0]*. The function will continue until a false statement is found, or the comparison completes. If the comparison completes with no false statements the output *A* will then be true. The mode could have also been *All* to execute all the comparisons in one scan, or *Increment* to update when the input to the function is true - in this case the input is a plain wire, so it will always be true.

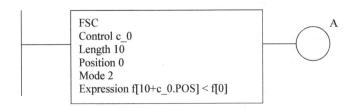

Figure 14.21   *File Comparison Using Expressions*

14.2.2 Boolean Functions

Figure 14.1 shows Boolean algebra functions. The function shown will obtain data words from bit memory, perform an and operation, and store the results in a new location in bit memory. These functions are all oriented to word level operations. The ability to perform Boolean operations allows logical operations on more than a single bit.

AND(value,value,destination) - Binary and function
OR(value,value,destination) - Binary or function
XOR(value,value,destination) - Binary exclusive or function
NOT(value,destination) - Binary not function

302

Figure 14.22   ***Boolean Functions***

The use of the Boolean functions is shown in Figure 14.1. The first three functions require two arguments, while the last function only requires one. The AND function will only turn on bits in the result that are true in both of the source words. The OR function will turn on a bit in the result word if either of the source word bits is on. The XOR function will only turn on a bit in the result word if the bit is on in only one of the source words. The NOT function reverses all of the bits in the source word.

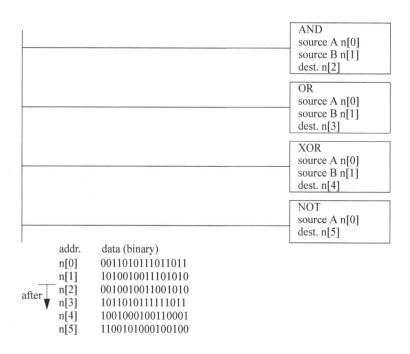

AND
source A n[0]
source B n[1]
dest. n[2]

OR
source A n[0]
source B n[1]
dest. n[3]

XOR
source A n[0]
source B n[1]
dest. n[4]

NOT
source A n[0]
dest. n[5]

addr.	data (binary)
n[0]	0011010111011011
n[1]	1010010011101010
n[2]	0010010011001010
n[3]	1011010111111011
n[4]	1001000100110001
n[5]	1100101000100100

after

Figure 14.23   ***Boolean Function Example***

14.3 DESIGN CASES

14.3.1 Simple Calculation

Problem: A switch will increment a counter on when engaged. This counter can be reset by a second switch. The value in the counter should be multiplied by 2, and then displayed as a BCD output using (O:0.0/0 - O:0.0/7)

303

Solution:

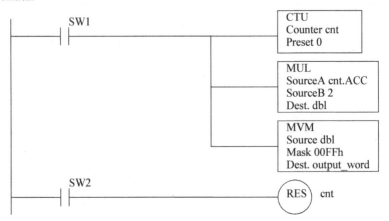

Figure 14.24   *A Simple Calculation Example*

14.3.2 For-Next

Problem: Design a for-next loop that is similar to ones found in traditional programming languages. When *A* is true the ladder logic should be active for 10 scans, and the scan number from 1 to 10 should be stored in n0.

Solution:

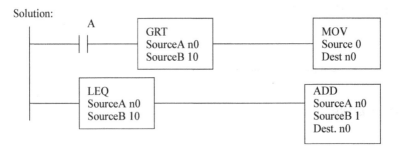

Figure 14.25   *A Simple Comparison Example*

As designed the program differs from traditional loops because it will only complete one 'loop' each time the logic is scanned.

14.3.3 Series Calculation

Problem: Create a ladder logic program that will start when input *A* is turned on and calculate the series below. The value of *n* will start at 1 and with each scan of the ladder logic *n* will increase until n=100. While the sequence is being incremented, any change in *A* will be ignored.

$$x = 2(n-1)$$

Solution:

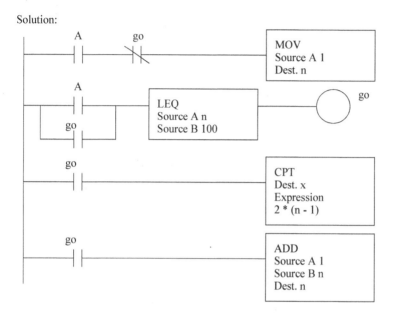

Figure 14.26   **A Series Calculation Example**

14.3.4 Flashing Lights

Problem: We are designing a movie theater marquee, and they want the traditional flashing lights. The lights have been connected to the outputs of the PLC from O[0] to O[17] - an INT. When the PLC is turned, every second light should be on. Every half second the lights should reverse. The result will be that in one second two lights side-by-side will be on half a second each.

Solution:

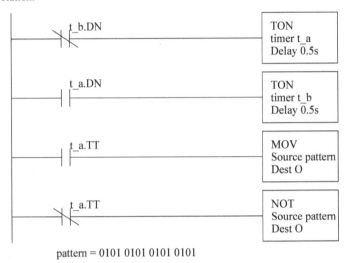

pattern = 0101 0101 0101 0101

Figure 14.27   *A Flashing Light Example*

14.4 SUMMARY

- Functions can get values from memory, do simple operations, and return the results to memory.
- Scientific and statistics math functions are available.
- Masked function allow operations that only change a few bits.
- Expressions can be used to perform more complex operations.
- Conversions are available for angles and BCD numbers.
- Array oriented file commands allow multiple operations in one scan.
- Values can be compared.to make decisions.
- Boolean functions allow bit level operations.
- Function change value in data memory immediately.

14.5 PRACTICE PROBLEMS

(Note: Problem solutions are available at http://sites.google.com/site/automatedmanufacturingsystems/)

1. Do the calculation below with ladder logic,

$$n\_2 = -(5 - n\_0 / n\_1)$$

2. Implement the following function,

$$x = \operatorname{atan}\left(y\left(\frac{y + \log(y)}{y + 1}\right)\right)$$

3. A switch will increment a counter on when engaged. This counter can be reset by a second switch. The value in the counter should be multiplied by 5, and then displayed as a binary output using output integer 'O_lights'.

4. Create a ladder logic program that will start when input A is turned on and calculate the series below. The value of n will start at 0 and with each scan of the ladder logic n will increase by 2 until n=20. While the sequence is being incremented, any change in A will be ignored.

$$x = 2(\log(n) - 1)$$

5. The following program uses indirect addressing. Indicate what the new values in memory will be when button A is pushed after the first and second instructions.

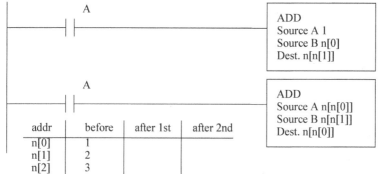

addr	before	after 1st	after 2nd
n[0]	1		
n[1]	2		
n[2]	3		

First instruction:
```
        A                          ADD
     ──┤ ├──────────────────       Source A 1
                                    Source B n[0]
                                    Dest. n[n[1]]
```

Second instruction:
```
        A                          ADD
     ──┤ ├──────────────────       Source A n[n[0]]
                                    Source B n[n[1]]
                                    Dest. n[n[0]]
```

6. A thumbwheel input card acquires a four digit BCD count. A sensor detects parts dropping down a chute. When the count matches the BCD value the chute is closed, and a light is turned on until a reset button is pushed. A start button must be pushed to start the part feeding. Develop the ladder logic for this controller. Use a structured design technique such as a state diagram.

Inputs

bcd_in - BCD input card
part_detect
start_button
reset_button

Outputs

chute_open
light

7. Describe the difference between incremental, all and a number for file oriented instruction, such as *FAL*.

8. What is the maximum number of elements that moved with a file instruction? What might happen if too many are transferred in one scan?

9. Write a ladder logic program to do the following calculation. If the result is greater than 20.0, then the output 'solenoid' will be turned on.

$$A = D - Be^{-\frac{T}{C}}$$

10. Write ladder logic to reset an RTO counter (timer) without using the RES instruction.

11. Write a program that will use Boolean operations and comparison functions to determine if bits 9, 4 and 2 are set in the input word *input_card*. If they are set, turn on output bit *match*.

12. Explain how the mask works in the following MVM function. Develop a Boolean equation.

```
        ────────────────────       MVM
                                    Source S
                                    Mask M
                                    Dest D
```

13. A machine is being designed for a foreign parts supplier. As part of the contractual agreement the logic will run until February 26, 2008. However, after that date the machine will enable a 'contract_expired' value and no lon-

ger run. Write the ladder logic.

14. Use an FAL instruction to average the values in n[0] to n[20] and store them in 'n_avg'.

15. The input bits from 'input_card_A' are to be read and XORed with the inputs from 'input_card_B'. The result is to be written to the output card 'output_card'. If the binary pattern of the least 16 output bits is 1010 0101 0111 0110 then the output 'match_bell' will be set. Write the ladder logic.

16. Write some simple ladder logic to change the preset value of a counter 'cnt'. When the input 'A' is active the preset should be 13, otherwise it will be 9.

17. A machine ejects parts into three chutes. Three optical sensors (A, B and C) are positioned in each of the slots to count the parts. The count should start when the reset (R) button is pushed. The count will stop, and an indicator light (L) turned on when the average number of parts counted is 100 or greater.

18. a) Write ladder logic to calculate and store the binary (geometric) sequence in 32 bit integer (DINT) memory starting at n[0] up to n[200] so that n[0] = 1, n[1] = 2, n[2] = 4, n[3] = 16, n[4] = 64, etc. b) Will the program operate as expected?

14.6 ASSIGNMENT PROBLEMS

1. Write a ladder logic program that will implement the function below, and if the result is greater than 100.5 then the output 'too_hot' will be turned on.

$$X = 6 + Ae^{B}\cos(C + 5)$$

2. Write ladder logic to calculate the average of the values from thickness[0] to thickness[99]. The operation should start after a momentary contact push button A is pushed. The result should be stored in 'thickness_avg'. If button B is pushed, all operations should be complete in a single scan. Otherwise, only ten values will be calculated each scan. (Note: this means that it will take 10 scans to complete the calculation if A is pushed.)

3. Write a ladder logic program that will calculate the standard deviation of numbers in the locations f[0] to f[29] without using the STD function.

4. A program is to perform the following actions for a self-service security check. The device will allow bags to be inserted to the test chamber through an entrance door. If the bag passes the check it can be removed through an exit door, otherwise an alarm is sounded. Create a state diagram using the steps below.
   1. The machine starts in an 'idle' state. The 'open_entry' output is activated to open the input door. The 'open_exit' output is deactivated to close the output door.
   2. When a bag is inserted the 'bag_detected' input goes high. The 'open_entry' input should be deactivated to close the door.
   3. When the 'entry_door_closed' and 'exit_door_closed' inputs are active then a 'test' output will be set high to start a scan of the bags.
   4. When the scan of the bags is complete a 'scan_done' input is set. The 'test' output should be turned off.
   5. The scan results in two real values 'nitrates' and 'mass'. The calculation below is performed. If the 'risk' is below 0.3, or above 23.5, then the machine enters an alarm state (step 8), otherwise it continues to step 6.

$$risk = 4^{nitrates} + sqrt(mass)nitrates$$

   6. The 'open_exit' output is activated to open the exit door. The machine waits until the 'bag_detected' input goes low.
   7. The 'open_exit' output is deactivated to close the door. The machine waits until the 'exit_door_closed' input is high before returning to the 'idle state.
   8. In the alarm state an operator input 'key' must be active to open the exit door. After this input is released the door will close and return to the 'idle' state.

14.1 PRACTICE PROBLEM SOLUTIONS

1.

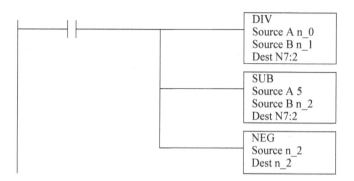

2.

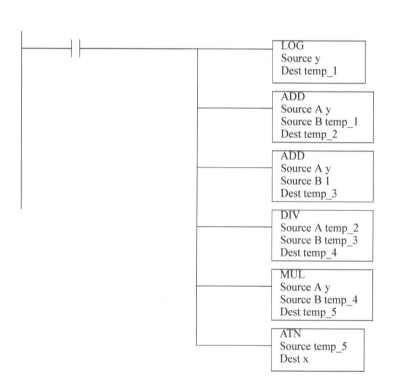

3.

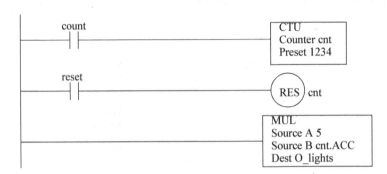

count
─┤ ├───
```
CTU
Counter cnt
Preset 1234
```

reset
─┤ ├───(RES) cnt

```
MUL
Source A 5
Source B cnt.ACC
Dest O_lights
```

4.

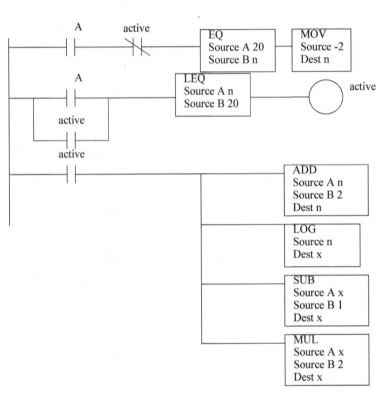

```
    A        active
──┤ ├────────┤/├──────
```
```
EQ
Source A 20
Source B n
```
```
MOV
Source -2
Dest n
```

```
    A
──┤ ├──────────
```
```
LEQ
Source A n
Source B 20
```
( ) active

```
 active
──┤ ├──
```

```
 active
──┤ ├──────────────────────
```
```
ADD
Source A n
Source B 2
Dest n
```
```
LOG
Source n
Dest x
```
```
SUB
Source A x
Source B 1
Dest x
```
```
MUL
Source A x
Source B 2
Dest x
```

5.

addr	before	after 1st	after 2nd
n[0]	1	1	1
n[1]	2	2	4
n[2]	3	2	2

6.

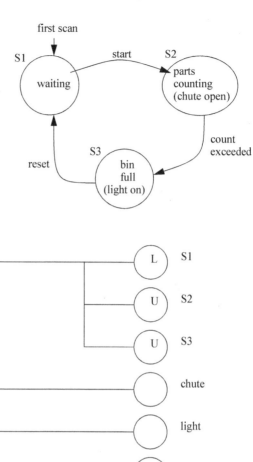

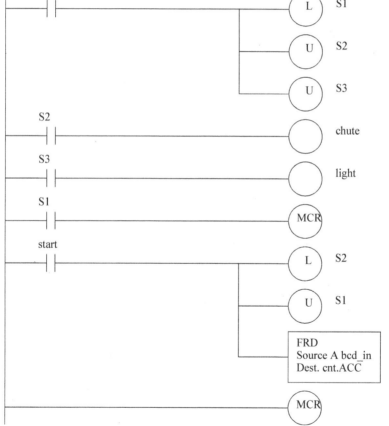

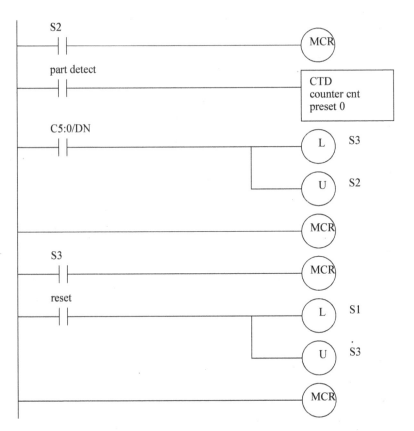

7. an incremental mode will do one calculation when the input to the function is a positive edge - goes from false to true. The all mode will attempt to complete the calculation in a single scan. If a number is used, the function will do that many calculations per scan while the input is true.

8. The maximum number is 1000. If the instruction takes too long the instruction may be paused and continued the next scan, or it may lead to a PLC fault because the scan takes too long.

9.

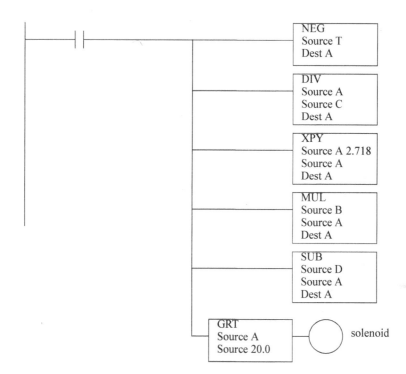

10.

11.

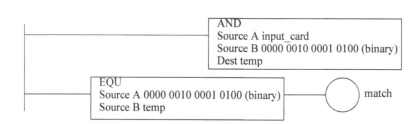

12.

    The data in the source location will be moved bit by bit to the destination for every bit that is set in the mask. Every other bit in the destination will be retain the previous value. The source address is not changed.

$$D = (S \,\&\, M) + (D \,\&\, \overline{M})$$

13.

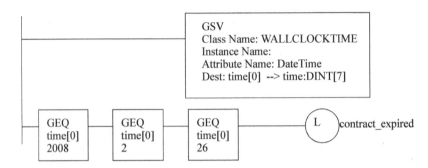

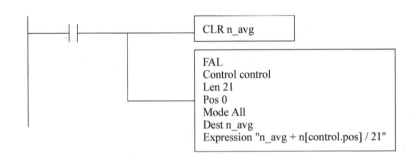

14.

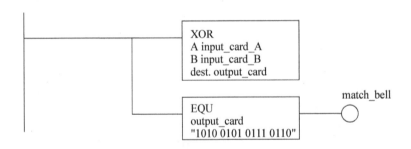

15.

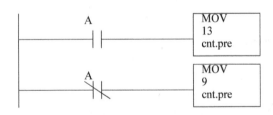

16.

314

17.

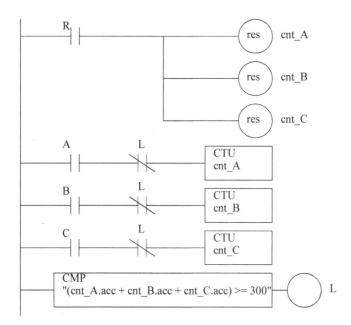

18.

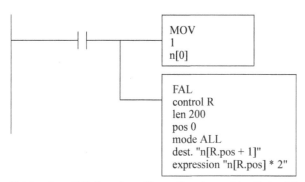

b) No, after n[31] the value will overflow the positive limit of the 32 bit 2's compliment integer and take on a large negative value.

15ADVANCED LADDER LOGIC FUNCTIONS

Topics:

- Shift registers, stacks and sequencers
- Program control; branching, looping, subroutines, temporary ends and one shots
- Interrupts; timed, fault and input driven
- Immediate inputs and outputs
- Block transfer
- Conversion of State diagrams using program subroutines
- Design examples

Objectives:

- To understand shift registers, stacks and sequencers.
- To understand program control statements.
- To understand the use of interrupts.
- To understand the operation of immediate input and output instructions.
- To be prepared to use the block transfer instruction later.
- Be able to apply the advanced function in ladder logic design.

This chapter covers *advanced* functions, but this definition is somewhat arbitrary. The array functions in the last chapter could be classified as advanced functions. The functions in this section tend to do things that are not oriented to simple data values. The list functions will allow storage and recovery of bits and words. These functions are useful when implementing buffered and queued systems. The program control functions will do things that don't follow the simple model of ladder logic execution - these functions recognize the program is executed left-to-right top-to-bottom. Finally, the input output functions will be discussed, and how they allow us to work around the normal input and output scans.

15.1 LIST FUNCTIONS

15.1.1 Shift Registers

Shift registers are oriented to single data bits. A shift register can only hold so many bits, so when a new bit is put in, one must be removed. An example of a shift register is given in Figure 15.1. The shift register is the word 'example', and it is 5 bits long. When A becomes true the bits all shift right to the least significant bit. When they shift a new bit is needed, and it is taken from *new_bit*. The bit that is shifted out, on the right hand side, is moved to the control word UL (unload) bit *c.UL*. This function will not complete in a single ladder logic scan, so the control word c is used. The function is edge triggered, so A would have to turn on 5 more times before the bit just loaded from *new_bit* would emerge to the unload bit. When A has a positive edge the 5 bits in *example* will be shifted in memory. In this case it is taking the value of bit *example.0* and putting it in the control word bit *c.UL*. It then shifts the bits once to the right, *example.0* = *example.1* then *example.1* = *example.2* then *example.2* = *example.3* then *example.3* = *example.4*. Then the input bit is put into the most significant bit *example.4* = *new_bit*. The bits in the shift register would be shifted to the left with the BSR function.

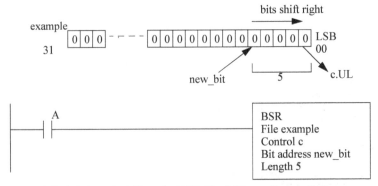

BSL - shifts left from the LSB to the MSB. The LSB must be supplied
BSR - similar to the BSL, except the bit is input to the MSB and shifted to the LSB

Figure 15.1   *Shift Register Functions*

There are other types of shift registers not implemented in the ControlLogix processors. These are shown in Figure 15.1. The primary difference is that the arithmetic shifts will put a zero into the shift register, instead of allowing an arbitrary bit. The rotate functions shift bits around in an endless circle. These functions can also be implemented using the BSR and BSL instructions when needed.

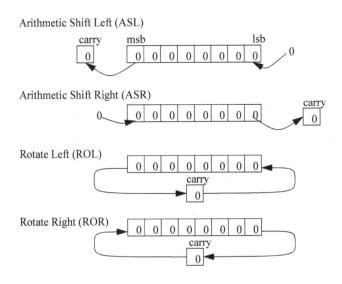

Figure 15.2   *Shift Register Variations*

15.1.2 Stacks

Stacks store integer words in a two ended buffer. There are two basic types of stacks; first-on-first-out (FIFO) and last-in-first-out (LIFO). As words are pushed on the stack it gets larger, when words are pulled off it gets smaller. When you retrieve a word from a LIFO stack you get the word that is the entry end of the stack. But, when you get a word from a FIFO stack you get the word from the exit end of the stack (it has also been there the longest). A useful analogy is a pile of work on your desk. As new work arrives you drop it on the top of the stack. If your stack is LIFO, you pick your next job from the top of the pile. If your stack is FIFO, you pick your work from the bottom of the pile. Stacks are very helpful when dealing with practical situations such as buffers in production lines. If the buffer is only a delay then a FIFO stack will keep the data in order. If product is buffered by piling it up then a LIFO stack works better, as shown in Figure 15.1. In a FIFO stack the parts pass through an entry gate, but are stopped by the exit gate. In the LIFO stack the parts enter the stack and lower the plate, when more parts are needed the plate is raised. In this arrangement the order of the parts in the stack will be reversed.

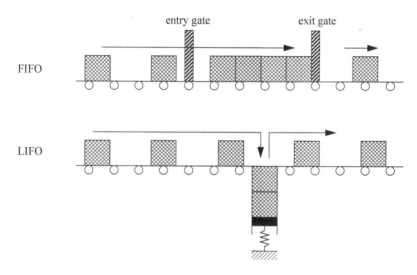

Figure 15.3   **Buffers and Stack Types**

The ladder logic functions are FFL to load the stack, and FFU to unload it. The example in Figure 15.1 shows two instructions to load and unload a FIFO stack. The first time this FFL is activated (edge triggered) it will grab the word (16 bits) from the input card *word_in* and store them on the stack, at *stack[0]*. The next value would be stored at *stack[1]*, and so on until the stack length is reached at *stack[4]*. When the FFU is activated the word at *stack[0]* will be moved to the output card *word_out*. The values on the stack will be shifted up so that the value previously in *stack[1]* moves to *stack[0]*, *stack[2]* moves to *stack[1]*, etc. If the stack is full or empty, an a load or unload occurs the error bit will be set *c.ER*.

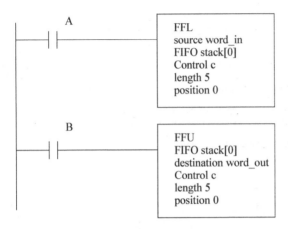

Figure 15.4   **FIFO Stack Instructions**

The LIFO stack commands are shown in Figure 15.1. As values are loaded on the stack the will be added sequentially stack[0], stack[1], stack[2], stack[3] then stack[4]. When values are unloaded they will be taken from the last loaded position, so if the stack is full the value of stack[4] will be removed first.

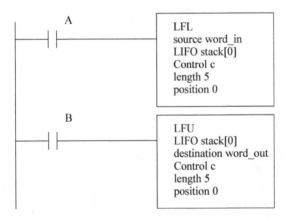

Figure 15.5   **LIFO Stack Commands**

15.1.3 Sequencers

A mechanical music box is a simple example of a sequencer. As the drum in the music box turns it has small pins that will sound different notes. The song sequence is fixed, and it always follows the same pattern. Traffic light controllers are now controlled with electronics, but previously they used sequencers that were based on a rotating drum with cams that would open and close relay terminals. One of these cams is shown in Figure 15.1. The cam rotates slowly, and the surfaces under the contacts will rise and fall to open and close contacts. For a traffic light controllers the speed of rotation would set the total cycle time for the traffic lights. Each cam will control one light, and by adjusting the circumferential length of rises and drops the on and off times can be adjusted.

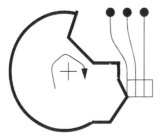

As the cam rotates it makes contact with none, one, or two terminals, as determined by the depressions and rises in the rotating cam.

Figure 15.6   ***A Single Cam in a Drum Sequencer***

A PLC sequencer uses a list of words in memory. It recalls the words one at a time and moves the words to another memory location or to outputs. When the end of the list is reached the sequencer will return to the first word and the process begins again. A sequencer is shown in Figure 15.1. The SQO instruction will retrieve words from bit memory starting at *sequence[0]*. The length is *4* so the end of the list will be at *sequence[0]+4* or *sequence[4]* (the total length of 'sequence' is actually 5). The sequencer is edge triggered, and each time *A* becomes true the retrieve a word from the list and move it to *output_lights*. When the sequencer reaches the end of the list the sequencer will return to the second position in the list *sequence[1]*. The first item in the list is *sequence[0]*, and it will only be sent to the output if the *SQO* instruction is active on the first scan of the PLC, otherwise the first word sent to the output is *sequence[1]*. The mask value is *000Fh*, or *0000000000001111b* so only the four least significant bits will be transferred to the output, the other output bits will not be changed. The other instructions allow words to be added or removed from the sequencer list.

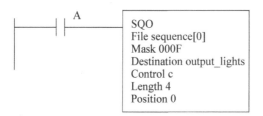

```
                      A         ┌──────────────────────────┐
────────────┤ ├────────┤ SQO                       │
                                │ File sequence[0]          │
                                │ Mask 000F                 │
                                │ Destination output_lights │
                                │ Control c                 │
                                │ Length 4                  │
                                │ Position 0                │
                                └──────────────────────────┘
```

SQO(start,mask,destination,control,length) - sequencer output from table to memory
SQI(start,mask,source,control,length) - sequencer input from memory address to table
SQL(start,source,control,length) - sequencer load to set up the sequencer parameters

Figure 15.7   ***The Basic Sequencer Instruction***

An example of a sequencer is given in Figure 15.1 for traffic light control. The light patterns are stored in memory (entered manually by the programmer). These are then moved out to the output card as the function is activated. The mask (003Fh = 0000000000111111b) is used so that only the 6 least significant bits are changed.

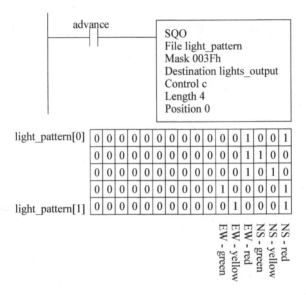

Figure 15.8   *A Sequencer For Traffic Light Control*

Figure 15.1 shows examples of the other sequencer functions. When *A* goes from false to true, the SQL function will move to the next position in the sequencer list, for example *sequence_rem[1]*, and load a value from *input_word*. If *A* then remains true the value in *sequence_rem[1]* will be overwritten each scan. When the end of the sequencer list is encountered, the position will reset to 1.

The sequencer input (SQI) function will compare values in the sequence list to the source *compare_word* while *B* is true. If the two values match *match_output* will stay on while *B* remains true. The mask value is *0005h* or *0000000000000101b*, so only the first and third bits will be compared. This instruction does not automatically change the position, so logic is shown that will increment the position every scan while *C* is true.

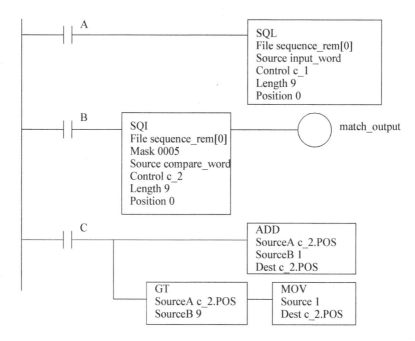

Figure 15.9   **Sequencer Instruction Examples**

These instructions are well suited to processes with a single flow of execution, such as traffic lights.

15.2 PROGRAM CONTROL

15.2.1 Branching and Looping

These functions allow parts of ladder logic programs to be included or excluded from each program scan. These functions are similar to functions in other programming languages such as C, C++, Java, Pascal, etc.

Entire sections of programs can be bypassed using the JMP instruction in Figure 15.1. If *A* is true the program will jump over the next three lines to the line with the *LBL Label_01*. If *A* is false the *JMP* statement will be ignored, and the program scan will continue normally. If *A* is false *X* will have the same value as *B*, and *Y* can be turned on by *C* and off by *D*. If *A* is true then *X* and *Y* will keep their previous values, unlike the *MCR* statement. Any instructions that follow the *LBL* statement will not be affected by the *JMP* so *Z* will always be equal to *E*. If a jump statement is true the program will run faster.

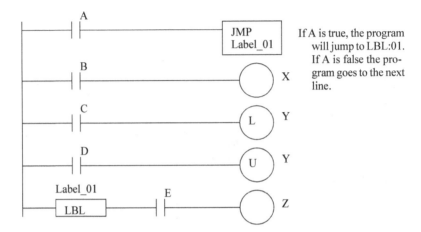

Figure 15.10   *A JMP Instruction*

Subroutines jump to other programs, as is shown in Figure 15.1. When *A* is true the *JSR* function will jump to the subroutine program in file 3. The *JSR* instruction two arguments are passed, *A* and *B*. The subroutine (SBR) function receives these two arguments and puts them in *X* and *Y*. When *B* is true the subroutine will end and return to program *file 2* where it was called (Note: a subroutine can have multiple returns). The *RET* function returns the value *Z* to the calling program where it is put in location *C*. By passing arguments (instead of having the subroutine use global memory locations) the subroutine can be used for more than one operation. For example, a subroutine could be given an angle in degrees and return a value in radians. A subroutine can be called more than once in a program, but if not called, it will be ignored.

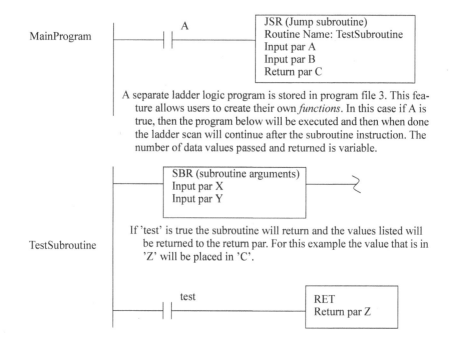

MainProgram

| A | JSR (Jump subroutine) |
| Routine Name: TestSubroutine |
| Input par A |
| Input par B |
| Return par C |

A separate ladder logic program is stored in program file 3. This feature allows users to create their own *functions*. In this case if A is true, then the program below will be executed and then when done the ladder scan will continue after the subroutine instruction. The number of data values passed and returned is variable.

SBR (subroutine arguments)
Input par X
Input par Y

TestSubroutine

If 'test' is true the subroutine will return and the values listed will be returned to the return par. For this example the value that is in 'Z' will be placed in 'C'.

test

RET
Return par Z

Figure 15.11   **Subroutines**

The 'FOR' function in Figure 15.1 will (within the same logic scan) call a subroutine 5 times (from 0 to 9 in steps of 2) when A is true. In this example the subroutine contains an *ADD* function that will add 1 to the value of *i*. So when this 'FOR' statement is complete the value of *j* will 5 larger. For-next loops can be put inside other for-next loops, this is called nesting. If A was false the program not call the subroutine. When A is true, all 5 loops will be completed in a single program scan. If B is true the *NXT* statement will return to the *FOR* instruction, and stop looping, even if the loop is not complete. Care must be used for this instruction so that the ladder logic does not get caught in an infinite, or long loop - if this happens the PLC will experience a fault and halt.

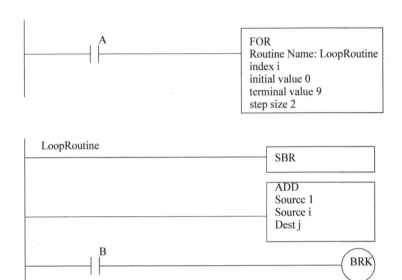

Note: if A is true then the loop will repeat 10 times, and the value of i will be increased by 10. If A is not true, then the subroutine will never be called.

Figure 15.12   *A For-Next Loop*

Ladder logic programs always have an end statement, as shown in Figure 15.1. Most modern software automatically inserts this. PLCs will experience faults if this is not present. The temporary end (TND) statement will skip the remaining portion of a program. If C is true then the program will end, and the next line with D and Y will be ignored. If C is false then the TND will have no effect and Y will be equal to D.

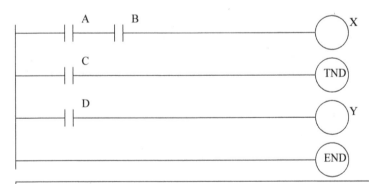

When the end (or End Of File) is encountered the PLC will stop scanning the ladder, and start updating the outputs. This will not be true if it is a subroutine or a step in an SFC.

Figure 15.13   End Statements

The one shot contact in Figure 15.1 can be used to turn on a ladder run for a single scan. When *A* has a positive edge the oneshot will turn on the run for a single scan. Bit *last_bit_value* is used here to track to rung status.

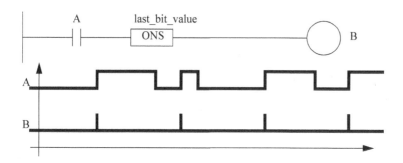

Figure 15.14   One Shot Instruction

15.2.2 Fault Handling

A fault condition can stop a PLC. If the PLC is controlling a dangerous process this could lead to significant damage to personnel and equipment. There are two types of faults that occur; terminal (major) and warnings (minor). A minor fault will normally set an error bit, but not stop the PLC. A major failure will normally stop the PLC, but an interrupt can be used to run a program that can reset the fault bit in memory and continue operation (or shut down safely). Not all major faults are recoverable. A complete list of these faults is available in PLC processor manuals.

The PLC can be set up to run a program when a fault occurs, such as a divide by zero. These routines are program files under 'Control Fault Handler'. These routines will be called when a fault occurs. Values are set in status memory to indicate the source of the faults.

Figure 15.1 shows two example programs. The default program 'MainProgram' will generate a fault, and the interrupt program called 'Recover' will detect the fault and fix it. When *A* is true a compute function will interpret the expression, using indirect addressing. If *B* becomes true then the value in *n[0]* will become negative. If *A* becomes true after this then the expression will become *n[10] +10*. The negative value for the address will cause a fault, and program file 'Recover' will be run.

In the fault program the fault values are read with an GSV function and the fault code is checked. In this case the error will result in a status error of 0x2104. When this is the case the n[0] is set back to zero, and the fault code in *fault_data[2]* is cleared. This value is then written back to the status memory using an SSV function. If the fault was not cleared the PLC would enter a fault state and stop (the fault light on the front of the PLC will turn on).

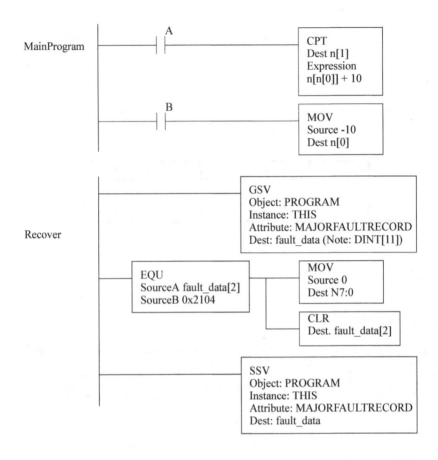

Figure 15.15   **A Fault Recovery Program**

15.2.3 Interrupts

The PLC can be set up to run programs automatically using interrupts. This is routinely done for a few reasons;

- to run a program at a regular timed interval (e.g. SPC calculations)
- to respond when a long instruction is complete (e.g. analog input)
- when a certain input changed (e.g. panic button)

Allen Bradley allows interrupts, but they are called periodic/event tasks. By default the main program is defined as a 'continuous' task, meaning that it runs as often as possible, typically 10-100 times per second. Only one continuos task is allowed. A 'periodic' task can be created that has a given update time. 'Event' tasks can be triggered by a variety of actions, including input changes, tag changes, EVENT instructions, and servo control changes.

A timed interrupt will run a program at regular intervals. To set a timed interrupt the program in file number should be put in S2:31. The program will be run every S2:30 times 1 milliseconds. In Figure 15.1 program 2 will set up an interrupt that will run program *3* every 5 seconds. Program 3 will add the value of *I:000* to *N7:10*. This type of timed interrupt is very useful when controlling processes where a constant time interval is important. The timed

328

interrupts are enabled by setting bit S2:2/1 in PLC-5s.

When activated, interrupt routines will stop the PLC, and the ladder logic is interpreted immediately. If multiple interrupts occur at the same time the ones with the higher priority will occur first. If the PLC is in the middle of a program scan when interrupted this can cause problems. To overcome this a program can disable interrupts temporarily using the UID and UIE functions. Figure 15.1 shows an example where the interrupts are disabled for a FAL instruction. Only the ladder logic between the *UID* and *UIE* will be disabled, the first line of ladder logic could be interrupted. This would be important if an interrupt routine could change a value between *n[0]* and *n[4]*. For example, an interrupt could occur while the FAL instruction was at *n[7]=n[2]+5*. The interrupt could change the values of *n[1]* and *n[4]*, and then end. The FAL instruction would then complete the calculations. But, the results would be based on the old value for *n[1]* and the new value for *n[4]*.

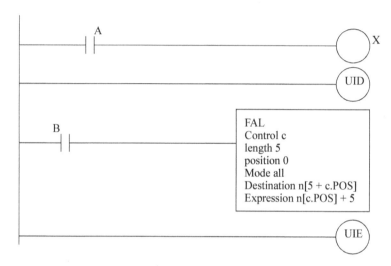

Figure 15.16   *Disabling Interrupts*

15.3 INPUT AND OUTPUT FUNCTIONS

15.3.1 Immediate I/O Instructions

The input scan normally records the inputs before the program scan, and the output scan normally updates the outputs after the program scan, as shown in Figure 15.1. Immediate input and output instructions can be used to update some of the inputs or outputs during the program scan.

• The normal operation of the PLC is

fast [input scan]

slow [ladder logic is checked]

fast [outputs updated]

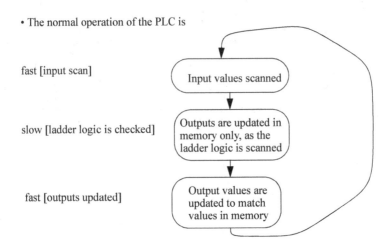

Figure 15.17   *Input, Program and Output Scan*

Figure 15.1 shows a segment within a program that will update the input word *input_value*, determine a new value for *output_value.1*, and update the output word *output_value* immediately. The process can be repeated many times during the program scan allowing faster than normal response times. These instructions are less useful on newer PLCs with networked hardware and software, so Allen Bradley does not support IIN for newer PLCs such as ControlLogix, even though the IOT is supported.

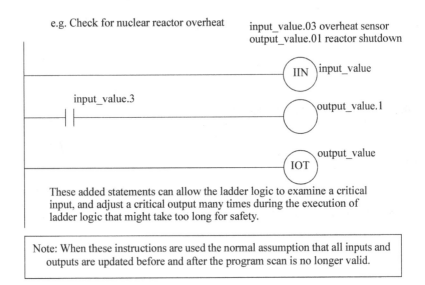

Figure 15.18   *Immediate Inputs and Outputs*

15.4 DESIGN TECHNIQUES

15.4.1 State Diagrams

The block logic method was introduced in chapter 8 to implement state diagrams using MCR blocks. A better implementation of this method is possible using subroutines in program files. The ladder logic for each state will be put in separate subroutines.

Consider the state diagram in Figure 15.1. This state diagram shows three states with four transitions. There is a potential conflict between transitions *A* and *C*.

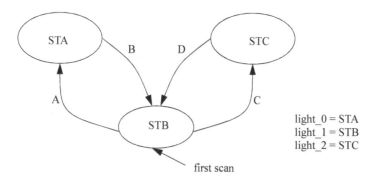

light_0 = STA
light_1 = STB
light_2 = STC

Figure 15.19   *A State Diagram*

The main program for the state diagram is shown in Figure 15.1. This program is stored in the MainProgram so that it is run by default. The first rung in the program resets the states so that the first scan state is on, while the other states are turned off. The following logic will call the subroutine for each state. The logic that uses the current state is placed in the main program. It is also possible to put this logic in the state subroutines.

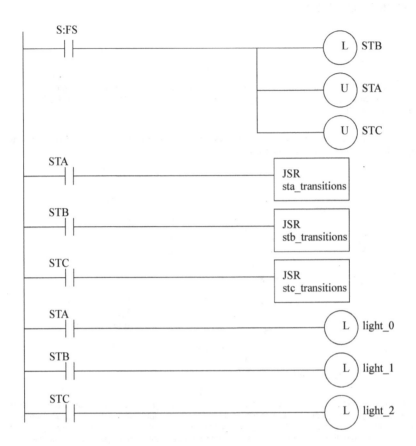

Figure 15.20   *The Main Program for the State Diagram (Program File 2)*

The ladder logic for each of the state subroutines is shown in Figure 15.1. These blocks of logic examine the transitions and change states as required. Note that state *STB* includes logic to give state *C* higher priority, by blocking *A* when *C* is active.

332

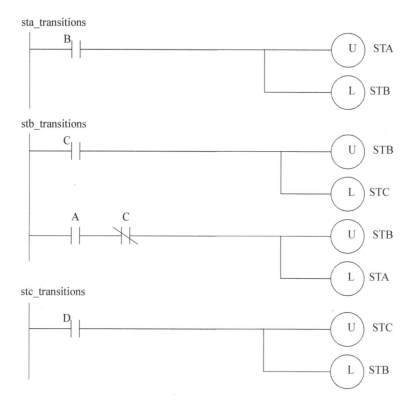

Figure 15.21   *Subroutines for the States*

The arrangement of the subroutines in Figure 15.1 and Figure 15.1 could experience problems with *racing* conditions. For example, if STA is active, and both *B* and *C* are true at the same time the main program would jump to subroutine 3 where STB would be turned on. then the main program would jump to subroutine 4 where STC would be turned on. For the output logic STB would never have been on. If this problem might occur, the state diagram can be modified to slow down these race conditions. Figure 15.1 shows a technique that blocks race conditions by blocking a transition out of a state until the transition into a state is finished. The solution may not always be appropriate.

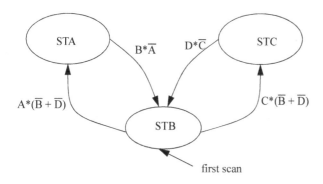

Figure 15.22   *A Modified State Diagram to Prevent Racing*

Another solution is to force the transition to wait for one scan as shown in Figure 15.1 for state *STA*. A wait bit is used to indicate when a delay of at least one scan has occurred since the transition out of the state *B* became true. The wait bit is set by having the exit transition *B* true. The *B3/0-STA* will turn off the wait *B3/10-wait* when the transition to state *B3/1-STB* has occurred. If the wait was not turned off, it would still be on the next time we return to this state.

Program 3 for STA

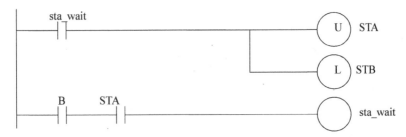

Figure 15.23   *Subroutines for State STA to Prevent Racing*

15.5 DESIGN CASES

15.5.1 If-Then

Problem: Convert the following C/Java program to ladder logic.

```
void main(){
        int A;
        for(A = 1; A < 10 ; A++){
        if (A >= 5) then A = add(A);
        }
}
int add(int x){
        x = x + 1;
        return x;
}
```

Solution:

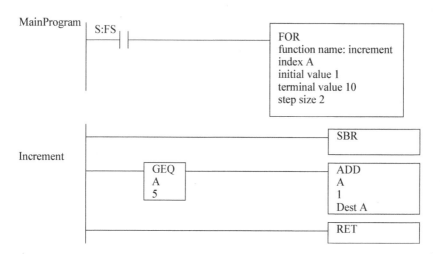

Figure 15.24   *C Program Implementation*

15.5.2 Traffic Light

Problem: Design and write ladder logic for a simple traffic light controller that has a single fixed sequence of 16 seconds for both green lights and 4 second for both yellow lights. Use either stacks or sequencers.

Solution: The sequencer is the best solution to this problem.

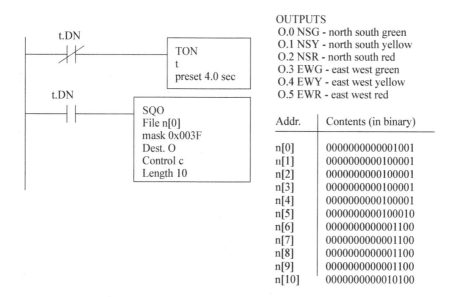

Figure 15.25   *An Example Traffic Light Controller*

15.6 SUMMARY

- Shift registers move bits through a queue.
- Stacks will create a variable length list of words.
- Sequencers allow a list of words to be stepped through.
- Parts of programs can be skipped with jump and MCR statements, but MCR statements shut off outputs.
- Subroutines can be called in other program files, and arguments can be passed.
- For-next loops allow parts of the ladder logic to be repeated.
- Interrupts allow parts to run automatically at fixed times, or when some event happens.
- Immediate inputs and outputs update I/O without waiting for the normal scans.

15.7 PRACTICE PROBLEMS

(Note: Problem solutions are available at http://sites.google.com/site/automatedmanufacturingsystems/)

1. Design and write ladder logic for a simple traffic light controller that has a single fixed sequence of 16 seconds for both green lights and 4 seconds for both yellow lights. Use shift registers to implement it.

2. A PLC is to be used to control a carillon (a bell tower). Each bell corresponds to a musical note and each has a pneumatic actuator that will ring it. The table below defines the tune to be programmed. Write a program that will run the tune once each time a start button is pushed. A stop button will stop the song.

time sequence in seconds

O:000/00	0	1	2	3	4	5	6	7	8	9	10	11	12	13	14	15	16
O:000/00	0	0	0	0	0	0	0	1	0	0	0	0	0	0	0	0	1
O:000/01	1	0	0	0	0	0	0	0	0	0	0	0	1	0	0	0	
O:000/02	1	0	0	1	0	0	0	0	0	1	1	0	0	0	1	0	0
O:000/03	0	0	0	0	1	0	0	0	0	0	1	0	1	0	0	1	0
O:000/04	0	1	1	0	0	0	0	0	0	0	0	0	0	0	0	0	
O:000/05	0	0	0	0	0	0	1	0	0	0	0	0	0	0	0	0	
O:000/06	0	0	0	0	0	1	1	0	0	0	0	0	1	0	0	0	0
O:000/07	0	0	0	0	0	0	0	0	1	0	0	0	0	0	0	0	0

3. Consider a conveyor where parts enter on one end. they will be checked to be in a left or right orientation with a vision system. If neither left nor right is found, the part will be placed in a reject bin. The conveyor layout is shown below.

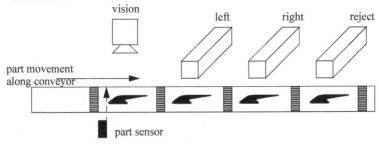

4. Why are MCR blocks different than JMP statements?

5. What is a suitable reason to use interrupts?

6. When would immediate inputs and outputs be used?

7. Explain the significant differences between shift registers, stacks and sequencers.

8. Design a ladder logic program that will run once every 30 seconds using interrupts. It will check to see if a water

tank is full with input tank_full. If it is full, then a shutdown value ('shutdown') will be latched on.

9. At MOdern Manufacturing (MOMs), pancakes are made by multiple machines in three flavors; chocolate, blue-berry and plain. When the pancakes are complete they travel along a single belt, in no specific order. They are buffered by putting them on the top of a stack. When they arrive at the stack the input 'detected' becomes true, and the stack is loaded by making output 'stack' high for one second. As the pancakes are put on the stack, a color detector is used to determine the pancakes type. A value is put in 'color_stack' (1=chocolate, 2=blueberry, 3=plain) and bit 'unload' is made true. A pancake can be requested by pushing a button ('chocolate', 'blueberry'; 'plain'). Pancakes are then unloaded from the stack, by making 'unload' high for 1 second, until the desired fla-vor is removed. Any pancakes removed aren't returned to the stack. Design a ladder logic program to control this stack.

10. a) What are the two fundamental types of interrupts?
    b) What are the advantages of interrupts in control programs?
    c) What potential problems can they create?
    d) Which instructions can prevent this problem?

11. Write a ladder logic program to drive a set of flashing lights. In total there are 10 lights connected to 'lights[0]' to 'lights[9]'. At any time every one out of three lights should be on. Every second the pattern on the lights should shift towards 'lights[9]'.

12. Implement the following state diagram using subroutines.

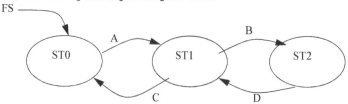

13. A SQO control word 'c' has a value of c.LEN = 5, but the array of values is 6 long. Why?

15.8 ASSIGNMENT PROBLEMS

1. Using 3 different methods write a program that will continuously cycle a pattern of 12 lights connected to a PLC output card. The pattern should have one out of every three lights set. The light patterns should appear to move endlessly in one direction.

2. Look at the manuals for the status memory in your PLC.
    a) Describe how to run program 'GetBetter' when a divide by zero error occurs.
    b) Write the ladder logic needed to clear a PLC fault.
    c) Describe how to set up a timed interrupt to run 'Slowly' every 2 seconds.

3. Write an interrupt driven program that will run once every 5 seconds and calculate the average of the numbers from 'f[0]' to 'f[19]', and store the result in 'f_avg'. It will also determine the median and store it in 'f_med'.

4. Write a program for SPC (Statistical Process Control) that will run once every 20 minutes using timed interrupts. When the program runs it will calculate the average of the data values in memory locations 'f[0]' to 'f[39]' (Note: these values are written into the PLC memory by another PLC using networking). The program will also find the range of the values by subtracting the maximum from the minimum value. The average will be compared to upper (f_ucl_x) and lower (f_lcl_x) limits. The range will also be compared to upper (f_ucl_r) and lower (f_lcl_r) limits. If the average, or range values are outside the limits, the process will stop, and an 'out of control' light will be turned on. The process will use start and stop buttons, and when running it will set memory bit 'in_control'.

5. Develop a ladder logic program to control a light display outside a theater. The display consists of a row of 8 lights. When a patron walks past an optical sensor the lights will turn on in sequence, moving in the same direc-tion. Initially all lights are off. Once triggered the lights turn on sequentially until all eight lights are on 1.6 sec-onds latter. After a delay of another 0.4 seconds the lights start to turn off until all are off, again moving in the same direction as the patron. The effect is a moving light pattern that follows the patron as they walk into the theater.

6. Write the ladder logic diagram that would be required to execute the following data manipulation for a preventative maintenance program.

        i) Keep track of the number of times a motor was started with toggle switch #1.
        ii) After 2000 motor starts turn on an indicator light on the operator panel.
        iii) Provide the capability to change the number of motor starts being tracked, prior to triggering of the indicator light. HINT: This capability will only require the change of a value in a compare statement rather than the addition of new lines of logic.
        iv) Keep track of the number of minutes that the motor has run.
        v) After 9000 minutes of operation turn the motor off automatically and also turn on an indicator light on the operator panel.

7. Parts arrive at an oven on a conveyor belt and pass a barcode scanner. When the barcode scanner reads a valid barcode it outputs the numeric code as 32 bits to 'scanner_value' and sets input 'scanner_value_valid'. The PLC must store this code until the parts pass through the oven. When the parts leave the oven they are detected by a proximity sensor connected to 'part_leaving'. The barcode value read before must be output to 'barcode_output'. Write the ladder logic for the process. There can be up to ten parts inside the oven at any time.

8. Write the ladder logic for the state diagram below using subroutines for the states.

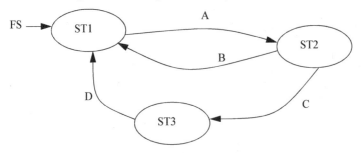

9. Convert the following state diagram to ladder logic using subroutines.

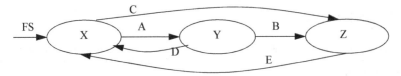

10. Implement the following state diagram using JMP statements.

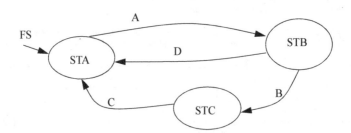

11. Write a traffic light program using a sequencer. Keep the program simple with a 4 second green and yellow in both directions. But, the traffic lights should only function when the system clock (WALLCLOCKTIME) is between 7am and 8pm. Other times the lights should be left green in one direction and red in the other.

15.1 PRACTICE PROBLEM SOLUTIONS

1.

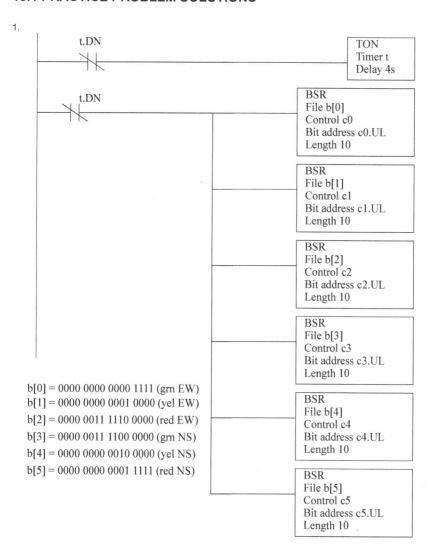

b[0] = 0000 0000 0000 1111 (grn EW)
b[1] = 0000 0000 0001 0000 (yel EW)
b[2] = 0000 0011 1110 0000 (red EW)
b[3] = 0000 0011 1100 0000 (grn NS)
b[4] = 0000 0000 0010 0000 (yel NS)
b[5] = 0000 0000 0001 1111 (red NS)

```
        b[0].0
    ────┤├──────────────────────────────────────( )  grn_EW

        b[1].0
    ────┤├──────────────────────────────────────( )  yel_EW

        b[2].0
    ────┤├──────────────────────────────────────( )  red_EW

        b[3].0
    ────┤├──────────────────────────────────────( )  grn_NS

        b[4].0
    ────┤├──────────────────────────────────────( )  yel_NS

        b[5].0
    ────┤├──────────────────────────────────────( )  red_NS
```

2.

n[0] = 0000 0000 0000 0000
n[1] = 0000 0000 0000 0110
n[2] = 0000 0000 0001 0000
n[3] = 0000 0000 0001 0000
n[4] = 0000 0000 0000 0100
n[5] = 0000 0000 0000 1000
n[6] = 0000 0000 0100 0000
n[7] = 0000 0000 0110 0000
n[8] = 0000 0000 0000 0001

n[9] = 0000 0000 1000 0000
n[10] = 0000 0000 0000 0100
n[11] = 0000 0000 0000 1100
n[12] = 0000 0000 0000 0000
n[13] = 0000 0000 0100 1000
n[14] = 0000 0000 0000 0010
n[15] = 0000 0000 0000 0100
n[16] = 0000 0000 0000 1000
n[17] = 0000 0000 0000 0001

```
     start                          stop
    ──┤├──────────────────────────┤/├──────────────────( )  play
                                        t.DN        ┌──────────────┐
     play   ┌──────────────────┐       ─┤/├─        │ TON          │
    ──┤├────┤ NEQ              ├┘                   │ Timer t      │
            │ Source A  c.POS   │                    │ Delay 4s     │
            │ Source B  17      │                    └──────────────┘
            └──────────────────┘

                                            ┌──────────────────────┐
     t.DN                                   │ SQO                   │
    ──┤├─────────────────────────────────── │ File n[0]             │
                                            │ Mask 0x00FF           │
                                            │ Destination lights    │
                                            │ Control c             │
                                            │ Length 17             │
                                            │ Position 0            │
                                            └──────────────────────┘
```

3.

assume:
    sensors.0 = left orientation
    sensors.1 = right orientation
    sensors.2 = reject
    sensors.3 = part sensor

```
  sensors.3
 ───┤ ├───┬─────────────────┌──────────────────────┐
           │                  │ BSR                  │
           │                  │ File b[0]            │
           │                  │ Control c0           │
           │                  │ Bit address sensors.0│
           │                  │ Length 4             │
           │                  └──────────────────────┘
           │                  ┌──────────────────────┐
           │                  │ BSR                  │
           │                  │ File b[1]            │
           │                  │ Control c1           │
           │                  │ Bit address sensors.1│
           │                  │ Length 4             │
           │                  └──────────────────────┘
           │                  ┌──────────────────────┐
           │                  │ BSR                  │
           │                  │ File b[2]            │
           │                  │ Control c2           │
           │                  │ Bit address sensors.2│
           │                  │ Length 4             │
           │                  └──────────────────────┘
           │    b[0].2
           ├────┤ ├────────────────( )  left
           │    b[1].1
           ├────┤ ├────────────────( )  right
           │    b[2].0
           └────┤ ├────────────────( )  reject
```

4. In MCR blocks the outputs will all be forced off. This is not a problem for outputs such as retentive timers and latches, but it will force off normal outputs. JMP statements will skip over logic and not examine it or force it off.

5. Timed interrupts are useful for processes that must happen at regular time intervals. Polled interrupts are useful to monitor inputs that must be checked more frequently than the ladder scan time will permit. Fault interrupts are important for processes where the complete failure of the PLC could be dangerous.

6. These can be used to update inputs and outputs more frequently than the normal scan time permits.

7. The main differences are: Shift registers focus on bits, stacks and sequencers on words Shift registers and sequencers are fixed length, stacks are variable lengths

8.

```
                    tank_full
Checker           ───┤ ├──────────────────────────────( L )  shutdown
configuration
periodic task
update 30000ms
```

9.

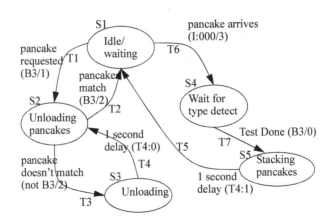

$T1 = S1 \bullet B3/1$

$T2 = S2 \bullet B3/2$

$T3 = S2 \bullet \overline{B3/2}$

$T4 = S3 \bullet T4:0/DN$

$T5 = S5 \bullet T4:1/DN$

$T6 = S1 \bullet I:000/3$

$T7 = S4 \bullet B3/0$

$S1 = (S1 + T2 + T5 + FS) \bullet \overline{T1} \bullet \overline{T6}$

$S2 = (S2 + T1 \bullet \overline{T6} + T4) \bullet \overline{T2} \bullet \overline{T3}$

$S3 = (S3 + T3) \bullet \overline{T4}$

$S4 = (S4 + T6) \bullet \overline{T7}$

$S5 = (S5 + T7) \bullet \overline{T5}$

342

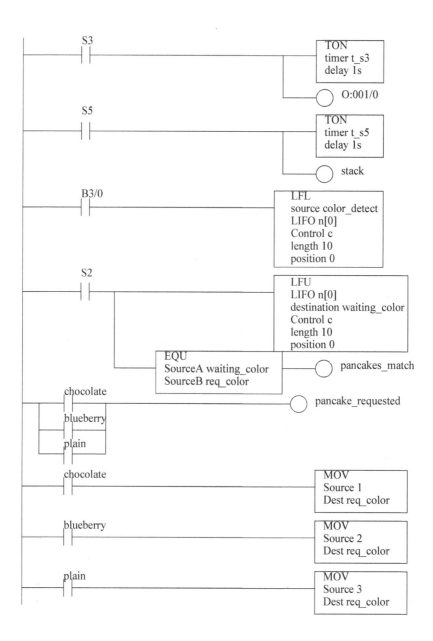

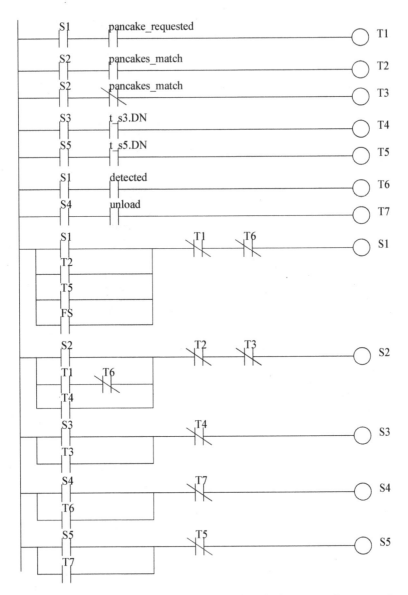

10. a) Timed, polled and fault, b) They remove the need to check for times or scan for memory changes, and they allow events to occur more often than the ladder logic is scanned. c) A few rungs of ladder logic might count on a value remaining constant, but an interrupt might change the memory, thereby corrupting the logic. d) The UID and UIE

11.

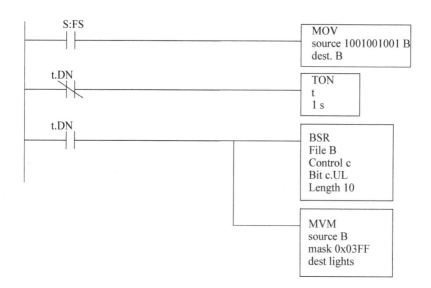

12.

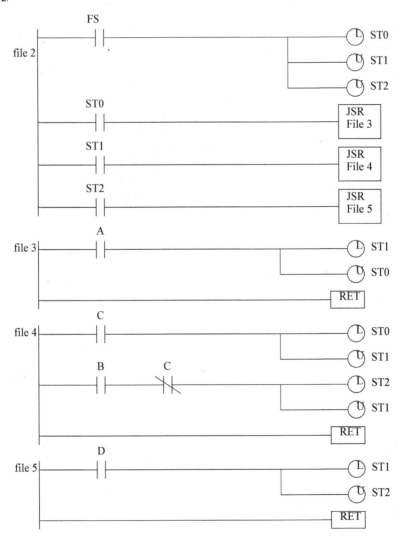

13. The first element of the array is loaded if the input to the SQO is true on the first scan, but after that it is never used again. So in this example the array[0] value will be used the first time, and the array[1] to array[5] values will be used for the normal sequence.

16 OPEN CONTROLLERS

Topics:

- Open systems
- IEC 61131 standards
- Open architecture controllers

Objectives:

- To understand the decision between choosing proprietary and public standards.
- To understand the basic concepts behind the IEC 61131 standards.

In previous decades (and now) PLC manufacturers favored "proprietary" or "closed" designs. This gave them control over the technology and customers. Essentially, a proprietary architecture kept some of the details of a system secret. This tended to limit customer choices and options. It was quite common to spend great sums of money to install a control system, and then be unable to perform some simple task because the manufacturer did not sell that type of solution. In these situations customers often had two choices; wait for the next release of the hardware/software and hope for a solution, or pay exorbitant fees to have custom work done by the manufacturer.

"Open" systems have been around for decades, but only recently has their value been recognized. The most significant step occurred in 1981 when IBM broke from it's corporate tradition and released a personal computer that could use hardware and software from other companies. Since that time IBM lost control of it's child, but it has now adopted the open system philosophy as a core business strategy. All of the details of an open system are available for users and developers to use and modify. This has produced very stable, flexible and inexpensive solutions. Controls manufacturers are also moving toward open systems. One such effort involves Devicenet, which is discussed in a later chapter.

A troubling trend that you should be aware of is that many manufacturers are mislabeling closed and semi-closed systems as open. An easy acid test for this type of system is the question "does the system allow me to choose alternate suppliers for all of the components?" If even one component can only be purchased from a single source, the system is not open. When you have a choice you should avoid "not-so-open" solutions.

16.1 IEC 61131

The IEC 1131 standards were developed to be a common and open framework for PLC architecture, agreed to by many standards groups and manufacturers. They were initially approved in 1992, and since then they have been reviewed as the IEC-61131 standards. The main components of the standard are;

IEC 61131-1 Overview
IEC 61131-2 Requirements and Test Procedures
IEC 61131-3 Data types and programming
IEC 61131-4 User Guidelines
IEC 61131-5 Communications
IEC 61131-7 Fuzzy control

This standard is defined loosely enough so that each manufacturer will be able to keep their own look-and-feel, but the core data representations should become similar. The programming models (IEC 61131-3) have the greatest impact on the user.

IL (Instruction List) - This is effectively mnemonic programming
ST (Structured Text) - A BASIC like programming language
LD (Ladder Diagram) - Relay logic diagram based programming
FBD (Function Block Diagram) - A graphical dataflow programming method
SFC (Sequential Function Charts) - A graphical method for structuring programs

Most manufacturers already support most of these models, except Function Block programming. The pro-

347

gramming model also describes standard functions and models. Most of the functions in the models are similar to the functions described in this book. The standard data types are shown in Figure 16.1.

Name	Type	Bits	Range
BOOL	boolean	1	0 to 1
SINT	short integer	8	-128 to 127
INT	integer	16	-32768 to 32767
DINT	double integer	32	-2.1e-9 to 2.1e9
LINT	long integer	64	-9.2e19 to 9.2e19
USINT	unsigned short integer	8	0 to 255
UINT	unsigned integer	16	0 to 65536
UDINT	unsigned double integer	32	0 to 4.3e9
ULINT	unsigned long integer	64	0 to 1.8e20
REAL	real numbers	32	
LREAL	long reals	64	
TIME	duration	not fixed	not fixed
DATE	date	not fixed	not fixed
TIME_OF_DAY, TOD	time	not fixed	not fixed
DATE_AND_TIME, DT	date and time	not fixed	not fixed
STRING	string	variable	variable
BYTE	8 bits	8	NA
WORD	16 bits	16	NA
DWORD	32 bits	32	NA
LWORD	64 bits	64	NA

Figure 16.1   *IEC 61131-3 Data Types*

Previous chapters have described Ladder Logic (LD) programming in detail, and Sequential Function Chart (SFC) programming briefly. Following chapters will discuss Instruction List (IL), Structured Test (ST) and Function Block Diagram (FBD) programming in greater detail.

16.2 OPEN ARCHITECTURE CONTROLLERS

Personal computers have been driving the open architecture revolution. A personal computer is capable of replacing a PLC, given the right input and output components. As a result there have been many companies developing products to do control using the personal computer architecture. Most of these devices use two basic variations;

- a standard personal computer with a normal operating system, such as Windows NT, runs a virtual PLC.
  - the computer is connected to a normal PLC rack
  - I/O cards are used in the computer to control input/output functions
  - the computer is networked to various sensors
- a miniaturized personal computer is put into a PLC rack running a virtual PLC.

In all cases the system is running a standard operating system, with some connection to rugged input and output cards. The PLC functions are performed by a virtual PLC that interprets the ladder logic and simulates a PLC. These can be fast, and more capable than a stand alone PLC, but also prone to the reliability problems of normal computers. For example, if an employee installs and runs a game on the control computer, the controller may act erratically, or stop working completely. Solutions to these problems are being developed, and the stability problem should be solved in the near future.

16.3 SUMMARY

- Open systems can be replaced with software or hardware from a third party.
- Some companies call products open incorrectly.
- The IEC 61131 standard encourages interchangeable systems.
- Open architecture controllers replace a PLC with a computer.

16.4 PRACTICE PROBLEMS

(Note: Problem solutions are available at http://sites.google.com/site/automatedmanufacturingsystems/)

1. Describe why traditional PLC racks are not 'open'.

2. Discuss why the IEC 61131 standards should lead to open architecture control systems.

16.5 ASSIGNMENT PROBLEMS

1. Write a ladder logic program to perform the function outlined below. (Hint: use a structured technique.)
  - i) when the input 'part' turns on, the value 'weight' should be added to an array in memory.
  - ii) if any 'weight' value is greater than 15, and output 'halt' should be turned on, and the process should stop. A 'reset' input will be turned on to clear the array and start the process again.
  - iii) when 'part' has been activated 10 times the median of the part weights should be found. If it is greater that 14 the process should be stopped as described in step ii).
  - iv) if the median is less than or equal to 14, then a 'dump' output should be turned on for 2 seconds. After that the matrix should be reset and the process should begin again.

16.1 PRACTICE PROBLEM SOLUTIONS

1. The hardware and software are only sold by Allen Bradley, and users are not given details to modify or change the hardware and software.

2. The IEC standards are a first step to make programming methods between PLCs the same. The standard does not make programming uniform across all programming platforms, so it is not yet ready to develop completely portable controller programs and hardware.

17 INSTRUCTION LIST PROGRAMMING

Topics:

- Instruction list (IL) opcodes and operations
- Converting from ladder logic to IL
- Stack oriented instruction delay
- The Allen Bradley version of IL

Note: Allen Bradley does not offer IL programming as a standard option so this chapter may be considered optional.

Objectives:

- Instruction list (IL) opcodes and operations
- Converting from ladder logic to IL
- Stack oriented instruction delay
- The Allen Bradley version of IL

Instruction list (IL) programming is defined as part of the IEC 61131 standard. It uses very simple instructions similar to the original mnemonic programming languages developed for PLCs. (Note: some readers will recognize the similarity to assembly language programming.) It is the most fundamental level of programming language - all other programming languages can be converted to IL programs. Most programmers do not use IL programming on a daily basis, unless they are using hand held programmers.

17.1 THE IEC 61131 VERSION

To ease understanding, this chapter will focus on the process of converting ladder logic to IL programs. A simple example is shown in Figure 17.1 using the definitions found in the IEC standard. The rung of ladder logic contains four inputs, and one output. It can be expressed in a Boolean equation using parentheses. The equation can then be directly converted to instructions. The beginning of the program begins at the *START* label. At this point the first value is loaded, and the rest of the expression is broken up into small segments. The only significant change is that *AND NOT* becomes *ANDN*.

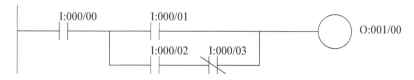

read as O:001/00 = I:000/00 AND (I:000/01 OR (I:000/02 AND NOT I:000/03))

Label	Opcode	Operand	Comment
START:	LD	%I:000/00	(* Load input bit 00 *)
	AND(	%I:000/01	(* Start a branch and load input bit 01 *)
	OR(	%I:000/02	(* Load input bit 02 *)
	ANDN	%I:000/03	(* Load input bit 03 and invert *)
	)		
	)		
	ST	%O:001/00	(* SET the output bit 00 *)

Figure 17.1   *An Instruction List Example*

An important concept in this programming language is the stack. (Note: if you use a calculator with RPN you are already familiar with this.) You can think of it as a do later list. With the equation in Figure 17.1 the first term in the expression is LD I:000/00, but the first calculation should be (I:000/02 AND NOT I:000/03). The instruction values are pushed on the stack until the most deeply nested term is found. Figure 17.1 illustrates how the expression is pushed on the stack. The LD instruction pushes the first value on the stack. The next instruction is an AND, but it is followed by a '(' so the stack must drop down. The OR(that follows also has the same effect. The ANDN instruction does not need to wait, so the calculation is done immediately and a result_1 remains. The next two ')' instructions remove the blocking '(' instruction from the stack, and allow the remaining OR I:000/1 and AND I:000/0 instructions to be done. The final result should be a single bit result_3. Two examples follow given different input conditions. If the final result in the stack is 0, then the output ST O:001/0 will set the output, otherwise it will turn it off.

LD I:000/0	AND(I:000/1	OR(I:000/2	ANDN I:000/3	)		)
I:000/0	I:000/1	I:000/2	result_1	result_2		result_3
	(	(	(	(		
	AND I:000/0	OR I:000/1	OR I:000/1	AND I:000/0		
		(	(			
		AND I:000/0	AND I:000/0			

Given:

I:000/0 = 1	1	0	1	1	1	1	1
I:000/1 = 0		(	(	(	(	AND 1	
I:000/2 = 1		AND 1	OR 0	OR 0	AND 1		
I:000/3 = 0			(	(			
			AND 1	AND 1			

Given:

I:000/0 = 0	0	1	0	0	0	0	0
I:000/1 = 1		(	(	(	(	AND 1	
I:000/2 = 0		AND 0	OR 1	OR 1	AND 1		
I:000/3 = 1			(	(			
			AND 0	AND 0			

Figure 17.2   ***Using a Stack for Instruction Lists***

A list of operations is given in Figure 17.1. The modifiers are;

> N - negates an input or output
> (- nests an operation and puts it on a stack to be pulled off by ')'
> C - forces a check for the currently evaluated results at the top of the stack

These operators can use multiple data types, as indicated in the data types column. This list should be supported by all vendors, but additional functions can be called using the CAL function.

Operator	Modifiers	Data Types	Description
LD	N	many	set current result to value
ST	N	many	store current result to location
S, R		BOOL	set or reset a value (latches or flip-flops)
AND, &	N, (	BOOL	boolean and
OR	N, (	BOOL	boolean or
XOR	N, (	BOOL	boolean exclusive or
ADD	(	many	mathematical add
SUB	(	many	mathematical subtraction
MUL	(	many	mathematical multiplication
DIV	(	many	mathematical division
GT	(	many	comparison greater than >
GE	(	many	comparison greater than or equal >=
EQ	(	many	comparison equals =
NE	(	many	comparison not equal <>
LE	(	many	comparison less than or equals <=
LT	(	many	comparison less than <
JMP	C, N	LABEL	jump to LABEL
CAL	C, N	NAME	call subroutine NAME
RET	C, N		return from subroutine call
)			get value from stack

Figure 17.3   *IL Operations*

17.2 THE ALLEN-BRADLEY VERSION

Allen Bradley only supports IL programming on the Micrologix 1000, and does not plan to support it in the future. Examples of the equivalent ladder logic and IL programs are shown in Figure 17.1 and Figure 17.1. The programs in Figure 17.1 show different variations when there is only a single output. Multiple IL programs are given where available. When looking at these examples recall the stack concept. When a *LD* or *LDN* instruction is encountered it will put a value on the top of the stack. The *ANB* and *ORB* instructions will remove the top two values from the stack, and replace them with a single value that is the result of an Boolean operation. The *AND* and *OR* functions take one value off the top of the stack, perform a Boolean operation and put the result on the top of the stack. The equivalent programs (to the right) are shorter and will run faster.

Ladder	Instruction List (IL)	
A —X	LD A	
	ST X	
A (NC) —X	LDN A	
	ST X	
A B —X	LD A	LD A
	LD B	AND B
	ANB	ST X
	ST X	
A B (NC) —X	LD A	LD A
	LDN B	ANDN B
	ANB	ST X
	ST X	
A C —X; B	LD A	LD A
	LD B	OR B
	ORB	AND C
	LD C	ST X
	ANB	
	ST X	
A B —X; C	LD A	LD A
	LD B	LD B
	LD C	OR C
	ORB	ANB
	ANB	ST X
	ST X	
A C —X; B D	LD A	LD A
	LD B	OR B
	ORB	LD C
	LD C	OR D
	LD D	ANB
	ORB	ST X
	ANB	
	ST X	

Figure 17.4   *IL Equivalents for Ladder Logic*

Figure 17.1 shows the IL programs that are generated when there are multiple outputs. This often requires that the stack be used to preserve values that would be lost normally using the *MPS*, *MPP* and *MRD* functions. The *MPS* instruction will store the current value of the top of the stack. Consider the first example with two outputs, the value of *A* is loaded on the stack with *LD A*. The instruction *ST X* examines the top of the stack, but does not remove the value, so it is still available for *ST Y*. In the third example the value of the top of the stack would not be correct when the second output rung was examined. So, when the output branch occurs the value at the top of the stack is copied using *MPS*, and pushed on the top of the stack. The copy is then ANDed with *B* and used to set *X*. After this the value at the top is pulled off with the *MPP* instruction, leaving the value at the top what is was before the first output rung. The last example shows multiple output rungs. Before the first rung the value is copied on the stack using MPS. Before the last rung the value at the top of the stack is discarded with the *MPP* instruction. But, the two center instructions use *MRD* to copy the right value to the top of the stack - it could be replaced with *MPP* then *MPS*.

356

Ladder	Instruction List (IL)	
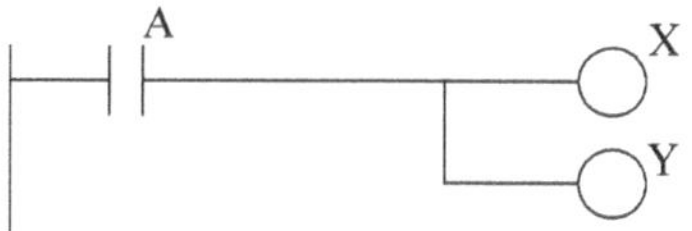	LD A ST X ST Y	
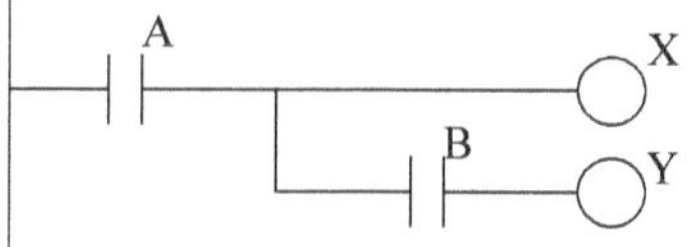	LD A ST X LD B ANB ST Y	LD A ST X AND B ST Y
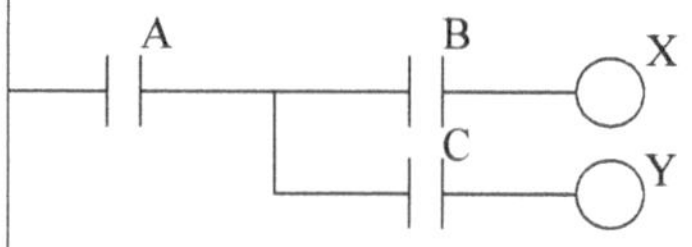	LD A MPS LD B ANB ST X MPP LD C ANB ST Y	LD A MPS AND B ST X MPP AND C ST Y
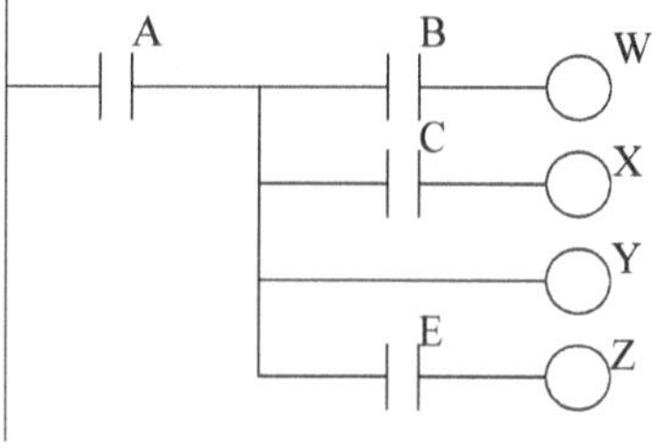	LD A MPS LD B ANB ST W MRD LD C ANB ST X MRD STY MPP LD E ANB ST Z	LD A MPS AND B ST W MRD AND C ST X MRD ST Y MPP AND E ST Z

Figure 17.5 ***IL Programs for Multiple Outputs***

Complex instructions can be represented in IL, as shown in Figure 17.1. Here the function are listed by their mnemonics, and this is followed by the arguments for the functions. The second line does not have any input contacts, so the stack is loaded with a true value.

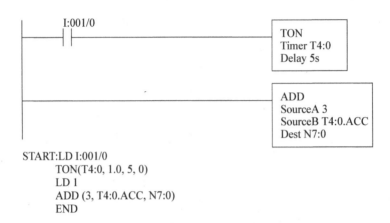

```
START:LD I:001/0
      TON(T4:0, 1.0, 5, 0)
      LD 1
      ADD (3, T4:0.ACC, N7:0)
      END
```

Figure 17.6   *A Complex Ladder Rung and Equivalent IL*

An example of an instruction language subroutine is shown in Figure 17.1. This program will examine a BCD input on card I:000, and if it becomes higher than 100 then 2 seconds later output O:001/00 will turn on.

Program File 2:

Label	Opcode	Operand	Comment
START:	CAL	3	(* Jump to program file 3 *)

Program File 3:

Label	Opcode	Operand	Comment
TEST:	LD	%I:000	(* Load the word from input card 000 *)
	BCD_TO_INT		(* Convert the BCD value to an integer *)
	ST	%N7:0	(* Store the value in N7:0 *)
	GT	100	(* Check for the stored value (N7:0) > 100 *)
	JMPC	ON	(* If true jump to ON *)
	CAL	RES(C5:0)	(* Reset the timer *)
ON:	LD	2	(* Load a value of 2 - for the preset *)
	ST	%C5:0.PR	(* Store 2 in the preset value *)
	CAL	TON(C5:0)	(* Update the timer *)
	LD	%C5:0.DN	(* Get the timer done condition bit *)
	ST	%O:001/00	(* Set the output bit *)
	RET		(* Return from the subroutine *)

358

Figure 17.7   *An Example of an IL Program*

17.3 SUMMARY

- Ladder logic can be converted to IL programs, but IL programs cannot always be converted to ladder logic.
- IL programs use a stack to delay operations indicated by parentheses.
- The Allen Bradley version is similar, but not identical to the IEC 61131 version of IL.

17.4 ASSIGNMENT PROBLEMS

1. Explain the operation of the stack.

2. Convert the following ladder logic to IL programs.

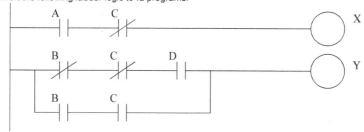

3. Write the ladder diagram programs that correspond to the following Boolean programs.

LD 001	LD 001	LD NOT 001
OR 003	AND 002	AND 002
LD 002	LD 004	LD 004
OR 004	AND 005	OR 007
AND LD	OR LD	AND 005
LD 005	OR 007	OR LD
OR 007	LD 003	LD 003
AND 006	OR NOT 006	OR NOT 006
OR LD	AND LD	AND LD
OUT 204		OR NOT 008
		OUT 204
		AND 009
		OUT 206
		AND NOT 010
		OUT 201

18STRUCTURED TEXT PROGRAMMING

Topics:

- Basic language structure and syntax
- Variables, functions, values
- Program flow commands and structures
- Function names
- Program Example

Objectives:

- To be able to write functions in Structured Text programs
- To understand the parallels between Ladder Logic and Structured Text
- To understand differences between Allen Bradley and the standard

If you know how to program in any high level language, such as Basic or C, you will be comfortable with Structured Text (ST) programming. ST programming is part of the IEC 61131 standard. An example program is shown in Figure 18.1. The program is called *main* and is defined between the statements *PROGRAM* and *END_PROGRAM*. Every program begins with statements the define the variables. In this case the variable *i* is defined to be an integer. The program follows the variable declarations. This program counts from 0 to 10 with a loop. When the example program starts the value of integer memory *i* will be set to zero. The *REPEAT* and *END_REPEAT* statements define the loop. The *UNTIL* statement defines when the loop must end. A line is present to increment the value of *i* for each loop.

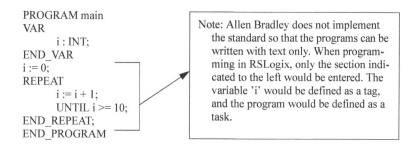

```
PROGRAM main
VAR
        i : INT;
END_VAR
i := 0;
REPEAT
        i := i + 1;
        UNTIL i >= 10;
END_REPEAT;
END_PROGRAM
```

Note: Allen Bradley does not implement the standard so that the programs can be written with text only. When programming in RSLogix, only the section indicated to the left would be entered. The variable 'i' would be defined as a tag, and the program would be defined as a task.

Figure 18.1   A Structured Text Example Program

One important difference between ST and traditional programming languages is the nature of program flow control. A ST program will be run from beginning to end many times each second. A traditional program should not reach the end until it is completely finished. In the previous example the loop could lead to a program that (with some modification) might go into an infinite loop. If this were to happen during a control application the controller would stop responding, the process might become dangerous, and the controller watchdog timer would force a fault.

ST has been designed to work with the other PLC programming languages. For example, a ladder logic program can call a structured text subroutine.

18.1 THE LANGUAGE

The language is composed of written statements separated by semicolons. The statements use predefined statements and program subroutines to change variables. The variables can be explicitly defined values, internally stored variables, or inputs and outputs. Spaces can be used to separate statements and variables, although they are not often necessary. Structured text is not case sensitive, but it can be useful to make variables lower case, and make statements upper case. Indenting and comments should also be used to increase readability and documents the program. Consider the example shown in Figure 18.1.

```
                FUNCTION sample
                    INPUT_VAR
GOOD
                                    start : BOOL;   (* a NO start input *)
                                    stop : BOOL;    (* a NC stop input *)
                    END_VAR
                    OUTPUT_VAR
                                    motor : BOOL;   (* a motor control relay
        *)
                    END_VAR
                    motor := (motor + start) * stop; (* get the motor output *)
                END_FUNCTION
```

```
                FUNCTION sample
                INPUT_VAR
BAD             START:BOOL;STOP:BOOL;
                END_VAR
                OUTPUT_VAR
                MOTOR:BOOL;
                END_VAR MOTOR:=(MOTOR+START)*STOP;END_FUNCTION
```

Figure 18.2   *A Syntax and Structured Programming Example*

18.1.1 Elements of the Language

ST programs allow named variables to be defined. This is similar to the use of symbols when programming in ladder logic. When selecting variable names they must begin with a letter, but after that they can include combinations of letters, numbers, and some symbols such as '_'. Variable names are not case sensitive and can include any combination of upper and lower case letters. Variable names must also be the same as other key words in the system as shown in Figure 18.1. In addition, these variable must not have the same name as predefined functions, or user defined functions.

Invalid variable names: START, DATA, PROJECT, SFC, SFC2, LADDER, I/O, ASCII, CAR, FORCE, PLC2, CONFIG, INC, ALL, YES, NO, STRUCTURED TEXT

Valid memory/variable name examples: TESTER, I, I:000, I:000/00, T4:0, T4:0/DN, T4:0.ACC

Figure 18.3   *Acceptable Variable Names*

When defining variables one of the declarations in Figure 18.1 can be used. These define the scope of the variables. The *VAR_INPUT*, *VAR_OUTPUT* and *VAR_IN_OUT* declarations are used for variables that are passed as arguments to the program or function. The *RETAIN* declaration is used to retain a variable value, even when the PLC power has been cycled. This is similar to a latch application. As mentioned before these are not used when writing Allen Bradley programs, but they are used when defining tags to be used by the structured programs.

Declaration	Description
VAR	the general variable declaration
VAR_INPUT	defines a variable list for a function
VAR_OUTPUT	defines output variables from a function
VAR_IN_OUT	defines variable that are both inputs and outputs from a function
VAR_EXTERNAL	
VAR_GLOBAL	a global variable
VAR_ACCESS	
RETAIN	a value will be retained when the power is cycled
CONSTANT	a value that cannot be changed
AT	can tie a variable to a specific location in memory (without this variable locations are chosen by the compiler
END_VAR	marks the end of a variable declaration

Figure 18.4   **Variable Declarations**

Examples of variable declarations are given in Figure 18.1.

Text Program Line	Description
VAR AT %B3:0 : WORD; END_VAR	a word in bit memory
VAR AT %N7:0 : INT; END_VAR	an integer in integer memory
VAR RETAIN AT %O:000 : WORD ; END_VAR	makes output bits retentive
VAR_GLOBAL A AT %I:000/00 : BOOL ; END_VAR	variable 'A' as input bit
VAR_GLOBAL A AT %N7:0 : INT ; END_VAR	variable 'A' as an integer
VAR A AT %F8:0 : ARRAY [0..14] OF REAL; END_VAR	an array 'A' of 15 real values
VAR A : BOOL; END_VAR	a boolean variable 'A'
VAR A, B, C : INT ; END_VAR	integers variables 'A', 'B', 'C'
VAR A : STRING[10] ; END_VAR	a string 'A' of length 10
VAR A : ARRAY[1..5,1..6,1..7] OF INT; END_VAR	a 5x6x7 array 'A' of integers
VAR RETAIN RTBT A : ARRAY[1..5,1..6] OF INT; END_VAR	a 5x6 array of integers, filled with zeros after power off
VAR A : B; END_VAR	'A' is data type 'B'
VAR CONSTANT A : REAL := 5.12345 ; END_VAR	a constant value 'A'
VAR A AT %N7:0 : INT := 55; END_VAR	'A' starts with 55
VAR A : ARRAY[1..5] OF INT := [5(3)]; END_VAR	'A' starts with 3 in all 5 spots
VAR A : STRING[10] := 'test'; END_VAR	'A' contains 'test' initially
VAR A : ARRAY[0..2] OF BOOL := [1,0,1]; END_VAR	an array of bits
VAR A : ARRAY[0..1,1..5] OF INT := [5(1),5(2)]; END_VAR	an array of integers filled with 1 for [0,x] and 2 for [1,x]

Figure 18.5   **Variable Declaration Examples**

Basic numbers are shown in Figure 18.1. Note the underline '_' can be ignored, it can be used to break up long numbers, ie. 10_000 = 10000. These are the literal values discussed for Ladder Logic.

number type	examples
integers	-100, 0, 100, 10_000
real numbers	-100.0, 0.0, 100.0, 10_000.0
real with exponents	-1.0E-2, -1.0e-2, 0.0e0, 1.0E2
binary numbers	2#111111111, 2#1111_1111, 2#1111_1101_0110_0101
octal numbers	8#123, 8#777, 8#14
hexadecimal numbers	16#FF, 16#ff, 16#9a, 16#01
boolean	0, FALSE, 1, TRUE

Figure 18.6   *Literal Number Examples*

Character strings defined as shown in Figure 18.1.

example	description
''	a zero length string
' ', 'a', '$"', '$$'	a single character, a space, or 'a', or a single quote, or a dollar sign $
'RL', 'rl','$0D$0A'	produces ASCII CR, LF combination - end of line characters
'$P', '$p'	form feed, will go to the top of the next page
'$T', '4t'	tab
'this%Tis a test$R$L'	a string that results in 'this\<TAB\>is a test\<NEXT LINE\>'

Figure 18.7   *Character String Data*

Basic time and date values are described in Figure 18.1 and Figure 18.1. Although it should be noted that for ControlLogix the GSV function is used to get the values.

Time Value	Examples
25ms	T#25ms, T#25.0ms, TIME#25.0ms, T#-25ms, t#25ms
5.5hours	TIME#5.3h, T#5.3h, T#5h_30m, T#5h30m
3days, 5hours, 6min, 36sec	TIME#3d5h6m36s, T#3d_5h_6m_36s

Figure 18.8   *Time Duration Examples*

description	examples
date values	DATE#1996-12-25, D#1996-12-25
time of day	TIME_OF_DAY#12:42:50.92, TOD#12:42:50.92
date and time	DATE_AND_TIME#1996-12-25-12:42:50.92, DT#1996-12-25-12:42:50.92

Figure 18.9   *Time and Date Examples*

The math functions available for structured text programs are listed in Figure 18.1. It is worth noting that these functions match the structure of those available for ladder logic. Other, more advanced, functions are also available - a general rule of thumb is if a function is available in one language, it is often available for others.

:=	assigns a value to a variable
+	addition
-	subtraction
/	division
*	multiplication
MOD(A,B)	modulo - this provides the remainder for an integer divide A/B
SQR(A)	square root of A
FRD(A)	from BCD to decimal
TOD(A)	to BCD from decimal
NEG(A)	reverse sign +/-
LN(A)	natural logarithm
LOG(A)	base 10 logarithm
DEG(A)	from radians to degrees
RAD(A)	to radians from degrees
SIN(A)	sine
COS(A)	cosine
TAN(A)	tangent
ASN(A)	arcsine, inverse sine
ACS(A)	arccosine - inverse cosine
ATN(A)	arctan - inverse tangent
XPY(A,B)	A to the power of B
A**B	A to the power of B

Figure 18.10   **Math Functions**

Functions for logical comparison are given in Figure 18.1. These will be used in expressions such as IF-THEN statements.

>	greater than
>=	greater than or equal
=	equal
<=	less than or equal
<	less than
<>	not equal

Figure 18.11   **Comparisons**

Boolean algebra functions are available, as shown in Figure 18.1. The can be applied to bits or integers.

AND(A,B)	logical and
OR(A,B)	logical or
XOR(A,B)	exclusive or
NOT(A)	logical not
!	logical not (note: not implemented on AB controllers)

Figure 18.12   **Boolean Functions**

The precedence of operations are listed in Figure 18.1 from highest to lowest. As normal expressions that are the most deeply nested between brackets will be solved first. (Note: when in doubt use brackets to ensure you get the sequence you expect.)

! - (Note: not available on AB controllers)
()
functions
XPY, **
negation
SQR, TOD, FRD, NOT, NEG, LN, LOG, DEG, RAD, SIN, COS, TAN, ASN, ACS, ATN
*, /, MOD
+, -
>, >=, =, <=, <, <>
AND (for word)
XOR (for word)
OR (for word)
AND (bit)
XOR (bit)
OR (bit)
ladder instructions

highest priority (vertical label with upward arrow on left side)

Figure 18.13   **Operator Precedence**

Common language structures include those listed in Figure 18.1.

IF-THEN-ELSIF-ELSE-END_IF;   normal if-then structure
CASE-value:-ELSE-END_CASE;   a case switching function
FOR-TO-BY-DO-END_FOR;   for-next loop
WHILE-DO-END_WHILE;

Figure 18.14   **Flow Control Functions**

Special instructions include those shown in Figure 18.1.

RETAIN();   causes a bit to be retentive
IIN();   immediate input update
EXIT;   will quit a FOR or WHILE loop
EMPTY

Figure 18.15   **Special Instructions**

18.1.2 Putting Things Together in a Program

Consider the program in Figure 18.1 to find the average of five values in a real array 'f[]'. The FOR loop in the example will loop five times adding the array values. After that the sum is divided to get the average.

```
avg := 0;
FOR (i := 0 TO 4) DO
        avg := avg + f[i];
END_FOR;
avg := avg / 5;
```

Figure 18.16   *A Program To Average Five Values In Memory With A For-Loop*

The previous example is implemented with a WHILE loop in Figure 18.1. The main differences is that the initial value and update for 'i' must be done manually.

```
avg := 0;
i := 0;
WHILE (i < 5) DO
        avg := avg + f[i];
        i := i + 1;
END_WHILE;
avg := avg / 5;
```

Figure 18.17   *A Program To Average Five Values In Memory With A While-Loop*

The example in Figure 18.1 shows the use of an IF statement. The example begins with a timer. These are handled slightly differently in ST programs. In this case if 'b' is true the timer will be active, if it is false the timer will reset. The second instruction calls 'TONR' to update the timer. (Note: ST programs use the FBD_TIMER type, instead of the TIMER type.) The IF statement works as normal, only one of the three cases will occur with the ELSE defining the default if the other two fail.

```
t.TimerEnable := b;
TONR(t);
IF (a = 1) THEN
        x := 1;
ELSIF (b = 1 AND t.DN = 1) THEN
        y := 1;
        IF (I:000/02 = 0) THEN
          z := 1;
        END_IF;
ELSE
        x := 0;
        y := 0;
        z := 0;
END_IF;
```

Figure 18.18   *Example With An If Statement*

Figure 18.1 shows the use of a CASE statement to set bits 0 to 3 of 'a' based upon the value of 'test'. In the event none of the values are matched, 'a' will be set to zero, turning off all bits.

```
CASE test OF
     0:
        a.0 := 1;
     1:
        a.1 := 1;
     2:
        a.2 := 1;
     3:
        a.3 := 1;
ELSE
     a := 0;
END_CASE;
```

Figure 18.19   *Use of a Case Statement*

The example in Figure 18.1 accepts a BCD input from 'bcd_input' and uses it to change the delay time for TON delay time. When the input 'test_input' is true the time will count. When the timer is done 'set' will become true.

```
FRD (bcd_input, delay_time);
t.PRE := delay_time;
IF (test_input) THEN
        t.EnableTimer := 1;
ELSE
        t.EnableTimer := 0;
END_IF;
TONR(t);
set := t.DN;
```

Figure 18.20   *Function Data Conversions*

Most of the IEC61131-3 defined functions with arguments are given in Figure 18.1. Some of the functions can be overloaded, for example ADD could have more than two values to add, and others have optional arguments. In most cases the optional arguments are things like preset values for timers. When arguments are left out they default to values, typically 0. ControlLogix uses many of the standard function names and arguments but does not support the overloading part of the standard.

Function	Description
ABS(A);	absolute value of A
ACOS(A);	the inverse cosine of A
ADD(A,B,...);	add A+B+...
AND(A,B,...);	logical and of inputs A,B,...
ASIN(A);	the inverse sine of A
ATAN(A);	the inverse tangent of A
BCD_TO_INT(A);	converts a BCD to an integer
CONCAT(A,B,...);	will return strings A,B,... joined together
COS(A);	finds the cosine of A
CTD(CD:=A,LD:=B,PV:=C);	down counter active <=0, A decreases, B loads preset
CTU(CU:=A,R:=B,PV:=C);	up counter active >=C, A decreases, B resets
CTUD(CU:=A,CD:=B,R:=C,LD:=D,PV:=E);	up/down counter combined functions of the up and down counters
DELETE(IN:=A,L:=B,P:=C);	will delete B characters at position C in string A
DIV(A,B);	A/B
EQ(A,B,C,...);	will compare A=B=C=...
EXP(A);	finds e**A where e is the natural number
EXPT(A,B);	A**B
FIND(IN1:=A,IN2:=B);	will find the start of string B in string A
F_TRIG(A);	a falling edge trigger
GE(A,B,C,...);	will compare A>=B, B>=C, C>=...
GT(A,B,C,...);	will compare A>B, B>C, C>...
INSERT(IN1:=A,IN2:=B,P:=C);	will insert string B into A at position C
INT_TO_BCD(A);	converts an integer to BCD
INT_TO_REAL(A);	converts A from integer to real
LE(A,B,C,...);	will compare A<=B, B<=C, C<=...
LEFT(IN:=A,L:=B);	will return the left B characters of string A
LEN(A);	will return the length of string A
LIMIT(MN:=A,IN:=B,MX:=C);	checks to see if B>=A and B<=C
LN(A);	natural log of A
LOG(A);	base 10 log of A
LT(A,B,C,...);	will compare A<B, B<C, C<...
MAX(A,B,...);	outputs the maximum of A,B,...
MID(IN:=A,L:=B,P:=C);	will return B characters starting at C of string A
MIN(A,B,...);	outputs the minimum of A,B,...
MOD(A,B);	the remainder or fractional part of A/B
MOVE(A);	outputs the input, the same as :=
MUL(A,B,...);	multiply values A*B*....
MUX(A,B,C,...);	the value of A will select output B,C,...
NE(A,B);	will compare A <> B
NOT(A);	logical not of A
OR(A,B,...);	logical or of inputs A,B,...

Function	Description
REAL_TO_INT(A);	converts A from real to integer
REPLACE(IN1:=A,IN2:=B,L:= C,P:=D);	will replace C characters at position D in string A with string B
RIGHT(IN:=A,L:=B);	will return the right A characters of string B
ROL(IN:=A,N:=B);	rolls left value A of length B bits
ROR(IN:=A,N:=B);	rolls right value A of length B bits
RS(A,B);	RS flip flop with input A and B
RTC(IN:=A,PDT:=B);	will set and/or return current system time
R_TRIG(A);	a rising edge trigger
SEL(A,B,C);	if a=0 output B if A=1 output C
SHL(IN:=A,N:=B);	shift left value A of length B bits
SHR(IN:=A,N:=B);	shift right value A of length B bits
SIN(A);	finds the sine of A
SQRT(A);	square root of A
SR(S1:=A,R:=B);	SR flipflop with inputs A and B
SUB(A,B);	A-B
TAN(A);	finds the tangent of A
TOF(IN:=A,PT:=B);	off delay timer
TON(IN:=A,PT:=B);	on delay timer
TP(IN:=A,PT:=B);	pulse timer - a rising edge fires a fixed period pulse
TRUNC(A);	converts a real to an integer, no rounding
XOR(A,B,...);	logical exclusive or of inputs A,B,...

Figure 18.21   **Structured Text Functions**

Control programs can become very large. When written in a single program these become confusing, and hard to write/debug. The best way to avoid the endless main program is to use subroutines to divide the main program. The IEC61131 standard allows the definition of subroutines/functions as shown in Figure 18.1. The function will accept up to three inputs and perform a simple calculation. It then returns one value. As mentioned before ControlLogix does not support overloading, so the function would not be able to have a variable size argument list.

```
....
D := TEST(1.3, 3.4); (* sample calling program, here C will default to 3.14 *)
E := TEST(1.3, 3.4, 6.28); (* here C will be given a new value *)
....

FUNCTION TEST : REAL
     VAR_INPUT A, B : REAL; C : REAL := 3.14159; END_VAR
     TEST := (A + B) / C;
END_FUNCTION
```

Figure 18.22   **Declaration of a Function**

18.2 AN EXAMPLE

The example beginning in Figure 18.1 shows a subroutine implementing traffic lights in ST for the ControlLogix processor. The variable 'state' is used to keep track of the current state of the lights. Timer enable bits are used to determine which transition should be checked. Finally the value of 'state' is used to set the outputs. (Note: this is possible because '=' and ':=' are not the same.) This subroutine would be stored under a name such as 'TrafficLights'. It would then be called from the main program as shown in Figure 18.1.

```
                                    ┌──────────────────────────────┐
────────────────────────────────────│ JSR                          │
                                   ──│ Function Name: TrafficLights  │
────────────────────────────────────└──────────────────────────────┘
```

Figure 18.23   *The Main Traffic Light Program*

```
    SBR();
          IF S:FS THEN
                                      state := 0;
                                      green_EW.TimerEnable := 1;
                                      yellow_EW.TimerEnable :=
 0;
                                      green_NS.TimerEnable := 0;
                                      yellow_NS.TimerEnable :=
 0;
          END_IF;

          TONR(green_EW); TONR(yellow_EW);
          TONR(green_NS); TONR(yellow_NS);

          CASE state OF
          0:                          IF green_EW.DN THEN
                                               state :=1;
 green_EW.TimerEnable := 0;

 yellow_EW.TimerEnable := 1;
                              END_IF
          1:                          IF yellow_EW.DN THEN
                                               state :=2;
 yellow_EW.TimerEnable := 0;

 green_NS.TimerEnable := 1;
                              END_IF
          2:                          IF green_NS.DN THEN
                                               state :=3;
 green_NS.TimerEnable := 0;

 yellow_NS.TimerEnable := 1;
                              END_IF
          3:                          IF yellow_NS.DN THEN
                                               state :=0;
 yellow_NS.TimerEnable := 0;

 green_EW.TimerEnable := 1;
```

┌──┐
│ Note: This example is for the AB   │
│ ControlLogix platform, so it   │
│ does not show the normal   │
│ function and tag definitions.   │
│ These are done separately in   │
│ the tag editor.   │
│   │
│ state : DINT   │
│ green_EW : FBD_TIMER   │
│ yellow_EW : FBD_TIMER   │
│ green_NS : FBD_TIMER   │
│ yellow_NS : FBD_TIMER   │
│ light_EW_green : BOOL alias =   │
│   rack:1:O.Data.0   │
│ light_EW_yellow : BOOL alias =   │
│   rack:1:O.Data.1   │
│ light_EW_red : BOOL alias =   │
│   rack:1:O.Data.2   │
│ light_NS_green : BOOL alias =   │
│   rack:1:O.Data.3   │
│ light_NS_yellow : BOOL alias =   │
│   rack:1:O.Data.4   │
│ light_NS_red : BOOL alias =   │
│   rack:1:O.Data.5   │
└──┘

Figure 18.24   *Traffic Light Subroutine*

18.3 SUMMARY

- Structured text programming variables, functions, syntax were discussed.
- The differences between the standard and the Allen Bradley implementation were indicated as appropriate.
- A traffic light example was used to illustrate a ControlLogix application

18.4 PRACTICE PROBLEMS

(Note: Problem solutions are available at http://sites.google.com/site/automatedmanufacturingsystems/)

1. Write a structured text program that will replace the following ladder logic.

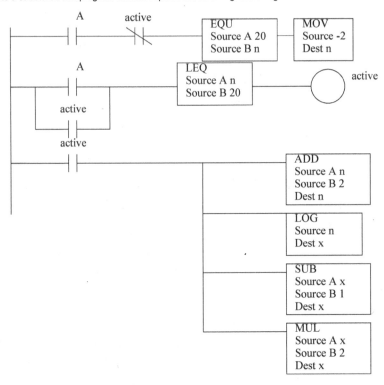

2. Implement the following Boolean equations in a Structured Text program. If the program was for a state machine what changes would be required to make it work?

$$light = (light + dark \bullet switch) \bullet \overline{switch} \bullet light$$
$$dark = (dark + light \bullet \overline{switch}) \bullet \overline{switch} \bullet dark$$

3. Convert the following state diagram to a Structured Text program.

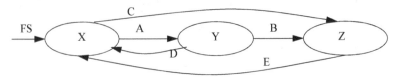

4. A temperature value is stored in F8:0. When it rises above 40 the following sequence should occur once. Write a ladder logic program that implement this function with a Structured Text program.

5. Write a structured text program that will replace the following ladder logic.

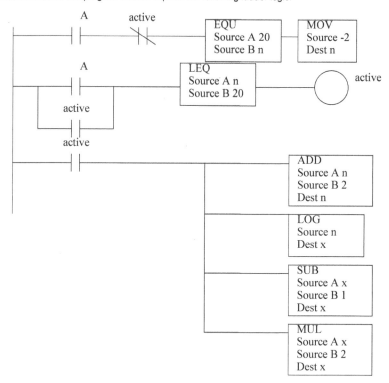

18.5 ASSIGNMENT PROBLEMS

1. Write logic for a traffic light controller using structured text.

2. Write a structured text program to control a press that has an advance and retract with limit switches. The press is started and stopped with start and stop buttons.

3. Write a structured text program to sort a set of ten integer numbers and then find the median value.

4. Write a ST program to accept an input string 'teststring' with the sample format of 'XWT55.36KG'. The program should extract the 5 digit number in the middle of the string starting at the 4th position. If the number is less than or equal to 20, or greater than 60, the output Y should be set.

5. Write a ST program to convert two string 'AA' and 'BB' to real numbers, add them, and convert the result back to a string 'CC'.

18.1 PRACTICE PROBLEM SOLUTIONS

1.

```
SBR();
        IF A AND NOT (active) AND (20 = n) THEN
        n := -2;
        END_IF
        active := (A OR active) AND (n <= 20);
        IF active THEN
        n := n + 2;
        x := LOG(n);
        x := x - 1;
        x := x * 2;
        END_IF
RET();
```

2.

Implemented Exactly

```
SBR()
        light := (light OR dark AND switch) AND NOT (NOT(switch) AND dark) OR S:FS;
        dark := (dark OR light AND NOT(switch)) AND NOT(switch AND dark);
RET();
```

Corrected for State Diagram

```
SBR()
        lightX := (light OR dark AND switch) AND NOT (NOT(switch) AND dark) OR S:FS;
        darkX := (dark OR light AND NOT(switch)) AND NOT(switch AND dark);
        light := lightX + S:FS;
        dark := darkX;
RET();
```

3.

```
SBR()
        T1 := X AND C;
        T2 := X AND A AND NOT (C);
        T3 := Y AND B AND NOT (D);
        T4 := Z AND E;
        T5 := Y AND D;

        X := (X OR T4 OR T5) AND NOT (T1 OR T2) OR S:FS
        Y := (Y OR T2) AND NOT (T3 OR T5)
        Z := (Z OR T1 OR T3) AND NOT T4;

RET();
```

4.

```
SBR()
        run := (run OR (F8:0 >= 40)) AND stop;

        t_2.TimerEnable := run;
        t_5.TimerEnable := run;
        t_11.TimerEnable := run;
        t_15.TimerEnable := run;
        TONR(t_2);
        TONR(t_5);
        TONR(t_11);
        TONR(t_15);

        horn := (t_2.DN AND t_5.TT) OR (t_11.DN AND t_15.TT);
RET();
```

5.

```
SBR();
        IF A AND NOT (active) AND (20 = n) THEN
        n := -2;
        END_IF
        active := (A OR active) AND (n <= 20);
        IF active THEN
        n := n + 2;
        x := LOG(n);
        x := x - 1;
        x := x * 2;
        END_IF
RET();
```

19SEQUENTIAL FUNCTION CHARTS

Topics:

- Describing process control SFCs
- Conversion of SFCs to ladder logic

Objectives:

- Learn to recognize parallel control problems.
- Be able to develop SFCs for a process.
- Be able to convert SFCs to ladder logic.

All of the previous methods are well suited to processes that have a single state active at any one time. This is adequate for simpler machines and processes, but more complex machines are designed perform simultaneous operations. This requires a controller that is capable of concurrent processing - this means more than one state will be active at any one time. This could be achieved with multiple state diagrams, or with more mature techniques such as Sequential Function Charts.

Sequential Function Charts (SFCs) are a graphical technique for writing concurrent control programs. (Note: They are also known as Grafcet or IEC 848.) SFCs are a subset of the more complex Petri net techniques that are discussed in another chapter. The basic elements of an SFC diagram are shown in Figure 19.1 and Figure 19.1.

flowlines - connects steps and transitions (these basically indicate sequence)
transition - causes a shift between steps, acts as a point of coordination

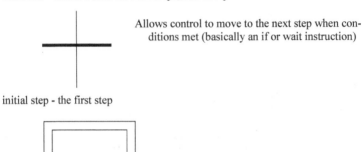

Allows control to move to the next step when conditions met (basically an if or wait instruction)

initial step - the first step

step - basically a state of operation. A state often has an associated action

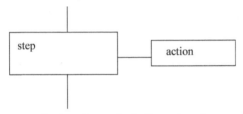

macrostep - a collection of steps (basically a subroutine

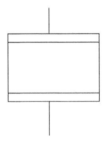

Figure 19.1   **Basic Elements in SFCs**

selection branch - an OR - only one path is followed

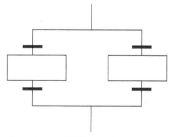

simultaneous branch - an AND - both (or more) paths are followed

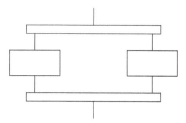

Figure 19.2   **Basic Elements in SFCs**

The example in Figure 19.1 shows a SFC for control of a two door security system. One door requires a two digit entry code, the second door requires a three digit entry code. The execution of the system starts at the top of the diagram at the *Start* block when the power is turned on. There is an action associated with the *Start* block that locks the doors. (Note: in practice the SFC uses ladder logic for inputs and outputs, but this is not shown on the diagram.) After the start block the diagram immediately splits the execution into two processes and both steps 1 and 6 are active. Steps are quite similar to states in state diagrams. The transitions are similar to transitions in state diagrams, but they are drawn with thick lines that cross the normal transition path. When the right logical conditions are satisfied the transition will stop one step and start the next. While step 1 is active there are two possible transitions that could occur. If the first combination digit is correct then step 1 will become inactive and step 2 will become active. If the digit is incorrect then the transition will then go on to wait for the later transition for the 5 second delay, and after that step 5 will be active. Step 1 does not have an action associated, so nothing should be done while waiting for either of the transitions. The logic for both of the doors will repeat once the cycle of combination-unlock-delay-lock has completed.

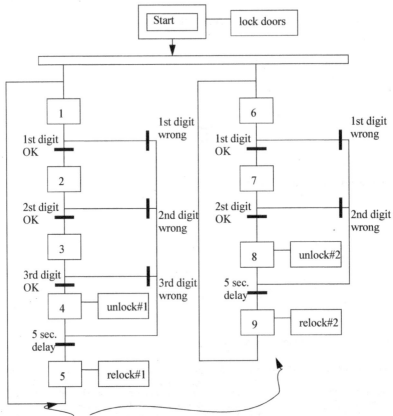

Parallel/Concurrent because things happen separately, but at same time
(this can also be done with state transition diagrams)

Figure 19.3   *SFC for Control of Two Doors with Security Codes*

A simple SFC for controlling a stamping press is shown in Figure 19.1. (Note: this controller only has a single thread of execution, so it could also be implemented with state diagrams, flowcharts, or other methods.) In the diagram the press starts in an idle state. when an *automatic* button is pushed the press will turn on the press power and lights. When a part is detected the press ram will advance down to the bottom limit switch. The press will then retract the ram until the top limit switch is contacted, and the ram will be stopped. A stop button can stop the press only when it is advancing. (Note: normal designs require that stops work all the time.) When the press is stopped a *reset* button must be pushed before the *automatic* button can be pushed again. After step 6 the press will wait until the part is not present before waiting for the next part. Without this logic the press would cycle continuously.

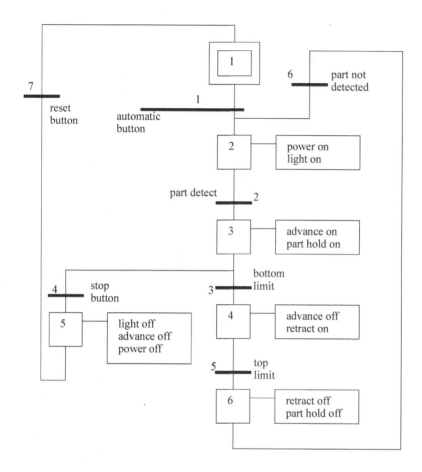

Figure 19.4   *SFC for Controlling a Stamping Press*

The SFC can be converted directly to ladder logic with methods very similar to those used for state diagrams as shown in Figure 19.1 to Figure 19.1. The method shown is patterned after the block logic method. One significant difference is that the transitions must now be considered separately. The ladder logic begins with a section to initialize the states and transitions to a single value. The next section of the ladder logic considers the transitions and then checks for transition conditions. If satisfied the following step or transition can be turned on, and the transition turned off. This is followed by ladder logic to turn on outputs as requires by the steps. This section of ladder logic corresponds to the actions for each step. After that the steps are considered, and the logic moves to the following transitions or steps. The sequence *examine transitions*, *do actions* then *do steps* is very important. If other sequences are used outputs may not be actuated, or steps missed entirely.

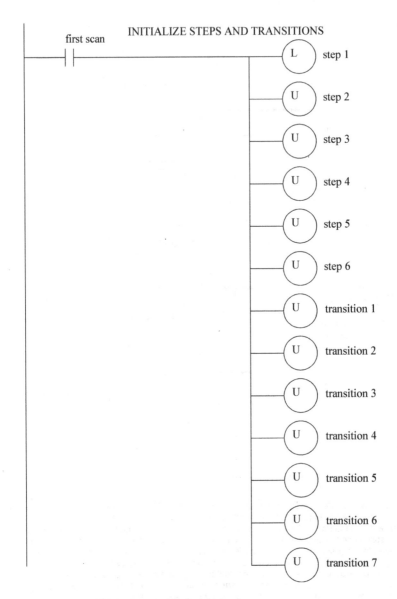

Figure 19.5   *SFC Implemented in Ladder Logic*

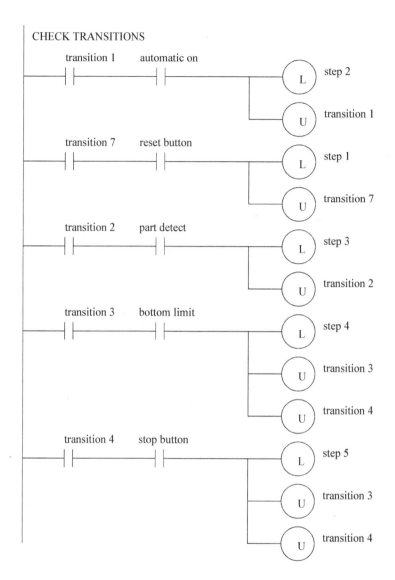

CHECK TRANSITIONS

transition 1	automatic on		step 2
transition 1	automatic on		transition 1
transition 7	reset button		step 1
transition 7	reset button		transition 7
transition 2	part detect		step 3
transition 2	part detect		transition 2
transition 3	bottom limit		step 4
transition 3	bottom limit		transition 3
			transition 4
transition 4	stop button		step 5
transition 4	stop button		transition 3
			transition 4

Figure 19.6   *SFC Implemented in Ladder Logic*

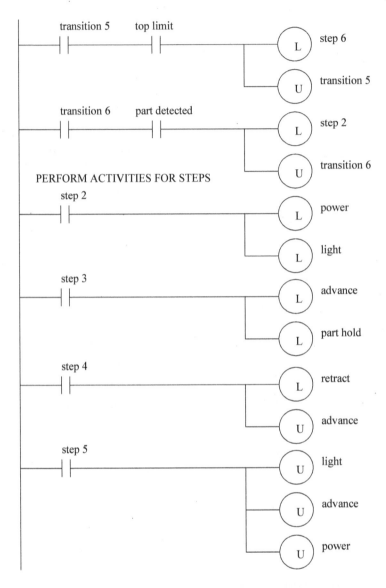

transition 5   top limit → (L) step 6

(U) transition 5

transition 6   part detected → (L) step 2

(U) transition 6

PERFORM ACTIVITIES FOR STEPS

step 2 → (L) power

(L) light

step 3 → (L) advance

(L) part hold

step 4 → (L) retract

(U) advance

step 5 → (U) light

(U) advance

(U) power

Figure 19.7   *SFC Implemented in Ladder Logic*

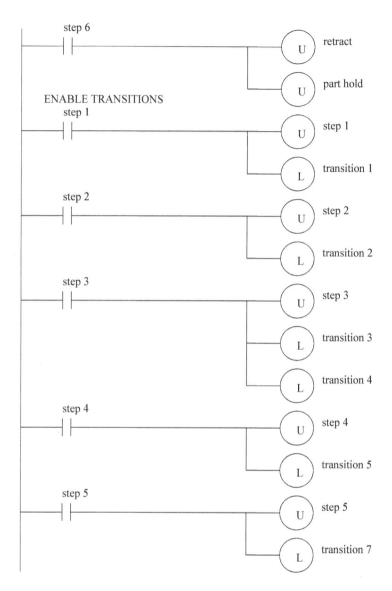

Figure 19.8   *SFC Implemented in Ladder Logic*

Figure 19.9   *SFC Implemented in Ladder Logic*

Many PLCs also allow SFCs to entered be as graphic diagrams. Small segments of ladder logic must then be entered for each transition and action. Each segment of ladder logic is kept in a separate program. If we consider the previous example the SFC diagram would be numbered as shown in Figure 19.1. The numbers are sequential and are for both transitions and steps.

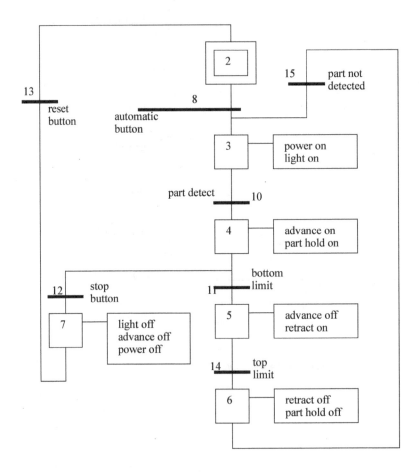

Figure 19.10   *SFC Renumbered*

Some of the ladder logic for the SFC is shown in Figure 19.1. Each program corresponds to the number on the diagram. The ladder logic includes a new instruction, EOT, that will tell the PLC when a transition has completed. When the rung of ladder logic with the EOT output becomes true the SFC will move to the next step or transition. when developing graphical SFCs the ladder logic becomes very simple, and the PLC deals with turning states on and off properly.

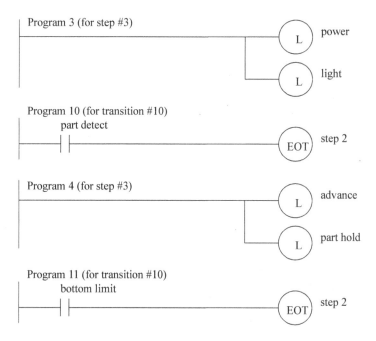

Figure 19.11   *Sample Ladder Logic for a Graphical SFC Program*

SFCs can also be implemented using ladder logic that is not based on latches, or built in SFC capabilities. The previous SFC example is implemented below. The first segment of ladder logic in Figure 19.1 is for the transitions. The logic for the steps is shown in Figure 19.1.

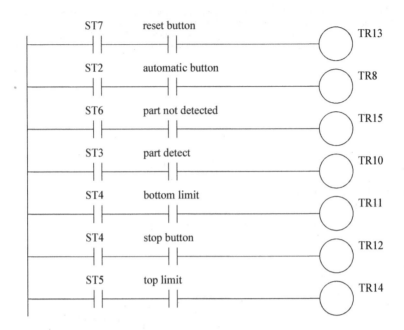

Figure 19.12   Ladder logic for transitions

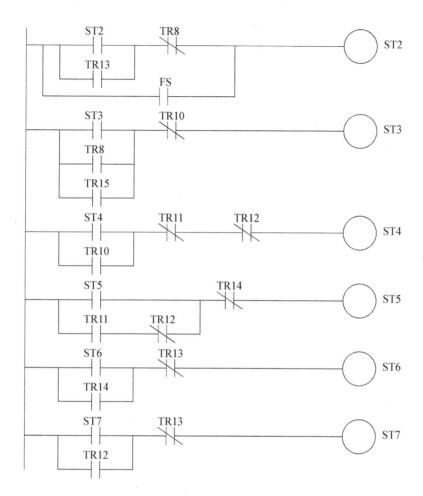

Figure 19.13   *Step logic*

Aside: The SFC approach can also be implemented with traditional programming languages. The example below shows the previous example implemented for a Basic Stamp II microcontroller.

```
autoon = 1; detect=2; bottom=3; top=4; stop=5;reset=6 'define input pins
input autoon; input detect; input button; input top; input stop; input reset
s1=1; s2=0; s3=0; s4=0; s5=0; s6=0 'set to initial step
advan=7;onlite=8; hold=9;retrac=10 'define outputs
output advan; output onlite; output hold; output retrac
step1: if s1<>1 then step2; s1=2
step2: if s2<>1 then step3; s2=2
step3: if s3<>1 then step4; s3=2
step4: if s4<>1 then step5; s4=2
step5: if s5<>1 then step6; s5=2
step6: if s6<>1 then trans1; s6=2
trans1: if (in1<>1 or s1<>2) then trans2;s1=0;s2=1
trans2: (if in2<>1 or s2<>2) then trans3;s2=0;s3=1
trans3: ...................
stepa1: if (st2<>1) then goto stepa2: high onlite
.................
goto step1
```

Figure 19.14   *Implementing SFCs with High Level Languages*

19.1 A COMPARISON OF METHODS

These methods are suited to different controller designs. The most basic controllers can be developed using process sequence bits and flowcharts. More complex control problems should be solved with state diagrams. If the controller needs to control concurrent processes the SFC methods could be used. It is also possible to mix methods together. For example, it is quite common to mix state based approaches with normal conditional logic. It is also possible to make a concurrent system using two or more state diagrams.

19.2 SUMMARY

• Sequential function charts are suited to processes with parallel operations
• Controller diagrams can be converted to ladder logic using MCR blocks
• The sequence of operations is important when converting SFCs to ladder logic.

19.3 PRACTICE PROBLEMS

(Note: Problem solutions are available at http://sites.google.com/site/automatedmanufacturingsystems/)

1. Develop an SFC for a two person assembly station. The station has two presses that may be used at the same time. Each press has a cycle button that will start the advance of the press. A bottom limit switch will stop the advance, and the cylinder must then be retracted until a top limit switch is hit.

2. Create an SFC for traffic light control. The lights should have cross walk buttons for both directions of traffic lights. A normal light sequence for both directions will be green 16 seconds and yellow 4 seconds. If the cross walk button has been pushed, a walk light will be on for 10 seconds, and the green light will be extended to 24 seconds.

3. Draw an SFC for a stamping press that can advance and retract when a cycle button is pushed, and then stop until the button is pushed again.

4. Design a garage door controller using an SFC. The behavior of the garage door controller is as follows,
- there is a single button in the garage, and a single button remote control.
- when the button is pushed the door will move up or down.
- if the button is pushed once while moving, the door will stop, a second push will start motion again in the opposite direction.
- there are top/bottom limit switches to stop the motion of the door.
- there is a light beam across the bottom of the door. If the beam is cut while the door is closing the door will stop and reverse.
- there is a garage light that will be on for 5 minutes after the door opens or closes.

19.4 ASSIGNMENT PROBLEMS

1. Develop an SFC for a vending machine and expand it into ladder logic.

19.1 PRACTICE PROBLEM SOLUTIONS

1.

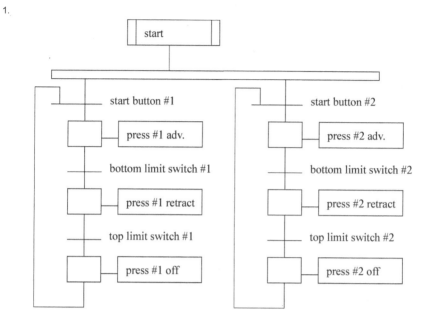

2.

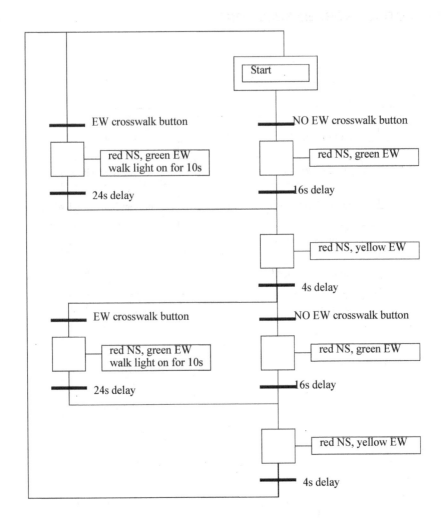

3.

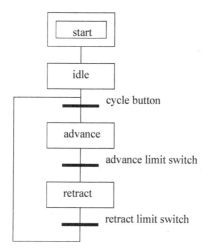

```
              ┌─────────────┐
              │  ┌───────┐  │
              │  │ start │  │
              │  └───────┘  │
              └──────┬──────┘
                     │
              ┌──────┴──────┐
              │    idle     │
              └──────┬──────┘
    ┌────────────────┤
    │          ━━━━━━━━━━━━  cycle button
    │                │
    │         ┌──────┴──────┐
    │         │   advance   │
    │         └──────┬──────┘
    │                │
    │          ━━━━━━━━━━━━  advance limit switch
    │                │
    │         ┌──────┴──────┐
    │         │   retract   │
    │         └──────┬──────┘
    │                │
    │          ━━━━━━━━━━━━  retract limit switch
    └────────────────┘
```

4.

```
    ┌──────────────────────────────────┐
    │                     ┌──────────┐  │
    │                     │ ┌──────┐ │  │
    │                     │ │step 1│ │  │
    │                     │ └──────┘ │  │
    │                     └────┬─────┘  │
    │                          │
    │                     ┌────┴─────┐
    │                     │  step 2  │
    │                     └────┬─────┘
    │                       T1 │
    │                    ━━━━━━━━━━━  button + remote
    │                          │
    │                     ┌────┴─────┐      ┌────────────┐
    │                     │  step 3  │──────│ close door │
    │                     └────┬─────┘      └────────────┘
    │              ┌───────────┤
    │          T3  │        T2 │
    │        ━━━━━━━━━      ━━━━━━━━━  button + remote + bottom limit
    │        light beam        │
    │              │      ┌────┴─────┐
    │              │      │  step 4  │
    │              │      └────┬─────┘
    │              │        T4 │
    │              │     ━━━━━━━━━━━  button + remote
    │              └───────────┤
    │                          │
    │                     ┌────┴─────┐      ┌────────────┐
    │                     │  step 5  │──────│ open door  │
    │                     └────┬─────┘      └────────────┘
    │                       T5 │
    │                    ━━━━━━━━━━━  button + remote + top limit
    └──────────────────────────┘
```

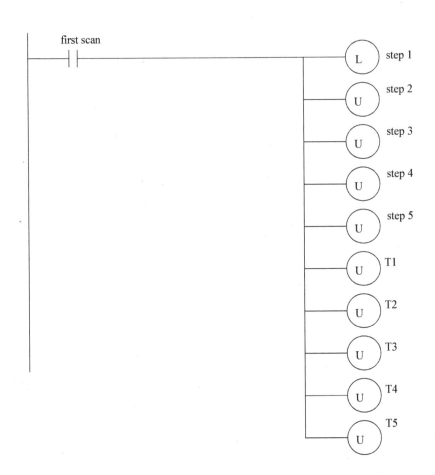

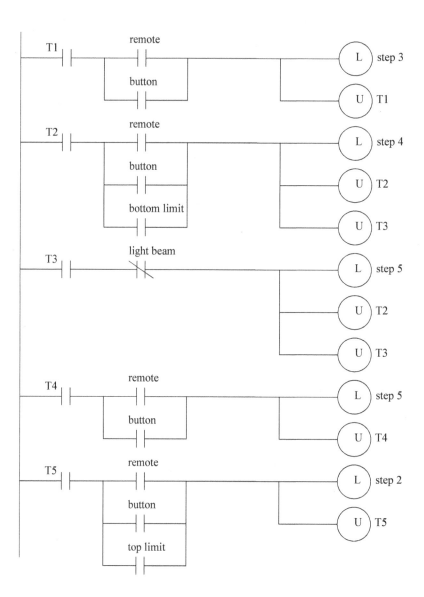

397

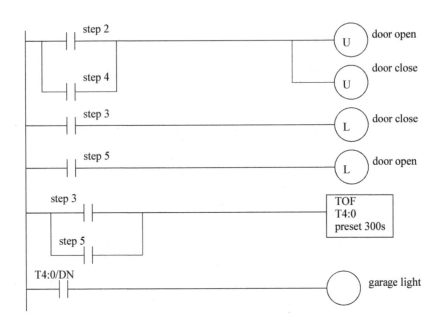

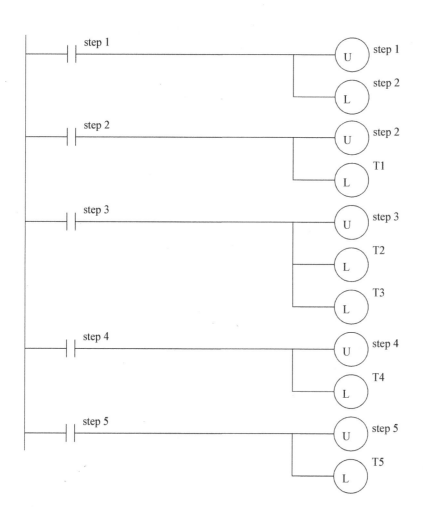

20 FUNCTION BLOCK PROGRAMMING

Topics:

- The basic construction of FBDs
- The relationship between ST and FBDs
- Constructing function blocks with structured text
- Design case

Objectives:

- To be able to write simple FBD programs

Function Block Diagrams (FBDs) are another part of the IEC 61131-3 standard. The primary concept behind a FBD is data flow. In these types of programs the values flow from the inputs to the outputs, through function blocks. A sample FBD is shown in Figure 20.1. In this program the inputs *A* and *B* are used to calculate a value *sin(A) * ln(B)*. The result of this calculation is compared to *C*. If the calculated value is less than *C* then the output *X* is turned on, otherwise it is turned off. Many readers will note the similarity of the program to block diagrams for control systems.

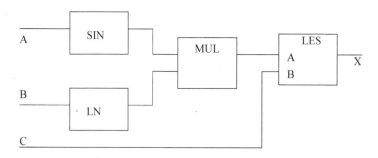

Figure 20.1   *A Simple Calculation and Comparison Program*

It is possible to disable part of the FBDs using enables. These are available for each function block but may not be displayed. Figure 20.2 shows an XOR calculation. Both of the Boolean AND functions have the enable inputs connected to 'enable'. If 'enable' is true, then the system works as expected and the output 'X' is the exclusive OR of 'A' and 'B'. However if 'enable' is off then the BAND functions will not operate. In this case the 'enable' input is not connected to the BOR function, but because it relies on the outputs from the BAND blocks, it will not function, and the output 'X' will not be changed.

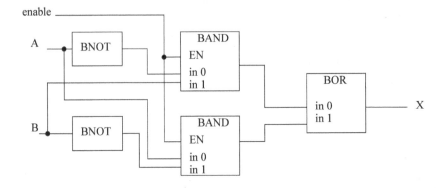

Figure 20.2   Using Enables in FBDs

A FBD program is constructed using function blocks that are connected together to define the data exchange. The connecting lines will have a data type that must be compatible on both ends. The inputs and outputs of function blocks can be inverted. This is normally shown with a small circle at the point where the line touches the function block, as shown in Figure 20.3. (Note: this is NOT available for Allen Bradley RSLogix, so BNOT functions should be used instead.)

Figure 20.3   Inverting Inputs and Outputs on Function Blocks

The basic functions used in FBD programs are equivalent to the basic set used in Structured Text (ST) programs. Consider the basic addition function shown in Figure 20.4. The ST function on the left adds *A* and *B*, and stores the result in *O*. The function block on the right is equivalent. By convention the inputs are on the left of the function blocks, and the outputs on the right.

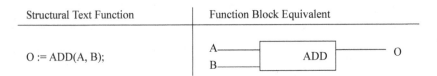

Structural Text Function	Function Block Equivalent
O := ADD(A, B);	A———[ADD]——— O B———

Figure 20.4   A Simple Function Block

Some functions allow a variable number of arguments. In Figure 20.5 there is a third value input to the *ADD* block. This is known as overloading.

Structural Text Function	Function Block Equivalent

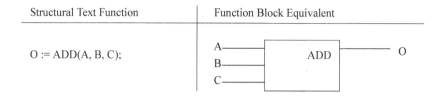

Figure 20.5   *A Function with A Variable Argument List*

The ADD function in the previous example will add all of the arguments in any order and get the same result, but other functions are more particular. Consider the circular limit function shown in Figure 20.6. In the first ST function the maximum *MX*, minimum *MN* and test *IN* values are all used. In the second function the *MX* value is not defined and will default to *0*. Both of the ST functions relate directly to the function blocks on the right side of the figure.

Structural Text Function	Function Block Equivalent

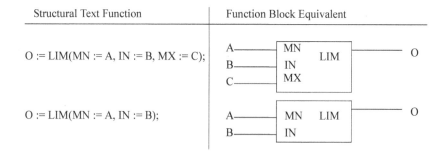

Figure 20.6   *Function Argument Lists*

20.1 CREATING FUNCTION BLOCKS

When developing a complex system it is desirable to create additional function blocks. This can be done with other FBDs, or using other IEC 61131-3 program types. Figure 20.7 shows a divide function block created using ST. In this example the first statement declares it as a *FUNCTION_BLOCK* called *divide*. The input variables *a* and *b*, and the output variable *c* are declared. In the function the denominator is checked to make sure it is not *0*. If not, the division will be performed, otherwise the output will be zero.

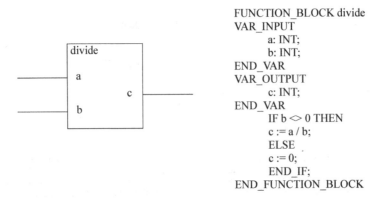

```
FUNCTION_BLOCK divide
VAR_INPUT
        a: INT;
        b: INT;
END_VAR
VAR_OUTPUT
        c: INT;
END_VAR
        IF b <> 0 THEN
        c := a / b;
        ELSE
        c := 0;
        END_IF;
END_FUNCTION_BLOCK
```

Figure 20.7   *Function Block Equivalencies*

20.2 DESIGN CASE

A simple state diagram is shown in Figure 20.8.

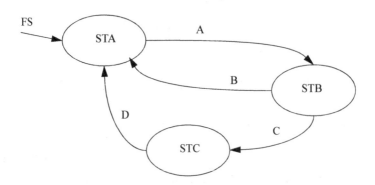

Figure 20.8   *An Example State Diagram*

The state diagram is implemented in FBD form in Figure 20.9. In this case the transition equations approach was used, although other methods are equally applicable. The transitions 'STA_TO_STB', "STB_TO_STA', 'STB_TO_STC', and 'STC_TO_STA' are calculated first. These are then used to update the states 'STA', 'STB', and 'STC'. Additional program steps could then be added to drive outputs.

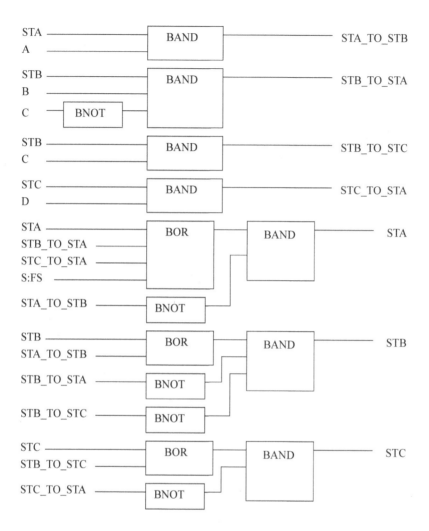

Figure 20.9   *An FBD Implementation of a State Diagram Using Transition Equations*

20.3 SUMMARY

- FBDs use data flow from left to right through function blocks
- Inputs and outputs can be inverted
- Function blocks can have variable argument list sizes
- When arguments are left off default values are used
- Function blocks can be created with ST

20.4 PRACTICE PROBLEMS

(Note: Problem solutions are available at http://sites.google.com/site/automatedmanufacturingsystems/)

1. Draw a timing diagram for the following FBD program.

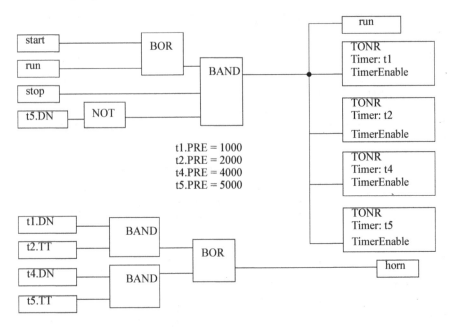

2. Write a function block diagram program that will replace the following ladder logic.

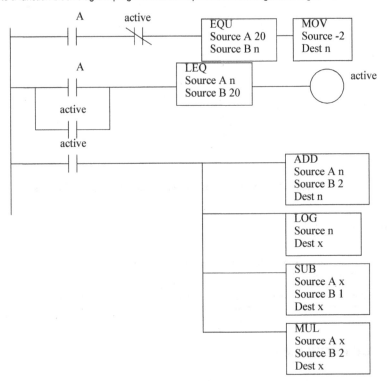

3. Write a Function Block Diagram program to implement the following timing diagram. The sequence should begin when a variable 'temp' rises above 80.

4. Develop a FBD for a system that will monitor a high temperature salt bath. The systems has *start* and *stop* buttons as normal. The temperature for the salt bath is available in *temp*. If the bath is above 250 C then the *heater* should be turned off. If the temperature is below 220 C then the *heater* should be turned on. Once the system has been in the acceptable range for 10 minutes the system should shut off.

5. Write a function block diagram program that will replace the following ladder logic.

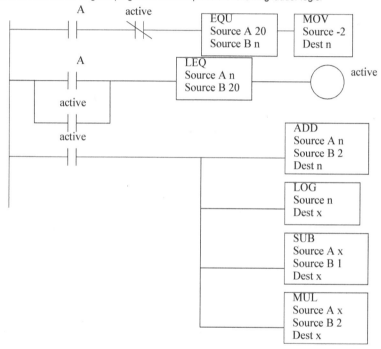

6. Write a structured text program that reads inputs from 'channel 0'. An input string of 'CLEAR' will clear a storage array. Up to 100 real values with the format 'XXX.XX' will arrive on 'channel 0' and are to be stored in the array. If the string 'AVG' is received, the average of the array contents will be calculated and written out 'Channel 0'.

20.5 ASSIGNMENT PROBLEMS

1. Convert the following state diagram to a Function Block Diagram.

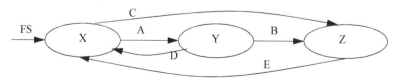

2. Write a FBD program that will be active when input 'X' is true. When the timer accumulator is between 5 and 10

seconds the output Y should be set.

20.1 PRACTICE PROBLEM SOLUTIONS

1.

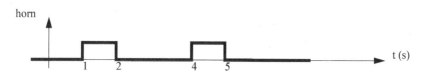

2.

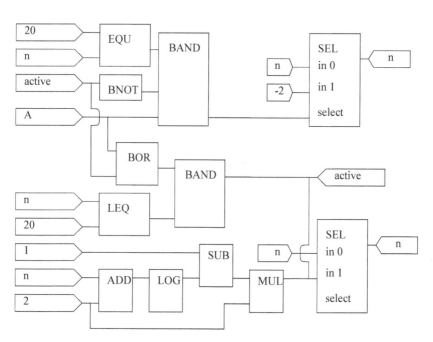

3.

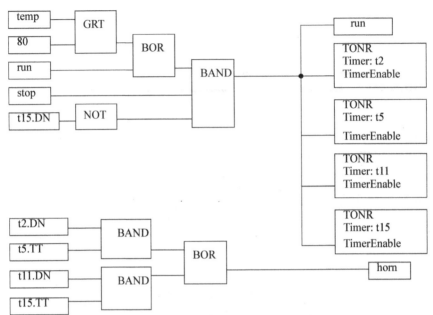

4.

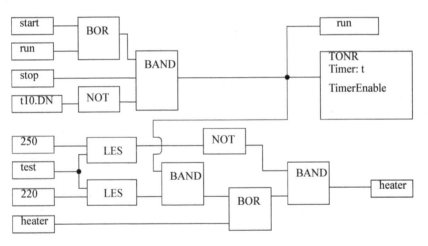

5.

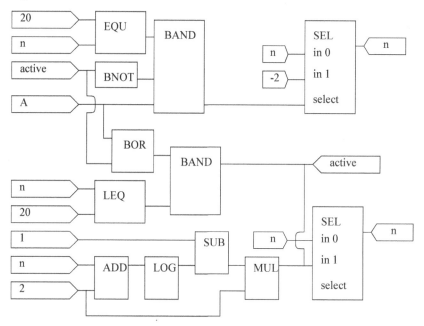

6.

```
SBR();
        IF S:FS THEN
        i = 0;
        END_IF;
        ACB(0, c);
        IF c.POS = 6 THEN
        ARL(0, str_in, s);
        IF i < 100 THEN
        r[i] = STOR(str_in);
        i := i + 1;
        END_IF;
        ELSE
        ARL(0, str_in, s);
        IF str_in = str_clear THEN
        i := 0;
        END_IF
        IF str_in = str_avg THEN
        sum := 0;
        FOR j = 0 to length-1 DO
        sum := sum + r[j];
        END_FOR;
        str_out := RTOS(sum / i);
        AWT(0, str_out, s);
        END_IF;
        END_IF;
RET();
```

Tags:

r:REAL[100]
i:INT
j:INT
sum:REAL
c:SerialPortControl
s:SerialPortControl
str_in:STRING
str_out:STRING
str_clear:STRING = "CLEAR"
str_avg:STRING = "AVG"

21 ANALOG INPUTS AND OUTPUTS

Topics:

- Analog inputs and outputs
- Sampling issues; aliasing, quantization error, resolution
- Analog I/O with a PLC

Objectives:

- To understand the basics of conversion to and from analog values.
- Be able to use analog I/O on a PLC.

An analog value is continuous, not discrete, as shown in Figure 21.1. In the previous chapters, techniques were discussed for designing logical control systems that had inputs and outputs that could only be on or off. These systems are less common than the logical control systems, but they are very important. In this chapter we will examine analog inputs and outputs so that we may design continuous control systems in a later chapter.

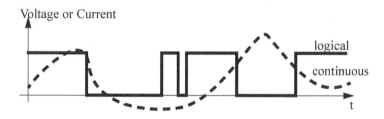

Figure 21.1   **Logical and Continuous Values**

Typical analog inputs and outputs for PLCs are listed below. Actuators and sensors that can be used with analog inputs and outputs will be discussed in later chapters.

Inputs:
- oven temperature
- fluid pressure
- fluid flow rate

Outputs:
- fluid valve position
- motor position
- motor velocity

This chapter will focus on the general principles behind digital-to-analog (D/A) and analog-to-digital (A/D) conversion. The chapter will show how to output and input analog values with a PLC.

21.1 ANALOG INPUTS

To input an analog voltage (into a PLC or any other computer) the continuous voltage value must be *sampled* and then converted to a numerical value by an A/D converter. Figure 21.1 shows a continuous voltage changing over time. There are three samples shown on the figure. The process of sampling the data is not instantaneous, so each sample has a start and stop time. The time required to acquire the sample is called the *sampling time*. A/D converters can only acquire a limited number of samples per second. The time between samples is called the sampling period T, and the inverse of the sampling period is the sampling frequency (also called sampling rate). The sampling time is often much smaller than the sampling period. The sampling frequency is specified when buying hardware, but for a PLC a maximum sampling rate might be 20Hz.

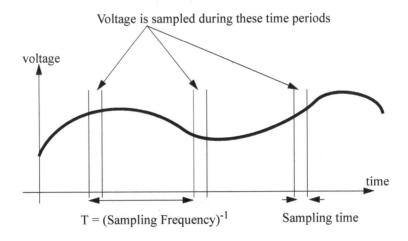

Figure 21.2   *Sampling an Analog Voltage*

A more realistic drawing of sampled data is shown in Figure 21.1. This data is noisier, and even between the start and end of the data sample there is a significant change in the voltage value. The data value sampled will be somewhere between the voltage at the start and end of the sample. The maximum (Vmax) and minimum (Vmin) voltages are a function of the control hardware. These are often specified when purchasing hardware, but reasonable ranges are;

        0V to 5V
        0V to 10V
        -5V to 5V
        -10V to 10V

The number of bits of the A/D converter is the number of bits in the result word. If the A/D converter is *8 bit* then the result can read up to 256 different voltage levels. Most A/D converters have 12 bits, 16 bit converters are used for precision measurements.

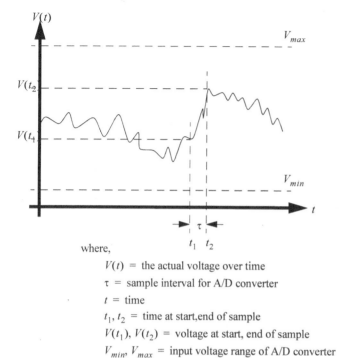

where,

$V(t)$ = the actual voltage over time

τ = sample interval for A/D converter

t = time

t_1, t_2 = time at start,end of sample

$V(t_1), V(t_2)$ = voltage at start, end of sample

V_{min}, V_{max} = input voltage range of A/D converter

N = number of bits in the A/D converter

Figure 21.3   *Parameters for an A/D Conversion*

The parameters defined in Figure 21.1 can be used to calculate values for A/D converters. These equations are summarized in Figure 21.1. Equation 1 relates the number of bits of an A/D converter to the resolution. In a normal A/D converter the minimum range value, Rmin, is zero, however some devices will provide 2's compliment negative numbers for negative voltages. Equation 2 gives the error that can be expected with an A/D converter given the range between the minimum and maximum voltages, and the resolution (this is commonly called the quantization error). Equation 3 relates the voltage range and resolution to the voltage input to estimate the integer that the A/D converter will record. Finally, equation 4 allows a conversion between the integer value from the A/D converter, and a voltage in the computer.

$$R = 2^N = R_{max} - R_{min} \tag{1}$$

$$V_{ERROR} = \left(\frac{V_{max} - V_{min}}{2R}\right) \tag{2}$$

$$V_I = INT\left[\left(\frac{V_{in} - V_{min}}{V_{max} - V_{min}}\right)(R-1) + R_{min}\right] \tag{3}$$

$$V_C = \left(\frac{V_I - R_{min}}{(R-1)}\right)(V_{max} - V_{min}) + V_{min} \tag{4}$$

where,

R, R_{min}, R_{max} = absolute and relative resolution of A/D converter

V_I = the integer value representing the input voltage

V_C = the voltage calculated from the integer value

V_{ERROR} = the maximum quantization error

Figure 21.4 **A/D Converter Equations**

Consider a simple example, a 10 bit A/D converter can read voltages between -10V and 10V. This gives a resolution of 1024, where 0 is -10V and 1023 is +10V. Because there are only 1024 steps there is a maximum error of ±9.8mV. If a voltage of 4.564V is input into the PLC, the A/D converter converts the voltage to an integer value of 745. When we convert this back to a voltage the result is 4.565V. The resulting quantization error is 4.565V-4.564V=+0.001V. This error can be reduced by selecting an A/D converter with more bits. Each bit halves the quantization error.

Given,

$$N = 10, R_{min} = 0$$
$$V_{max} = 10V$$
$$V_{min} = -10V$$
$$V_{in} = 4.564V$$

Calculate,

$$R = R_{max} = 2^N = 1024$$

$$V_{ERROR} = \left(\frac{V_{max} - V_{min}}{2R}\right) = 0.0098V$$

$$V_I = INT\left[\left(\frac{V_{in} - V_{min}}{V_{max} - V_{min}}\right)(R-1) + 0\right] = 745$$

$$V_C = \left(\frac{V_I - 0}{R-1}\right)(V_{max} - V_{min}) + V_{min} = 4.565V$$

Figure 21.5 **Sample Calculation of A/D Values**

If the voltage being sampled is changing too fast we may get false readings, as shown in Figure 21.1. In the upper graph the waveform completes seven cycles, and 9 samples are taken. The bottom graph plots out the values read. The sampling frequency was too low, so the signal read appears to be different that it actually is, this is called aliasing.

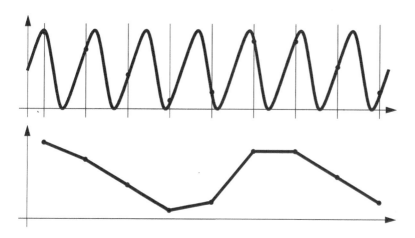

Figure 21.6   *Low Sampling Frequencies Cause Aliasing*

The Nyquist criterion specifies that sampling frequencies should be at least twice the frequency of the signal being measured, otherwise aliasing will occur. The example in Figure 21.1 violated this principle, so the signal was aliased. If this happens in real applications the process will appear to operate erratically. In practice the sample frequency should be 4 or more times faster than the system frequency.

$$f_{AD} > 2f_{signal} \quad \text{where,}$$
$$f_{AD} = \text{sampling frequency}$$
$$f_{signal} = \text{maximum frequency of the input}$$

There are other practical details that should be considered when designing applications with analog inputs;

- Noise - Since the sampling window for a signal is short, noise will have added effect on the signal read. For example, a momentary voltage spike might result in a higher than normal reading. Shielded data cables are commonly used to reduce the noise levels.
- Delay - When the sample is requested, a short period of time passes before the final sample value is obtained.
- Multiplexing - Most analog input cards allow multiple inputs. These may share the A/D converter using a technique called multiplexing. If there are 4 channels using an A/D converter with a maximum sampling rate of 100Hz, the maximum sampling rate per channel is 25Hz.
- Signal Conditioners - Signal conditioners are used to amplify, or filter signals coming from transducers, before they are read by the A/D converter.
- Resistance - A/D converters normally have high input impedance (resistance), so they do not affect circuits they are measuring.
- Single Ended Inputs - Voltage inputs to a PLC can use a single common for multiple inputs, these types of inputs are called *single* ended inputs. These tend to be more prone to noise.
- Double Ended Inputs - Each double ended input has its own common. This reduces problems with electrical noise, but also tends to reduce the number of inputs by half.

ASIDE: This device is an 8 bit A/D converter. The main concept behind this is the successive approximation logic. Once the reset is toggled the converter will start by setting the most significant bit of the 8 bit number. This will be converted to a voltage *Ve* that is a function of the +/-*Vref* values. The value of *Ve* is compared to *Vin* and a simple logic check determines which is larger. If the value of *Ve* is larger the bit is turned off. The logic then repeats similar steps from the most to least significant bits. Once the last bit has been set on/off and checked the conversion will be complete, and a done bit can be set to indicate a valid conversion value.

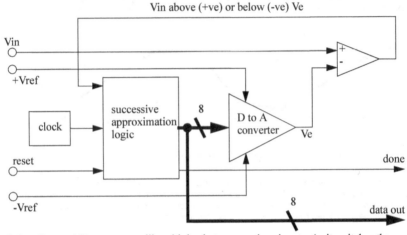

Quite often an A/D converter will multiplex between various inputs. As it switches the voltage will be sampled by a *sample and hold circuit*. This will then be converted to a digital value. The sample and hold circuits can be used before the multiplexer to collect data values at the same instant in time.

Figure 21.7　　**A Successive Approximation A/D Converter**

21.2 ANALOG OUTPUTS

Analog outputs are much simpler than analog inputs. To set an analog output an integer is converted to a voltage. This process is very fast, and does not experience the timing problems with analog inputs. But, analog outputs are subject to quantization errors. Figure 21.1 gives a summary of the important relationships. These relationships are almost identical to those of the A/D converter.

$$R = 2^N = R_{max} - R_{min} \qquad (5)$$

$$V_{ERROR} = \left(\frac{V_{max} - V_{min}}{2R}\right) \qquad (6)$$

$$V_I = INT\left[\left(\frac{V_{desired} - V_{min}}{V_{max} - V_{min}}\right)(R-1) + R_{min}\right] \qquad (7)$$

$$V_{output} = \left(\frac{V_I - R_{min}}{(R-1)}\right)(V_{max} - V_{min}) + V_{min} \qquad (8)$$

where,

R, R_{min}, R_{max} = absolute and relative resolution of A/D converter

V_{ERROR} = the maximum quantization error

V_I = the integer value representing the desired voltage

V_{output} = the voltage output using the integer value

$V_{desired}$ = the desired analog output value

Figure 21.8   *Analog Output Relationships*

Assume we are using an 8 bit D/A converter that outputs values between 0V and 10V. We have a resolution of 256, where 0 results in an output of 0V and 255 results in 10V. The quantization error will be 20mV. If we want to output a voltage of 6.234V, we would specify an output integer of 159, this would result in an output voltage of 6.235V. The quantization error would be 6.235V-6.234V=0.001V.

Given,

$N = 8, R_{min} = 0$

$V_{max} = 10V$

$V_{min} = 0V$

$V_{desired} = 6.234V$

Calculate,

$$R = R_{max} = 2^N = 256$$

$$V_{ERROR} = \left(\frac{V_{max} - V_{min}}{2R}\right) = 0.020V$$

$$V_I = INT\left[\left(\frac{V_{in} - V_{min}}{V_{max} - V_{min}}\right)(R-1) + 0\right] = 159$$

$$V_C = \left(\frac{V_I - 0}{R-1}\right)(V_{max} - V_{min}) + V_{min} = 6.235V$$

The current output from a D/A converter is normally limited to a small value, typically less than 20mA. This is enough for instrumentation, but for high current loads, such as motors, a current amplifier is needed. This type of interface will be discussed later. If the current limit is exceeded for 5V output, the voltage will decrease (so don't exceed the rated voltage). If the current limit is exceeded for long periods of time the D/A output may be damaged.

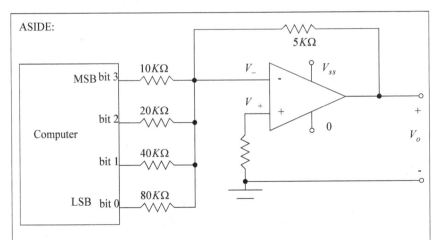

ASIDE:

$5K\Omega$

MSB bit 3 $10K\Omega$

V_- V_{ss}

bit 2 $20K\Omega$

V_+

Computer

bit 1 $40K\Omega$

0

V_o

LSB bit 0 $80K\Omega$

First we write the obvious,

$$V_+ = 0 = V_-$$

Next, sum the currents into the inverting input as a function of the output voltage and the input voltages from the computer,

$$\frac{V_{b_3}}{10K\Omega} + \frac{V_{b_2}}{20K\Omega} + \frac{V_{b_1}}{40K\Omega} + \frac{V_{b_0}}{80K\Omega} = \frac{V_o}{5K\Omega}$$

$$\therefore V_o = 0.5V_{b_3} + 0.25V_{b_2} + 0.125V_{b_1} + 0.0625V_{b_0}$$

Consider an example where the binary output is 1110, with 5V for on,

$$\therefore V_o = 0.5(5V) + 0.25(5V) + 0.125(5V) + 0.625(0V) = 4.375V$$

Figure 21.9   *A Digital-To-Analog Converter*

21.3 CONTROLLOGIX HARDWARE

In this section analog I/O will be discussed using a 1794-IE4XOE2/B 4 Input/2Output 24V DC Non-Isolated Analog module. The card has a 12 bit resolution. To use this module it is defined under the 'I/O Configuration'. While configuring the module the following options are available.

- Update rate (Requested Packet Interval) 2-750ms
- Input channel ranges for channels 0 to 3
    - 4 to 20mA
    - 0 to 10V/0 to 20mV
    - -10 to 10V
- Output channel ranges for channel 0 and 1
    - 4 to 20mA
    - 0 to 10V/0 to 20mV
    - -10 to 10V

After the card is configured the configuration words are available in the 'controller scooped tags'. These are listed below with descriptions assuming the card is in 'rack:2:'. The configuration words may also be used to update the card during operation. To do this the values are changed using normal program statements to read or write to values.

rack:2:C.Ch0SafeStateConfig - Sets the safe state when the module update fails
rack:2:C.Ch1SafeStateConfig - Sets the safe state when the module update fails
rack:2:C.Ch0InputFullRange - 0 = 4-20mA; 1 = -10-10V,0-10V,0-20mA
rack:2:C.Ch1InputFullRange - 0 = 4-20mA; 1 = -10-10V,0-10V,0-20mA
rack:2:C.Ch2InputFullRange - 0 = 4-20mA; 1 = -10-10V,0-10V,0-20mA
rack:2:C.Ch3InputFullRange - 0 = 4-20mA; 1 = -10-10V,0-10V,0-20mA
rack:2:C.Ch0OututFullRange - 0 = 4-20mA; 1 = -10-10V,0-10V,0-20mA
rack:2:C.Ch1OutputFullRange - 0 = 4-20mA; 1 = -10-10V,0-10V,0-20mA
rack:2:C.Ch0InputConfigSelect - 0 = 0-10V, 0-20mA; 1 = 4-20mA, -10V-10V
rack:2:C.Ch1InputConfigSelect - 0 = 0-10V, 0-20mA; 1 = 4-20mA, -10V-10V
rack:2:C.Ch2InputConfigSelect - 0 = 0-10V, 0-20mA; 1 = 4-20mA, -10V-10V
rack:2:C.Ch3InputConfigSelect - 0 = 0-10V, 0-20mA; 1 = 4-20mA, -10V-10V
rack:2:C.Ch0OutputConfigSelect - 0 = 0-10V, 0-20mA; 1 = 4-20mA, -10V-10V
rack:2:C.Ch1OutputConfigSelect - 0 = 0-10V, 0-20mA; 1 = 4-20mA, -10V-10V
rack:2:C.SSCh0OutputData - A safe output value for channel 0
rack:2:C.SSCh1OutputData - A safe output value for channel 1

rack:2:I.Fault - Returns a fault code for the module
rack:2:I.Ch0InputData - The analog input value read on channel 0
rack:2:I.Ch1InputData - The analog input value read on channel 1
rack:2:I.Ch2InputData - The analog input value read on channel 2
rack:2:I.Ch3InputData - The analog input value read on channel 3
rack:2:I.Ch0InputUnderrange - The channel 0 current is below 4mA
rack:2:I.Ch1InputUnderrange - The channel 1 current is below 4mA
rack:2:I.Ch2InputUnderrange - The channel 2 current is below 4mA
rack:2:I.Ch3InputUnderrange - The channel 3 current is below 4mA
rack:2:I.Ch0OutputOpenWire - The output current is zero - indicates broken wire
rack:2:I.Ch1OutputOpenWire - The output current is zero - indicates broken wire
rack:2:I.PowerUp - The module is configured and running normally

rack:2:O.Ch0OutputData - The analog output voltage for channel 0
rack:2:O.Ch1OutputData - The analog output voltage for channel 1

Figure 21.1 shows a simple analog IO example with some error checking. The system uses start and stop buttons to operate, along with a check for module errors. If the system is running the input voltage from input channel 0 will be divided by two and then set as the output voltage for output channel 0. If the system is not running the output voltage on channel zero is set to 0 (0V).

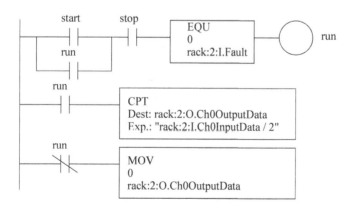

Figure 21.10   **A Voltage Divide by Two Example**

Although the card is a 12 bit card, it uses data values as if it has 15 digits of accuracy. Hence the valid range for the card is -32,768 to 32,767.

21.4 PLC-5 HARDWARE

(NOTE: This section is optional but is included for historical perspective.)

The PLC 5 ladder logic in Figure 21.1 will control an analog input card. The Block Transfer Write (BTW) statement will send configuration data from integer memory to the analog card in rack 0, slot 0. The data from *N7:30* to *N7:66* describes the configuration for different input channels. Once the analog input card receives this it will start doing analog conversions. The instruction is edge triggered, so it is run with the first scan, but the input is turned off while it is active, *BT10:0/EN*. This instruction will require multiple scans before all of the data has been written to the card. The *update* input is only needed if the configuration for the input changes, but this would be unusual. The Block Transfer Read (BTR) will retrieve data from the card and store it in memory *N7:10* to *N7:29*. This data will contain the analog input values. The function is edge triggered, so the enable bits prevent it from trying to read data before the card is configured *BT10:0/EN*. The *BT10:1/EN* bit will prevent if from starting another read until the previous one is complete. Without these the instructions experience continuous errors. The *MOV* instruction will move the data value from one analog input to another memory location when the BTR instruction is done.

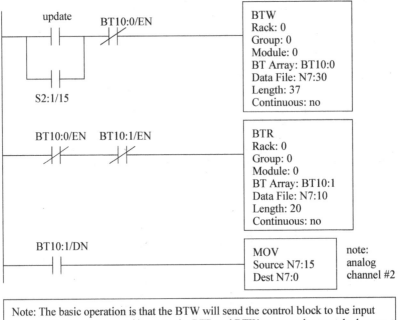

Note: The basic operation is that the BTW will send the control block to the input card. The inputs are used because the BTR and BTW commands may take longer than one scan.

Figure 21.11   **Ladder Logic to Control an Analog Input Card**

The data to configure a *1771-IFE Analog Input Card* is shown in Figure 21.1. (Note: each type of card will be different, and you need to refer to the manuals for this information.) The 1771-IFE is a 12 bit card, so the range will have up to 2**12 = 4096 values. The card can have 8 double ended inputs, or 16 single ended inputs (these are set with jumpers on the board). To configure the card a total of 37 data words are needed. The voltage range of different inputs are set using the bits in word 0 (N7:30) and 1 (N7:31). For example, to set the voltage range on channel 10 to -5V to 5V we would need to set the bits, N7:31/3 = 1 and N7:31/2 = 0. Bits in data word 2 (N7:32) are set to determine the general configuration of the card. For example, if word 2 was *0001 0100 0000 0000b* the card would be set for; a delay of *00010* between samples, to return 2s compliment results, using single ended inputs, and no filtering. The remaining data words, from 3 to 36, allow data values to be scaled to a new range. Words 3 and 4 are for channel 1, words 5 and 6 are for channels 2 and so on. To scale the data, the new minimum value is put in the first word (word 3 for channel 1), and the maximum value is put in the second word (word 4 for channel 1). The card then automatically converts the actual data reading between 0 and 4095 to the new data range indicated in word 3 and 4. One oddity of this card is that the data values for scaling must always be BCD, regardless of the data type setting. The manual for this card claims that putting zeros in the scaling values will cause the card to leave the data unscaled, but in practice it is better to enter values of 0 for the minimum and 4095 for the maximum.

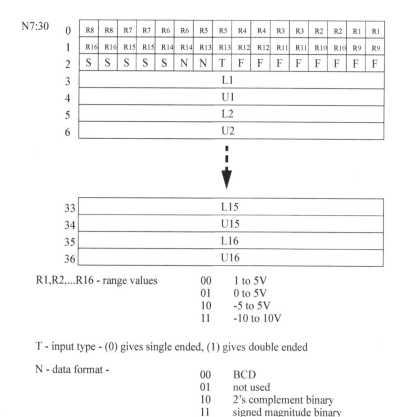

R1,R2,...R16 - range values

00	1 to 5V
01	0 to 5V
10	-5 to 5V
11	-10 to 10V

T - input type - (0) gives single ended, (1) gives double ended

N - data format -

00	BCD
01	not used
10	2's complement binary
11	signed magnitude binary

F - filter function - a value of (0) will result in no filtering, up to a value of (99BCD)

S - real time sampling mode - (0) samples always, (11111binary) gives long delays.

L1,L2,...L16 - lower input scaling word values

U1,U2,...,U16 - upper input scaling word values

Figure 21.12   *Configuration Data for an 1771-IFE Analog Input Card*

The block of data returned by the BTR statement is shown in Figure 21.1. Bits 0-2 in word 0 (N7:10) will indicate the status of the card, such as error conditions. Words 1 to 4 will reflect status values for each channel. Words 1 and 2 indicate if the input voltage is outside the set range (e.g., -5V to 5V). Word 3 gives the sign of the data, which is important if the data is not in 2s compliment form. Word 4 indicates when data has been read from a channel. The data values for the analog inputs are stored in words from 5 to 19. In this example, the status for channel 9 are N7:11/8 (under range), N7:12/8 (over range), N7:13/8 (sign) and N7:14/8 (data read). The data value for channel 9 is in N7:13.

N7:10																
0														D	D	D
1	u16	u15	u14	u13	u12	u11	u10	u9	u8	u7	u6	u5	u4	u3	u2	u1
2	v16	v15	v14	v13	v12	v11	v10	v9	v8	v7	v6	v5	v4	v3	v2	v1
3	s16	s15	s14	s13	s12	s11	s10	s9	s8	s7	s6	s5	s4	s3	s2	s1
4	d1	d1	d1	d1	d1	d1	d1	d1	d1	d1	d1	d1	d1	d1	d1	d1
19	d16	d16	d16	d16	d16	d16	d16	d16	d16	d16	d16	d16	d16	d16	d16	d16

D - diagnostics
u - under range for input channels
v - over range for input channels
s - sign of data
d - data values read from inputs

Figure 21.13   *Data Returned by the 1771-IFE Analog Input Card*

Most new PLC programming software provides tools, such as dialog boxes to help set up the data parameters for the card. If these aids are not available, the values can be set manually in the PLC memory.

The PLC-5 ladder logic in Figure 21.1 can be used to set analog output voltages with a 1771-OFE Analog Output Card. The BTW instruction will write configuration memory to the card (the contents are described later). Values can also be read back from the card using a BTR, but this is only valuable when checking the status of the card and detecting errors. The BTW is edge triggered, so the *BT10:0/EN* input prevents the BTW from restarting the instruction until the previous block has been sent. The MOV instruction will change the output value for channel 1 on the card.

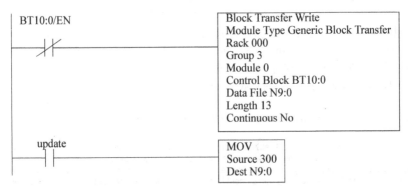

Figure 21.14   *Controlling a 1771-OFE Analog Output Card*

The configuration memory structure for the 1771-OFE Analog Output Card is shown in Figure 21.1. The card has four 12 bit output channels. The first four words set the output values for the card. Word 0 (N9:0) sets the value for channel 1, word 1 (N9:1) sets the value for channel 2, etc. Word 4 configures the card. Bit 16 (N9:4/15) will set the data format, bits 5 to 12 (/4 to /11) will enable scaling factors for channels, and bits 1 to 4 (/0 to /3) will provide signs for the data in words 0 to 3. The words from 5 to 13 allow scaling factors, so that the values in words 0 to 3 can be provided in another range of values, and then converted to the appropriate values. Good default values for the scaling factors are 0 for the lower limit and 4095 for the upper limit.

N9:0	0	D1
	1	D2
	2	D3
	3	D4
	4	f / / / s s s s s s s s p4 p3 p2 p1
	5	L1
	6	U1
	7	L2
	8	U2
	9	L3
	10	U3
	11	L4
	12	U4

D - data value words for channels 1, 2, 3 or 4
f - data format bit (1) binary, (0) BCD
s - scaling factor bits
p - data sign bits for the four output channels
L - lower scaling limit words for output channels 1, 2, 3 or 4
U - upper scaling limit words for output channels 1, 2, 3 or 4

Figure 21.15   *Configuration Data for a 1771-OFE Output Card*

21.4.1 PULSE WIDTH MODULATED (PWM) OUTPUTS

An equivalent analog output voltage can be generated using pulse width modulation, as shown in Figure 21.1. In this method the output circuitry is only capable of outputing a fixed voltage (in the figure 'A') or 0V. To obtain an analog voltage between the maximum and minimum the voltage is turned on and off quickly to reduce the effective voltage. The output is a square wave voltage at a high frequency, typically over 20Khz, above the hearing range. The duty cycle of the wave determines the effective voltage of the output. It is the percentage of time the output is on relative to the time it is off. If the duty cycle is 100% the output is always on. If the wave is on for the same time it is off the duty cycle is 50%. If the wave is always off, the duty cycle is 0%.

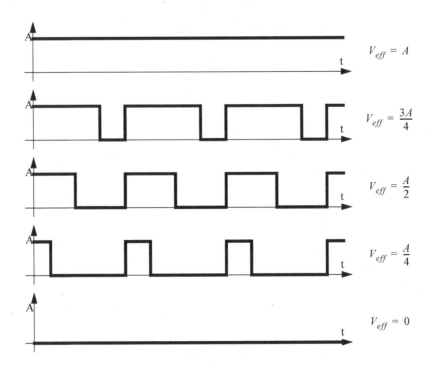

Figure 21.16   *Pulse Width Modulated (PWM) Signals*

PWM is commonly used in power electronics, such as servo motor control systems. In this case the response time of the motor is slow enough that the motor effectively filters the high frequency of the signal. The PWM signal can also be put through a low pass filter to produce an analog DC voltage.

Aside: A basic low pass RC filter is shown below. This circuit is suitable for an analog output that does not draw much current. (drawing too much current will result in large losses across the resistor.) The corner frequency can be easily found by looking at the circuit as a voltage divider.

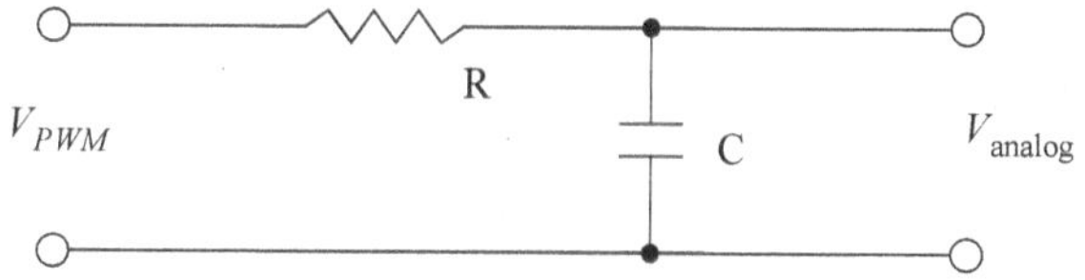

$$V_{analog} = V_{PWM}\left(\frac{\frac{1}{j\omega C}}{R + \frac{1}{j\omega C}}\right) = V_{PWM}\left(\frac{1}{j\omega CR + 1}\right)$$

$$\frac{V_{analog}}{V_{PWM}} = \frac{1}{j\omega CR + 1}$$

$$\omega = \frac{1}{CR}$$ ← corner frequency

As an example consider that the PWM signal is used at a frequency of 100KHz, an it is to be used with a system that has a response time (time constant) of 0.1seconds. Therefore the corner frequency should be between 10Hz (1/0.1s) and 100KHz. This can be put at the mid point of 1000Hz, or 6.2Krad/s. This system also requires the arbitrary selection of a resistor or capacitor value. We will pick the capacitor value to be 0.1uF so that we don't need an electrolytic.

$$R = \frac{1}{C\omega} = \frac{1}{10^{-7}2\pi 10^{3}} = \frac{10^{4}}{2\pi} = 1.59K\Omega$$

Figure 21.17 ***Converting a PWM Signal to an Analog Voltage***

In some cases the frequency of the output is not fixed, but the duty cycle of the output is maintained.

21.4.2 SHIELDING

When a changing magnetic field cuts across a conductor, it will induce a current flow. The resistance in the circuits will convert this to a voltage. These unwanted voltages result in erroneous readings from sensors, and signal to outputs. Shielding will reduce the effects of the interference. When shielding and grounding are done properly, the effects of electrical noise will be negligible. Shielding is normally used for; all logical signals in noisy environments, high speed counters or high speed circuitry, and all analog signals.

There are two major approaches to reducing noise; shielding and twisted pairs. Shielding involves encasing conductors and electrical equipment with metal. As a result electrical equipment is normally housed in metal cases. Wires are normally put in cables with a metal sheath surrounding both wires. The metal sheath may be a thin film, or a woven metal mesh. Shielded wires are connected at one end to "drain" the unwanted signals into the cases of the instruments. Figure 21.1 shows a thermocouple connected with a thermocouple. The cross section of the wire contains two insulated conductors. Both of the wires are covered with a metal foil, and final covering of insulation finishes the cable. The wires are connected to the thermocouple as expected, but the shield is only connected on the amplifier end to the case. The case is then connected to the shielding ground, shown here as three diagonal lines.

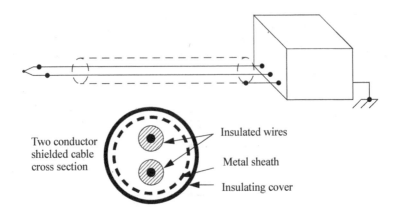

Two conductor
shielded cable
cross section

Insulated wires

Metal sheath

Insulating cover

Figure 21.18   **Shielding for a Thermocouple**

A twisted pair is shown in Figure 21.1. The two wires are twisted at regular intervals, effectively forming small loops. In this case the small loops reverse every twist, so any induced currents are cancel out for every two twists.

|1" or less typical|

Figure 21.19   **A Twisted Pair**

When designing shielding, the following design points will reduce the effects of electromagnetic interference.

- Avoid "noisy" equipment when possible.
- Choose a metal cabinet that will shield the control electronics.
- Use shielded cables and twisted pair wires.
- Separate high current, and AC/DC wires from each other when possible.
- Use current oriented methods such as sourcing and sinking for logical I/O.
- Use high frequency filters to eliminate high frequency noise.
- Use power line filters to eliminate noise from the power supply.

21.5 DESIGN CASES

21.5.1 Process Monitor

Problem: Design ladder logic that will monitor the dimension of a part in a die. If the

Solution:

21.6 SUMMARY

- A/D conversion will convert a continuous value to an integer value.
- D/A conversion is easier and faster and will convert a digital value to an analog value.
- Resolution limits the accuracy of A/D and D/A converters.
- Sampling too slowly will alias the real signal.
- Analog inputs are sensitive to noise.
- The analog I/O cards are configured with a few words of memory.
- BTW and BTR functions are needed to communicate with the analog I/O cards for older PLCs such as the PLC-5s.
- Analog shielding should be used to improve the quality of electrical signals.

21.7 PRACTICE PROBLEMS

(Note: Problem solutions are available at http://sites.google.com/site/automatedmanufacturingsystems/)

1. Analog inputs require:
    a) A Digital to Analog conversion at the PLC input interface module
    b) Analog to Digital conversion at the PLC input interface module
    c) No conversion is required
    d) None of the above

2. You need to read an analog voltage that has a range of -10V to 10V to a precision of +/-0.05V. What resolution of A/D converter is needed?

3. We are given a 12 bit analog input with a range of -10V to 10V. If we put in 2.735V, what will the integer value be after the A/D conversion? What is the error? What voltage can we calculate?

4. Use manuals on the web for a 1794 analog input card, and describe the process that would be needed to set up the card to read an input voltage between -2V and 7V. This description should include jumper settings, configuration memory and ladder logic.

5. We need to select a digital to analog converter for an application. The output will vary from -5V to 10V DC, and we need to be able to specify the voltage to within 50mV. What resolution will be required? How many bits will this D/A converter need? What will the accuracy be?

6. Write a program that will input an analog voltage, do the calculation below, and output an analog voltage.

$$V_{out} = \ln(V_{in})$$

7. The following calculation will be made when input A is true. If the result x is between 1 and 10 then the output B will be turned on. The value of x will be output as an analog voltage. Create a ladder logic program to perform these tasks.

$$x = 5^y \sqrt{1 + \sin y}$$

8. You are developing a controller for a game that measures hand strength. To do this a START button is pushed, 3 seconds later a LIGHT is turned on for one second to let the user know when to start squeezing. The analog value is read at 0.3s after the light is on. The value is converted to a force F with the equation below. The force is displayed by converting it to BCD and writing it to an output card (force_display). If the value exceeds 100 then a BIG_LIGHT and SIREN are turned on for 5sec. Use a structured design technique to develop ladder logic..

$$F = \frac{V_{in}}{6}$$

9. A machine is connected to a load cell that outputs a voltage proportional to the mass on a platform. When unloaded the cell outputs a voltage of 1V. A mass of 500Kg results in a 6V output. Write a program that will measure the mass when an input sensor (M) becomes true. If the mass is not between 300Kg and 400Kg and

alarm output (A) will be turned on. Write a program and indicate the general settings for the analog IO.

21.8 ASSIGNMENT PROBLEMS

1 In detail, describe the process of setting up analog inputs and outputs for a range of -10V to 10V in 2s compliment in realtime sampling mode.

2. Develop a program to sample analog data values and calculate the average, standard deviation, and the control limits. The general steps are listed below.
   1. Read 'm' sampled inputs.
   2. Randomly select values and calculate the average and store in memory. Calculate the standard deviation of the 'n' stored values.
   3. Compare the inputs to the standard deviation. If it is larger than 3 deviations from the mean, halt the process.
   4. If it is larger than 2 then increase a counter A, or if it is larger than 1 increase a second counter B. If it is less than 1 reset the counters.
   5. If counter A is =3 or B is =5 then shut down.
   6. Goto 1.

$$\bar{X}_j = \frac{\sum\limits_{i=1}^{m} X_i}{n} \qquad \bar{\bar{X}} = \sum\limits_{j=1}^{n} \bar{X}_j \qquad \sigma_{\bar{X}} = \sqrt{\frac{\sum\limits_{i=1}^{m} (X_i - \bar{X}_j)}{n-1}} \qquad \begin{array}{l} UCL = \bar{\bar{X}} + 3\sigma_{\bar{X}} \\[2mm] LCL = \bar{\bar{X}} - 3\sigma_{\bar{X}} \end{array}$$

21.1 PRACTICE PROBLEM SOLUTIONS

1. b)

2.

$$R = \frac{10V - (-10V)}{0.1V} = 200 \qquad \begin{array}{l} 7\text{ bits} = 128 \\ 8\text{ bits} = 256 \end{array}$$

The minimum number of bits is 8.

3.

$$N = 12 \quad R = 4096 \quad V_{min} = -10V \qquad V_{max} = 10V \qquad V_{in} = 2.735V$$

$$V_I = INT\left[\left(\frac{V_{in} - V_{min}}{V_{max} - V_{min}}\right)R\right] = 2608$$

$$V_C = \left(\frac{V_I}{R}\right)(V_{max} - V_{min}) + V_{min} = 2.734V$$

4. For the 1794-IE4XOE2/B card you would turn the key on the terminal block to match the back of the module. The card can then be installed in the terminal block. After the programming software is running the card is added to the IO configuration, and automatic settings can be used - these change the memory values to set values in integer memory. The values chosen would include a range of -10 to 10V.

5.

A card with a voltage range from -10V to +10V will be selected to cover the entire range.

$$R = \frac{10V - (-10V)}{0.050V} = 400 \qquad \text{minimum resolution}$$

$$\begin{array}{l} 8\text{ bits} = 256 \\ 9\text{ bits} = 512 \\ 10\text{ bits} = 1024 \end{array}$$

The A/D converter needs a minimum of 9 bits, but this number of bits is not commonly available, but 10 bits is, so that will be selected.

$$V_{ERROR} = \left(\frac{V_{max} - V_{min}}{2R}\right) = \frac{10V - (-10V)}{2(1024)} = \pm 0.00976V$$

6.

```
      BT9:1/DN                    ┌─────────────────────────────────────────────┐
───────┤ ├──────────────────────│ CPT                                         │
                                  │ Dest rack:2:Ch0OutputData                   │
                                  │ Expression "LN (rack:2:I.Ch0InputData)"     │
                                  └─────────────────────────────────────────────┘
```

7.

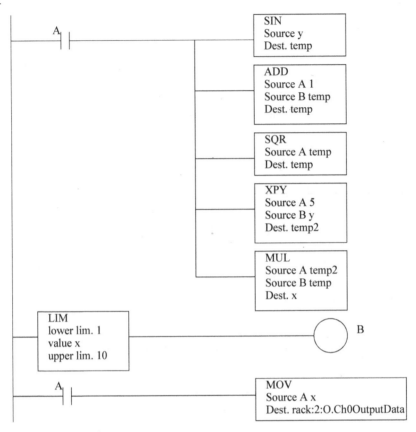

	SIN Source y Dest. temp
	ADD Source A 1 Source B temp Dest. temp
	SQR Source A temp Dest. temp
	XPY Source A 5 Source B y Dest. temp2
	MUL Source A temp2 Source B temp Dest. x

A

LIM
lower lim. 1
value x
upper lim. 10

B

A

MOV
Source A x
Dest. rack:2:O.Ch0OutputData

8.

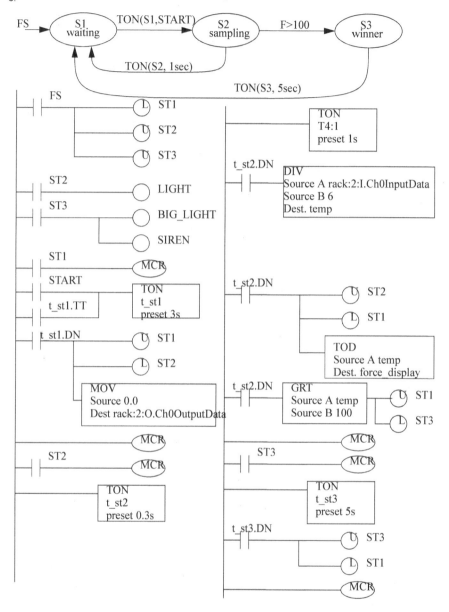

9.

```
SBR();
        v_in := rack:2:I.Ch0InputData / 32767; /* range is 0=0V,32767=10V */
        mass := 500.0 * (v_in - 1.0) / (6.0 - 1.0); /* convert to Kg */
        A := 0;
        IF M THEN
                IF NOT ((mass >= 300) AND (mass <= 400)) THEN
                        A := 1;
                END_IF
        END_IF
RET();
```

22CONTINUOUS SENSORS

Topics:

- Continuous sensor issues; accuracy, resolution, etc.
- Angular measurement; potentiometers, encoders and tachometers
- Linear measurement; potentiometers, LVDTs, Moire fringes and accelerometers
- Force measurement; strain gages and piezoelectric
- Liquid and fluid measurement; pressure and flow
- Temperature measurement; RTDs, thermocouples and thermistors
- Other sensors
- Continuous signal inputs and wiring
- Glossary

Objectives:

- To understand the common continuous sensor types.
- To understand interfacing issues.

Continuous sensors convert physical phenomena to measurable signals, typically voltages or currents. Consider a simple temperature measuring device, there will be an increase in output voltage proportional to a temperature rise. A computer could measure the voltage, and convert it to a temperature. The basic physical phenomena typically measured with sensors include;

- angular or linear position
- acceleration
- temperature
- pressure or flow rates
- stress, strain or force
- light intensity
- sound

Most of these sensors are based on subtle electrical properties of materials and devices. As a result the signals often require *signal conditioners*. These are often amplifiers that boost currents and voltages to larger voltages.

Sensors are also called transducers. This is because they convert an input phenomena to an output in a different form. This transformation relies upon a manufactured device with limitations and imperfection. As a result sensor limitations are often characterized with;

Accuracy - This is the maximum difference between the indicated and actual reading. For example, if a sensor reads a force of 100N with a ±1% accuracy, then the force could be anywhere from 99N to 101N.

Resolution - Used for systems that *step* through readings. This is the smallest increment that the sensor can detect, this may also be incorporated into the accuracy value. For example if a sensor measures up to 10 inches of linear displacements, and it outputs a number between 0 and 100, then the resolution of the device is 0.1 inches.

Repeatability - When a single sensor condition is made and repeated, there will be a small variation for that particular reading. If we take a statistical range for repeated readings (e.g., ±3 standard deviations) this will be the repeatability. For example, if a flow rate sensor has a repeatability of 0.5cfm, readings for an actual flow of 100cfm should rarely be outside 99.5cfm to 100.5cfm.

Linearity - In a linear sensor the input phenomenon has a linear relationship with the output signal. In most sensors this is a desirable feature. When the relationship is not linear, the conversion from the sensor output (e.g., voltage) to a calculated quantity (e.g., force) becomes more complex.

Precision - This considers accuracy, resolution and repeatability or one device relative to another.

Range - Natural limits for the sensor. For example, a sensor for reading angular rotation may only rotate 200 degrees.

Dynamic Response - The frequency range for regular operation of the sensor. Typically sensors will have an upper operation frequency, occasionally there will be lower frequency limits. For example, our ears hear best between 10Hz and 16KHz.

Environmental - Sensors all have some limitations over factors such as temperature, humidity, dirt/oil, corrosives and pressures. For example many sensors will work in relative humidities (RH) from 10% to 80%.

Calibration - When manufactured or installed, many sensors will need some calibration to determine or set the relationship between the input phenomena, and output. For example, a temperature reading sensor may need to be *zeroed* or adjusted so that the measured temperature matches the actual temperature. This may require special equipment, and need to be performed frequently.

Cost - Generally more precision costs more. Some sensors are very inexpensive, but the signal conditioning equipment costs are significant.

22.1 INDUSTRIAL SENSORS

This section describes sensors that will be of use for industrial measurements. The sections have been divided by the phenomena to be measured. Where possible details are provided.

22.1.1 Angular Displacement

22.1.1.1 - Potentiometers

Potentiometers measure the angular position of a shaft using a variable resistor. A potentiometer is shown in Figure 22.1. The potentiometer is resistor, normally made with a thin film of resistive material. A wiper can be moved along the surface of the resistive film. As the wiper moves toward one end there will be a change in resistance proportional to the distance moved. If a voltage is applied across the resistor, the voltage at the wiper interpolate the voltages at the ends of the resistor.

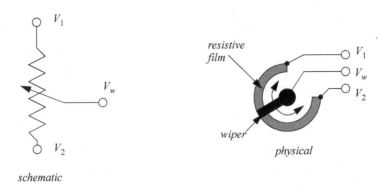

schematic

Figure 22.1 **A Potentiometer**

The potentiometer in Figure 22.1 is being used as a voltage divider. As the wiper rotates the output voltage will be proportional to the angle of rotation.

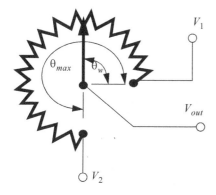

$$V_{out} = (V_2 - V_1)\left(\frac{\theta_w}{\theta_{max}}\right) + V_1$$

Figure 22.2　　*A Potentiometer as a Voltage Divider*

Potentiometers are popular because they are inexpensive, and don't require special signal conditioners. But, they have limited accuracy, normally in the range of 1% and they are subject to mechanical wear.

Potentiometers measure absolute position, and they are calibrated by rotating them in their mounting brackets, and then tightening them in place. The range of rotation is normally limited to less than 360 degrees or multiples of 360 degrees. Some potentiometers can rotate without limits, and the wiper will jump from one end of the resistor to the other.

Faults in potentiometers can be detected by designing the potentiometer to never reach the ends of the range of motion. If an output voltage from the potentiometer ever reaches either end of the range, then a problem has occurred, and the machine can be shut down. Two examples of problems that might cause this are wires that fall off, or the potentiometer rotates in its mounting.

22.1.2 Encoders

Encoders use rotating disks with optical windows, as shown in Figure 22.1. The encoder contains an optical disk with fine windows etched into it. Light from emitters passes through the openings in the disk to detectors. As the encoder shaft is rotated, the light beams are broken. The encoder shown here is a quadrature encode, and it will be discussed later.

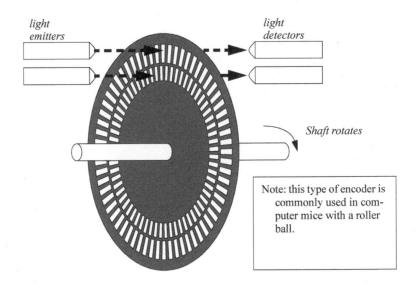

light emitters

light detectors

Shaft rotates

Note: this type of encoder is commonly used in computer mice with a roller ball.

Figure 22.3   **An Encoder Disk**

There are two fundamental types of encoders; absolute and incremental. An absolute encoder will measure the position of the shaft for a single rotation. The same shaft angle will always produce the same reading. The output is normally a binary or grey code number. An incremental (or relative) encoder will output two pulses that can be used to determine displacement. Logic circuits or software is used to determine the direction of rotation, and count pulses to determine the displacement. The velocity can be determined by measuring the time between pulses.

Encoder disks are shown in Figure 22.1. The absolute encoder has two rings, the outer ring is the most significant digit of the encoder, the inner ring is the least significant digit. The relative encoder has two rings, with one ring rotated a few degrees ahead of the other, but otherwise the same. Both rings detect position to a quarter of the disk. To add accuracy to the absolute encoder more rings must be added to the disk, and more emitters and detectors. To add accuracy to the relative encoder we only need to add more windows to the existing two rings. Typical encoders will have from 2 to thousands of windows per ring.

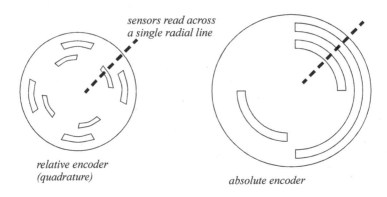

sensors read across a single radial line

relative encoder (quadrature)

absolute encoder

Figure 22.4   **Encoder Disks**

When using absolute encoders, the position during a single rotation is measured directly. If the encoder rotates multiple times then the total number of rotations must be counted separately.

When using a relative encoder, the distance of rotation is determined by counting the pulses from one of the rings. If the encoder only rotates in one direction then a simple count of pulses from one ring will determine the total distance. If the encoder can rotate both directions a second ring must be used to determine when to subtract pulses. The quadrature scheme, using two rings, is shown in Figure 22.1. The signals are set up so that one is out of phase with the other. Notice that for different directions of rotation, input B either leads or lags A.

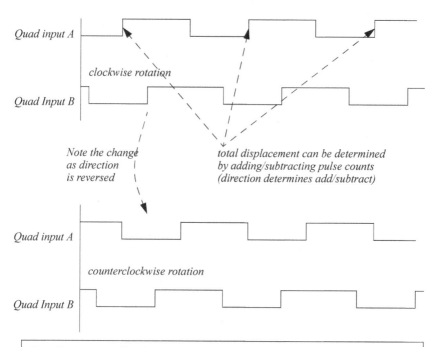

Note: To determine direction we can do a simple check. If both are off or on, the first to change state determines direction. Consider a point in the graphs above where both A and B are off. If A is the first input to turn on the encoder is rotating clockwise. If B is the first to turn on the rotation is counterclockwise.

Aside: A circuit (or program) can be built for this circuit using an up/down counter. If the positive edge of input A is used to trigger the clock, and input B is used to drive the up/down count, the counter will keep track of the encoder position.

Figure 22.5　　　**Quadrature Encoders**

Interfaces for encoders are commonly available for PLCs and as purchased units. Newer PLCs will also allow two normal inputs to be used to decode encoder inputs.

Normally absolute and relative encoders require a calibration phase when a controller is turned on. This normally involves moving an axis until it reaches a logical sensor that marks the end of the range. The end of range is then used as the zero position. Machines using encoders, and other relative sensors, are noticeable in that they normally move to some extreme position before use.

22.1.2.1 - Tachometers

Tachometers measure the velocity of a rotating shaft. A common technique is to mount a magnet to a rotating shaft. When the magnetic moves past a stationary pick-up coil, current is induced. For each rotation of the shaft there is a pulse in the coil, as shown in Figure 22.1. When the time between the pulses is measured the period for one rotation can be found, and the frequency calculated. This technique often requires some signal conditioning circuitry.

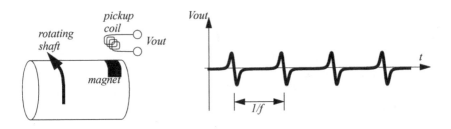

Figure 22.6   A Magnetic Tachometer

Another common technique uses a simple permanent magnet DC generator (note: you can also use a small DC motor). The generator is hooked to the rotating shaft. The rotation of a shaft will induce a voltage proportional to the angular velocity. This technique will introduce some drag into the system, and is used where efficiency is not an issue.

Both of these techniques are common, and inexpensive.

22.1.3 Linear Position

22.1.3.1 - Potentiometers

Rotational potentiometers were discussed before, but potentiometers are also available in linear/sliding form. These are capable of measuring linear displacement over long distances. Figure 22.1 shows the output voltage when using the potentiometer as a voltage divider.

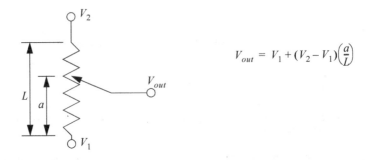

$$V_{out} = V_1 + (V_2 - V_1)\left(\frac{a}{L}\right)$$

Figure 22.7   Linear Potentiometer

Linear/sliding potentiometers have the same general advantages and disadvantages of rotating potentiometers.

22.1.3.2 - Linear Variable Differential Transformers (LVDT)

Linear Variable Differential Transformers (LVDTs) measure linear displacements over a limited range. The basic device is shown in Figure 22.1. It consists of outer coils with an inner moving magnetic core. High frequency alternating current (AC) is applied to the center coil. This generates a magnetic field that induces a current in the two outside coils. The core will pull the magnetic field towards it, so in the figure more current will be induced in the left hand coil. The outside coils are wound in opposite directions so that when the core is in the center the induced currents cancel, and the signal out is zero (0Vac). The magnitude of the *signal out* voltage on either line indicates the position of the core. Near the center of motion the change in voltage is proportional to the displacement. But, further from the center the relationship becomes nonlinear.

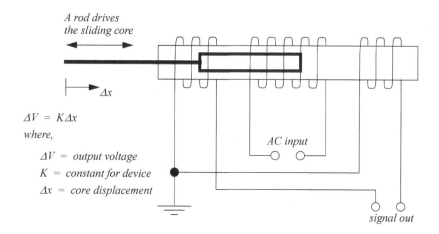

*A rod drives
the sliding core*

Δx

$\Delta V = K\Delta x$

where,

ΔV = *output voltage*

K = *constant for device*

Δx = *core displacement*

AC input

signal out

Figure 22.8   **An LVDT**

Aside: The circuit below can be used to produce a voltage that is proportional to position. The two diodes convert the AC wave to a half wave DC wave. The capacitor and resistor values can be selected to act as a low pass filter. The final capacitor should be large enough to smooth out the voltage ripple on the output.

Vac in   *LVDT*   *Vac out*   *Vdc out*

Figure 22.9   **A Simple Signal Conditioner for an LVDT**

These devices are more accurate than linear potentiometers, and have less friction. Typical applications for these devices include measuring dimensions on parts for quality control. They are often used for pressure measurements with Bourdon tubes and bellows/diaphragms. A major disadvantage of these sensors is the high cost, often in the thousands.

22.1.3.3 - Moire Fringes

High precision linear displacement measurements can be made with Moire Fringes, as shown in Figure 22.1. Both of the strips are transparent (or reflective), with black lines at measured intervals. The spacing of the lines determines the accuracy of the position measurements. The stationary strip is offset at an angle so that the strips interfere to give irregular patterns. As the moving strip travels by a stationary strip the patterns will move up, or down, depending upon the speed and direction of motion.

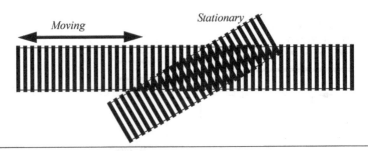

Note: you can recreate this effect with the strips below. Photocopy the pattern twice, overlay the sheets and hold them up to the light. You will notice that shifting one sheet will cause the stripes to move up or down.

Figure 22.10   **The Moire Fringe Effect**

A device to measure the motion of the moire fringes is shown in Figure 22.1. A light source is collimated by passing it through a narrow slit to make it one slit width. This is then passed through the fringes to be detected by light sensors. At least two light sensors are needed to detect the bright and dark locations. Two sensors, close enough, can act as a quadrature pair, and the same method used for quadrature encoders can be used to determine direction and distance of motion.

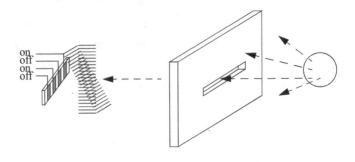

Figure 22.11   *Measuring Motion with Moire Fringes*

These are used in high precision applications over long distances, often meters. They can be purchased from a number of suppliers, but the cost will be high. Typical applications include Coordinate Measuring Machines (CMMs).

22.1.3.4 - Accelerometers

Accelerometers measure acceleration using a mass suspended on a force sensor, as shown in Figure 22.1. When the sensor accelerates, the inertial resistance of the mass will cause the force sensor to deflect. By measuring the deflection the acceleration can be determined. In this case the mass is cantilevered on the force sensor. A base and housing enclose the sensor. A small mounting stud (a threaded shaft) is used to mount the accelerometer.

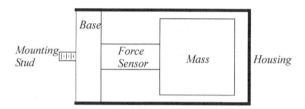

Figure 22.12   *A Cross Section of an Accelerometer*

Accelerometers are dynamic sensors, typically used for measuring vibrations between 10Hz to 10KHz. Temperature variations will affect the accuracy of the sensors. Standard accelerometers can be linear up to 100,000 m/s**2: high shock designs can be used up to 1,000,000 m/s**2. There is often a trade-off between a wide frequency range and device sensitivity (note: higher sensitivity requires a larger mass). Figure 22.1 shows the sensitivity of two accelerometers with different resonant frequencies. A smaller resonant frequency limits the maximum frequency for the reading. The smaller frequency results in a smaller sensitivity. The units for sensitivity is charge per m/s**2.

resonant freq. (Hz)	sensitivity
22 KHz	4.5 pC/(m/s**2)
180KHz	.004

Figure 22.13   *Piezoelectric Accelerometer Sensitivities*

The force sensor is often a small piece of piezoelectric material (discussed later in this chapter). The piezoelectic material can be used to measure the force in shear or compression. Piezoelectric based accelerometers typically have parameters such as,

        -100 to 250°C operating range
        1mV/g to 30V/g sensitivity
        operate well below one forth of the natural frequency

The accelerometer is mounted on the vibration source as shown in Figure 22.1. The accelerometer is electrically isolated from the vibration source so that the sensor may be grounded at the amplifier (to reduce electrical noise). Cables are fixed to the surface of the vibration source, close to the accelerometer, and are fixed to the surface as often as possible to prevent noise from the cable striking the surface. Background vibrations can be detected by attaching control electrodes to *non-vibrating* surfaces. Each accelerometer is different, but some general application guidelines are;

- The control vibrations should be less than 1/3 of the signal for the error to be less than 12%).
- Mass of the accelerometers should be less than a tenth of the measurement mass.
- These devices can be calibrated with shakers, for example a 1g shaker will hit a peak velocity of 9.81 m/s**2.

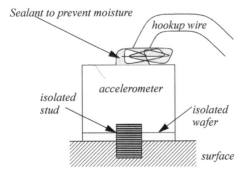

Figure 22.14   ***Mounting an Accelerometer***

Equipment normally used when doing vibration testing is shown in Figure 22.1. The sensor needs to be mounted on the equipment to be tested. A pre-amplifier normally converts the charge generated by the accelerometer to a voltage. The voltage can then be analyzed to determine the vibration frequencies.

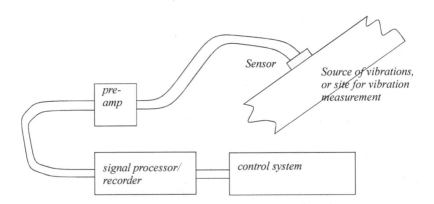

Figure 22.15   ***Typical Connection for Accelerometers***

Accelerometers are commonly used for control systems that adjust speeds to reduce vibration and noise. Computer Controlled Milling machines now use these sensors to actively eliminate chatter, and detect tool failure. The signal from accelerometers can be integrated to find velocity and acceleration.

Currently accelerometers cost hundreds or thousands per channel. But, advances in micromachining are already beginning to provide integrated circuit accelerometers at a low cost. Their current use is for airbag deployment systems in automobiles.

22.1.4 Forces and Moments

22.1.4.1 - Strain Gages

Strain gages measure strain in materials using the change in resistance of a wire. The wire is glued to the surface of a part, so that it undergoes the same strain as the part (at the mount point). Figure 22.1 shows the basic properties of the undeformed wire. Basically, the resistance of the wire is a function of the resistivity, length, and cross sectional area.

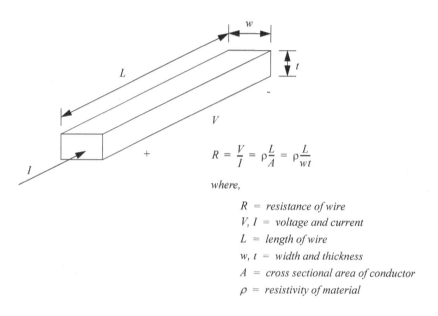

$$R = \frac{V}{I} = \rho \frac{L}{A} = \rho \frac{L}{wt}$$

where,

R = *resistance of wire*

V, I = *voltage and current*

L = *length of wire*

w, t = *width and thickness*

A = *cross sectional area of conductor*

ρ = *resistivity of material*

Figure 22.16   ***The Electrical Properties of a Wire***

After the wire in Figure 22.1 has been deformed it will take on the new dimensions and resistance shown in Figure 22.1. If a force is applied as shown, the wire will become longer, as predicted by Young's modulus. But, the cross sectional area will decrease, as predicted by Poison's ratio. The new length and cross sectional area can then be used to find a new resistance.

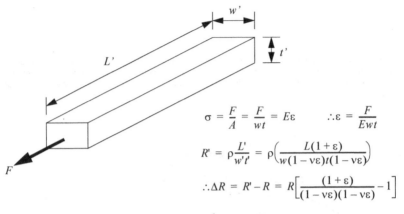

$$\sigma = \frac{F}{A} = \frac{F}{wt} = E\varepsilon \qquad \therefore \varepsilon = \frac{F}{Ewt}$$

$$R' = \rho\frac{L'}{w't'} = \rho\left(\frac{L(1+\varepsilon)}{w(1-\nu\varepsilon)t(1-\nu\varepsilon)}\right)$$

$$\therefore \Delta R = R' - R = R\left[\frac{(1+\varepsilon)}{(1-\nu\varepsilon)(1-\nu\varepsilon)} - 1\right]$$

where,

ν = poissons ratio for the material
F = applied force
E = Youngs modulus for the material
σ, ε = stress and strain of material

Aside: Gauge factor, as defined below, is a commonly used measure of stain gauge sensitivity.

$$GF = \frac{\left(\frac{\Delta R}{R}\right)}{\varepsilon}$$

Figure 22.17   *The Electrical and Mechanical Properties of the Deformed Wire*

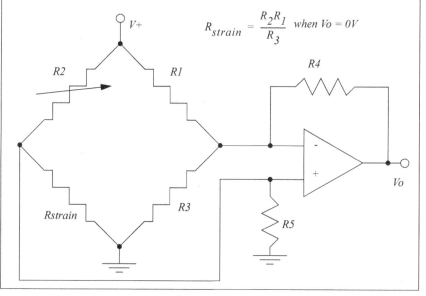

Aside: Changes in strain gauge resistance are typically small (large values would require strains that would cause the gauges to plastically deform). As a result, Wheatstone bridges are used to amplify the small change. In this circuit the variable resistor R2 would be tuned until Vo = 0V. Then the resistance of the strain gage can be calculated using the given equation.

$$R_{strain} = \frac{R_2 R_1}{R_3} \quad when \ Vo = 0V$$

Figure 22.18   *Measuring Strain with a Wheatstone Bridge*

A strain gage must be small for accurate readings, so the wire is actually wound in a uniaxial or rosette pattern, as shown in Figure 22.1. When using uniaxial gages the direction is important, it must be placed in the direction of the normal stress. (Note: the gages cannot read shear stress.) Rosette gages are less sensitive to direction, and if a shear force is present the gage will measure the resulting normal force at 45 degrees. These gauges are sold on thin films that are glued to the surface of a part. The process of mounting strain gages involves surface cleaning. application of adhesives, and soldering leads to the strain gages.

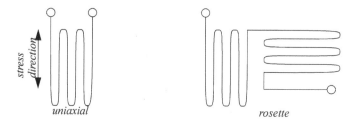

Figure 22.19   *Wire Arrangements in Strain Gages*

A design techniques using strain gages is to design a part with a narrowed neck to mount the strain gage on, as shown in Figure 22.1. In the narrow neck the strain is proportional to the load on the member, so it may be used to measure force. These parts are often called *load cells*.

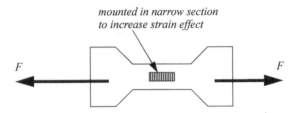

mounted in narrow section
to increase strain effect

F F

Figure 22.20   ***Using a Narrow to Increase Strain***

Strain gauges are inexpensive, and can be used to measure a wide range of stresses with accuracies under 1%. Gages require calibration before each use. This often involves making a reading with no load, or a known load applied. An example application includes using strain gages to measure die forces during stamping to estimate when maintenance is needed.

22.1.4.2 - Piezoelectric

When a crystal undergoes strain it displaces a small amount of charge. In other words, when the distance between atoms in the crystal lattice changes some electrons are forced out or drawn in. This also changes the capacitance of the crystal. This is known as the Piezoelectric effect. Figure 22.1 shows the relationships for a crystal undergoing a linear deformation. The charge generated is a function of the force applied, the strain in the material, and a constant specific to the material. The change in capacitance is proportional to the change in the thickness.

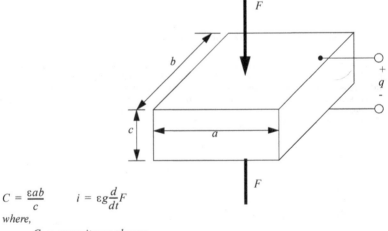

$$C = \frac{\varepsilon ab}{c} \qquad i = \varepsilon g \frac{d}{dt} F$$

where,

     C = *capacitance change*

     a, b, c = *geometry of material*

     ε = *dielectric constant (quartz typ. 4.06*10**-11 F/m)*

     i = *current generated*

     F = *force applied*

     g = *constant for material (quartz typ. 50*10**-3 Vm/N)*

     E = *Youngs modulus (quartz typ. 8.6*10**10 N/m**2)*

Figure 22.21   **The Piezoelectric Effect**

These crystals are used for force sensors, but they are also used for applications such as microphones and pressure sensors. Applying an electrical charge can induce strain, allowing them to be used as actuators, such as audio speakers.

When using piezoelectric sensors charge amplifiers are needed to convert the small amount of charge to a larger voltage. These sensors are best suited to dynamic measurements, when used for static measurements they tend to *drift* or slowly lose charge, and the signal value will change.

22.1.5 Liquids and Gases

There are a number of factors to be considered when examining liquids and gasses.

- Flow velocity
- Density
- Viscosity
- Pressure

There are a number of differences factors to be considered when dealing with fluids and gases. Normally a fluid is considered incompressible, while a gas normally follows the ideal gas law. Also, given sufficiently high enough temperatures, or low enough pressures a fluid can be come a gas.

$$PV = nRT$$

where,

P = *the gas pressure*

V = *the volume of the gas*

n = *the number of moles of the gas*

R = *the ideal gas constant* =

T = *the gas temperature*

When flowing, the flow may be smooth, or laminar. In case of high flow rates or unrestricted flow, turbulence may result. The Reynold's number is used to determine the transition to turbulence. The equation below is for calculation the Reynold's number for fluid flow in a pipe. A value below 2000 will result in laminar flow. At a value of about 3000 the fluid flow will become uneven. At a value between 7000 and 8000 the flow will become turbulent.

$$R = \frac{VD\rho}{u}$$

where,

R = *Reynolds number*

V = *velocity*

D = *pipe diameter*

ρ = *fluid density*

u = *viscosity*

22.1.5.1 - Pressure

Figure 22.1 shows different two mechanisms for pressure measurement. The Bourdon tube uses a circular pressure tube. When the pressure inside is higher than the surrounding air pressure (14.7psi approx.) the tube will straighten. A position sensor, connected to the end of the tube, will be elongated when the pressure increases.

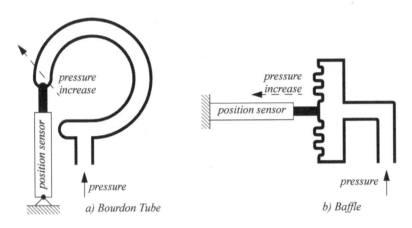

Figure 22.22   **Pressure Transducers**

These sensors are very common and have typical accuracies of 0.5%.

22.1.5.2 - Venturi Valves

When a flowing fluid or gas passes through a narrow pipe section (neck) the pressure drops. If there is no flow the pressure before and after the neck will be the same. The faster the fluid flow, the greater the pressure difference before and after the neck. This is known as a Venturi valve. Figure 22.1 shows a Venturi valve being used to measure a fluid flow rate. The fluid flow rate will be proportional to the pressure difference before and at the neck (or after the neck) of the valve.

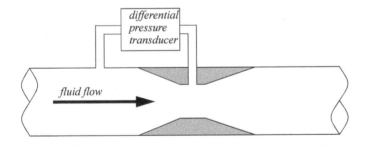

Figure 22.23   **A Venturi Valve**

Aside: Bernoulli's equation can be used to relate the pressure drop in a venturi valve.

$$\frac{p}{\rho} + \frac{v^2}{2} + gz = C$$

where,

p = pressure
ρ = density
v = velocity
g = gravitational constant
z = height above a reference
C = constant

Consider the centerline of the fluid flow through the valve. Assume the fluid is incompressible, so the density does not change. And, assume that the center line of the valve does not change. This gives us a simpler equation, as shown below, that relates the velocity and pressure before and after it is compressed.

$$\frac{p_{before}}{\rho} + \frac{v_{before}^2}{2} + gz = C = \frac{p_{after}}{\rho} + \frac{v_{after}^2}{2} + gz$$

$$\frac{p_{before}}{\rho} + \frac{v_{before}^2}{2} = \frac{p_{after}}{\rho} + \frac{v_{after}^2}{2}$$

$$p_{before} - p_{after} = \rho \left(\frac{v_{after}^2}{2} - \frac{v_{before}^2}{2} \right)$$

The flow velocity v in the valve will be larger than the velocity in the larger pipe section before. So, the right hand side of the expression will be positive. This will mean that the pressure before will always be higher than the pressure after, and the difference will be proportional to the velocity squared.

Figure 22.24   **The Pressure Relationship for a Venturi Valve**

Venturi valves allow pressures to be read without moving parts, which makes them very reliable and durable. They work well for both fluids and gases. It is also common to use Venturi valves to generate vacuums for actuators, such as suction cups.

22.1.5.3 - Coriolis Flow Meter

Fluid passes through thin tubes, causing them to vibrate. As the fluid approaches the point of maximum vibration it accelerates. When leaving the point it decelerates. The result is a distributed force that causes a bending moment, and hence twisting of the pipe. The amount of bending is proportional to the velocity of the fluid flow. These devices typically have a large constriction on the flow, and result is significant loses. Some of the devices also use bent tubes to increase the sensitivity, but this also increases the flow resistance. The typical accuracy for a Coriolis flowmeter is 0.1%.

22.1.5.4 - Magnetic Flow Meter

A magnetic sensor applies a magnetic field perpendicular to the flow of a conductive fluid. As the fluid moves, the electrons in the fluid experience an electromotive force. The result is that a potential (voltage) can be measured perpendicular to the direction of the flow and the magnetic field. The higher the flow rate, the greater the voltage. The typical accuracy for these sensors is 0.5%.

These flowmeters don't oppose fluid flow, and so they don't result in pressure drops.

22.1.5.5 - Ultrasonic Flow Meter

A transmitter emits a high frequency sound at point on a tube. The signal must then pass through the fluid to a detector where it is picked up. If the fluid is flowing in the same direction as the sound it will arrive sooner. If the sound is against the flow it will take longer to arrive. In a transit time flow meter two sounds are used, one traveling forward, and the other in the opposite direction. The difference in travel time for the sounds is used to determine the flow velocity.

A doppler flowmeter bounces a soundwave off particle in a flow. If the particle is moving away from the emitter and detector pair, then the detected frequency will be lowered, if it is moving towards them the frequency will be higher.

The transmitter and receiver have a minimal impact on the fluid flow, and therefore don't result in pressure drops.

22.1.5.6 - Vortex Flow Meter

Fluid flowing past a large (typically flat) obstacle will shed vortices. The frequency of the vortices will be proportional to the flow rate. Measuring the frequency allows an estimate of the flow rate. These sensors tend be low cost and are popular for low accuracy applications.

22.1.5.7 - Positive Displacement Meters

In some cases more precise readings of flow rates and volumes may be required. These can be obtained by using a positive displacement meter. In effect these meters are like pumps run in reverse. As the fluid is pushed through the meter it produces a measurable output, normally on a rotating shaft.

22.1.5.8 - Pitot Tubes

Gas flow rates can be measured using Pitot tubes, as shown in Figure 22.1. These are small tubes that project into a flow. The diameter of the tube is small (typically less than 1/8") so that it doesn't affect the flow.

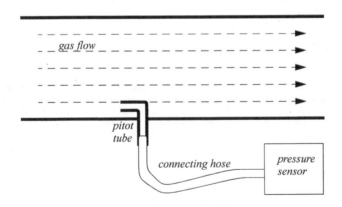

Figure 22.25   **Pitot Tubes for Measuring Gas Flow Rates**

22.1.6 Temperature

Temperature measurements are very common with control systems. The temperature ranges are normally described with the following classifications.

very low temperatures <-60 deg C - e.g. superconductors in MRI units
low temperature measurement -60 to 0 deg C - e.g. freezer controls
fine temperature measurements 0 to 100 deg C - e.g. environmental controls
high temperature measurements <3000 deg F - e.g. metal refining/processing
very high temperatures > 2000 deg C - e.g. plasma systems

22.1.6.1 - Resistive Temperature Detectors (RTDs)

When a metal wire is heated the resistance increases. So, a temperature can be measured using the resistance of a wire. Resistive Temperature Detectors (RTDs) normally use a wire or film of platinum, nickel, copper or nickel-iron alloys. The metals are wound or wrapped over an insulator, and covered for protection. The resistances of these alloys are shown in Figure 22.1.

Material	Temperature Range C (F)	Typical Resistance (ohms)
Platinum	-200 - 850 (-328 - 1562)	100
Nickel	-80 - 300 (-112 - 572)	120
Copper	-200 - 260 (-328 - 500)	10

Figure 22.26   **RTD Properties**

These devices have positive temperature coefficients that cause resistance to increase linearly with temperature. A platinum RTD might have a resistance of 100 ohms at 0C, that will increase by 0.4 ohms/°C. The total resistance of an RTD might double over the temperature range.

A current must be passed through the RTD to measure the resistance. (Note: a voltage divider can be used to convert the resistance to a voltage.) The current through the RTD should be kept to a minimum to prevent self heating. These devices are more linear than thermocouples, and can have accuracies of 0.05%. But, they can be expensive

22.1.6.2 - Thermocouples

Each metal has a natural potential level, and when two different metals touch there is a small potential difference, a voltage. (Note: when designing assemblies, dissimilar metals should not touch, this will lead to corrosion.) Thermocouples use a junction of dissimilar metals to generate a voltage proportional to temperature. This principle was discovered by T.J. Seebeck.

The basic calculations for thermocouples are shown in Figure 22.1. This calculation provides the measured voltage using a reference temperature and a constant specific to the device. The equation can also be rearranged to provide a temperature given a voltage.

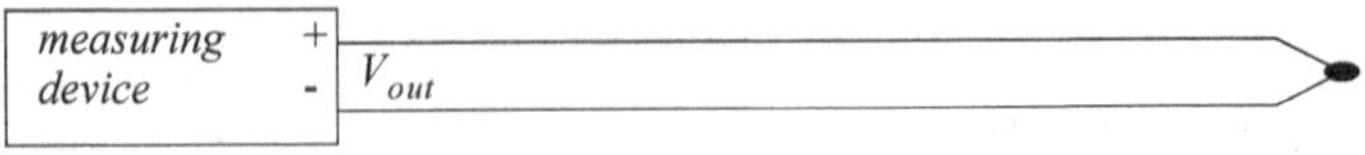

$$V_{out} = \alpha(T - T_{ref})$$

$$\therefore T = \frac{V_{out}}{\alpha} + T_{ref}$$

where,

$\alpha = constant\ (V/C)$ $50\frac{\mu V}{°C}$ *(typical)*

$T, T_{ref} = current\ and\ reference\ temperatures$

Figure 22.27   **Thermocouple Calculations**

The list in Table 1 shows different junction types, and the normal temperature ranges. Both thermocouples, and signal conditioners are commonly available, and relatively inexpensive. For example, most PLC vendors sell thermocouple input cards that will allow multiple inputs into the PLC.

Table 1: Thermocouple Types

ANSI Type	Materials	Temperature Range (°F)	Voltage Range (mV)
T	copper/constantan	-200 to 400	-5.60 to 17.82
J	iron/constantan	0 to 870	0 to 42.28
E	chromel/constantan	-200 to 900	-8.82 to 68.78
K	chromel/aluminum	-200 to 1250	-5.97 to 50.63
R	platinum-13%rhodium/platinum	0 to 1450	0 to 16.74
S	platinum-10%rhodium/platinum	0 to 1450	0 to 14.97
C	tungsten-5%rhenium/tungsten-26%rhenium	0 to 2760	0 to 37.07

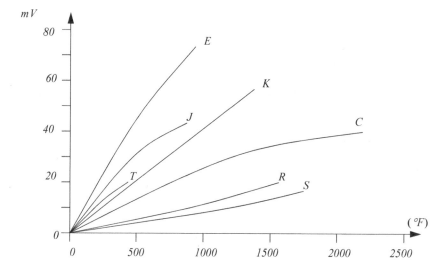

Figure 22.28 *Thermocouple Temperature Voltage Relationships (Approximate)*

The junction where the thermocouple is connected to the measurement instrument is normally cooled to reduce the thermocouple effects at those junctions. When using a thermocouple for precision measurement, a second thermocouple can be kept at a known temperature for reference. A series of thermocouples connected together in series produces a higher voltage and is called a thermopile. Readings can approach an accuracy of 0.5%.

22.1.6.3 - Thermistors

Thermistors are non-linear devices, their resistance will decrease with an increase in temperature. (Note: this is because the extra heat reduces electron mobility in the semiconductor.) The resistance can change by more than 1000 times. The basic calculation is shown in Figure 22.1.

often metal oxide semiconductors The calculation uses a reference temperature and resistance, with a constant for the device, to predict the resistance at another temperature. The expression can be rearranged to calculate the temperature given the resistance.

$$R_t = R_o e^{\beta\left(\frac{1}{T} - \frac{1}{T_o}\right)}$$

$$\therefore T = \frac{\beta T_o}{T_o \ln\left(\frac{R_t}{R_o}\right) + \beta}$$

where,

R_o, R_t = resistances at reference and measured temps.
T_o, T = reference and actual temperatures
β = constant for device

Figure 22.29 *Thermistor Calculations*

Aside: The circuit below can be used to convert the resistance of the thermistor to a voltage using a Wheatstone bridge and an inverting amplifier.

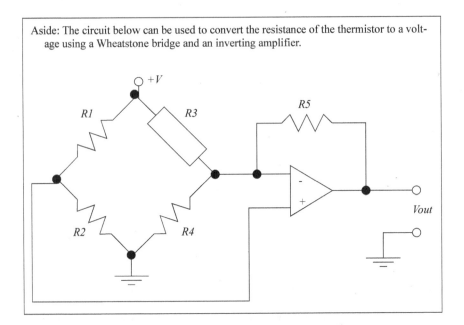

Figure 22.30   ***Thermistor Signal Conditioning Circuit***

Thermistors are small, inexpensive devices that are often made as beads, or metallized surfaces. The devices respond quickly to temperature changes, and they have a higher resistance, so junction effects are not an issue. Typical accuracies are 1%, but the devices are not linear, have a limited temperature/resistance range and can be self heating.

22.1.6.4 - Other Sensors

IC sensors are becoming more popular. They output a digital reading and can have accuracies better than 0.01%. But, they have limited temperature ranges, and require some knowledge of interfacing methods for serial or parallel data.

Pyrometers are non-contact temperature measuring devices that use radiated heat. These are normally used for high temperature applications, or for production lines where it is not possible to mount other sensors to the material.

22.1.7 Light

22.1.7.1 - Light Dependant Resistors (LDR)

Light dependant resistors (LDRs) change from high resistance (>Mohms) in bright light to low resistance (<Kohms) in the dark. The change in resistance is non-linear, and is also relatively slow (ms).

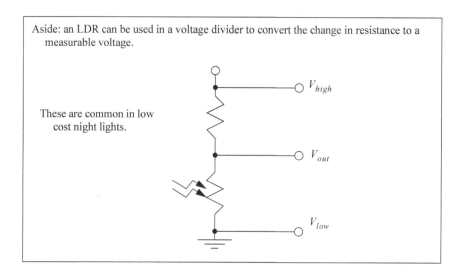

Aside: an LDR can be used in a voltage divider to convert the change in resistance to a measurable voltage.

These are common in low cost night lights.

Figure 22.31   *A Light Level Detector Circuit*

22.1.8 Chemical

22.1.8.1 - pH

The pH of an ionic fluid can be measured over the range from a strong base (alkaline) with pH=14, to a neutral value, pH=7, to a strong acid, pH=0. These measurements are normally made with electrodes that are in direct contact with the fluids.

22.1.8.2 - Conductivity

Conductivity of a material, often a liquid is often used to detect impurities. This can be measured directly be applying a voltage across two plates submerged in the liquid and measuring the current. High frequency inductive fields is another alternative.

22.1.9 Others

A number of other detectors/sensors are listed below,

Combustion - gases such as CO_2 can be an indicator of combustion
Humidity - normally in gases
Dew Point - to determine when condensation will form

22.2 INPUT ISSUES

Signals from transducers are typically too small to be read by a normal analog input card. Amplifiers are used to increase the magnitude of these signals. An example of a single ended signal amplifier is shown in Figure

22.1. The amplifier is in an inverting configuration, so the output will have an opposite sign from the input. Adjustments are provided for *gain* and *offset* adjustments.

> Note: op-amps are used in this section to implement the amplifiers because they are inexpensive, common, and well suited to simple design and construction projects. When purchasing a commercial signal conditioner, the circuitry will be more complex, and include other circuitry for other factors such as temperature compensation.

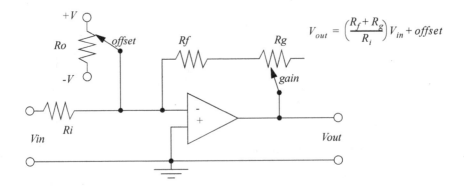

$$V_{out} = \left(\frac{R_f + R_g}{R_i}\right) V_{in} + offset$$

Figure 22.32   *A Single Ended Signal Amplifier*

A differential amplifier with a current input is shown in Figure 22.1. Note that Rc converts a current to a voltage. The voltage is then amplified to a larger voltage.

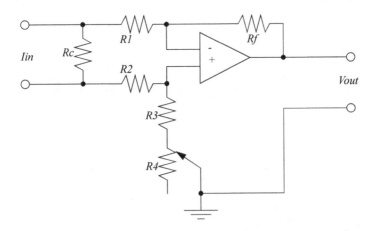

Figure 22.33   *A Current Amplifier*

The circuit in Figure 22.1 will convert a differential (double ended) signal to a single ended signal. The two input op-amps are used as unity gain followers, to create a high input impedance. The following amplifier amplifies the voltage difference.

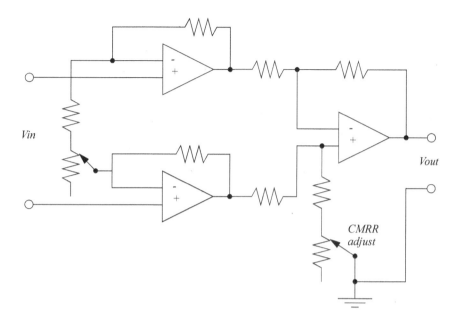

Figure 22.34   A Differential Input to Single Ended Output Amplifier

The Wheatstone bridge can be used to convert a resistance to a voltage output, as shown in Figure 22.1. If the resistor values are all made the same (and close to the value of R3) then the equation can be simplified.

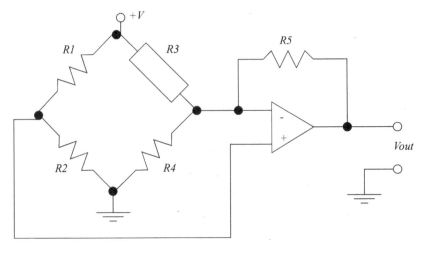

$$V_{out} = V(R_5)\left[\left(\frac{R_2}{R_1+R_2}\right)\left(\frac{1}{R_3}+\frac{1}{R_4}+\frac{1}{R_5}\right)-\frac{1}{R_3}\right]$$

or if $R = R_1 = R_2 = R_4 = R_5$

$$V_{out} = V\left(\frac{R}{2R_3}\right)$$

Figure 22.35   ***A Resistance to Voltage Amplifier***

22.3 SENSOR GLOSSARY

Ammeter - A meter to indicate electrical current. It is normally part of a DMM

Bellows - This is a flexible volumed that will expand or contract with a pressure change. This often looks like a cylinder with a large radius (typ. 2") but it is very thin (type 1/4"). It can be set up so that when pressure changes, the displacement of one side can be measured to determine pressure.

Bourdon tube - Widely used industrial gage to measure pressure and vacuum. It resembles a crescent moon. When the pressure inside changes the moon shape will tend to straighten out. By measuring the displacement of the tip the pressure can be measured.

Chromatographic instruments - laboratory-type instruments used to analyze chemical compounds and gases.

Inductance-coil pulse generator - transducer used to measure rotational speed. Output is pulse train.

Interferometers - These use the interference of light waves 180 degrees out of phase to determine distances. Typical sources of the monochromatic light required are lasers.

Linear-Variable-Differential transformer (LVDT) electromechanical transducer used to measure angular or linear displacement. Output is Voltage

Manometer - liquid column gage used widely in industry to measure pressure.

Ohmmeter - meter to indicate electrical resistance

Optical Pyrometer - device to measure temperature of an object at high temperatures by sensing the brightness of an objects surface.

Orifice Plate - widely used flowmeter to indicate fluid flow rates

Photometric Transducers - a class of transducers used to sense light, including phototubes, photodiodes, phototransistors, and photoconductors.

Piezoelectric Accelerometer - Transducer used to measure vibration. Output is emf.

Pitot Tube - Laboratory device used to measure flow.

Positive displacement Flowmeter - Variety of transducers used to measure flow. Typical output is pulse train.

Potentiometer - instrument used to measure voltage

Pressure Transducers - A class of transducers used to measure pressure. Typical output is voltage. Operation of the transducer can be based on strain gages or other devices.

Radiation pyrometer - device to measure temperature by sensing the thermal radiation emitted from the object.

Resolver - this device is similar to an incremental encoder, except that it uses coils to generate magnetic fields. This is like a rotary transformer.

Strain Gage - Widely used to indicate torque, force, pressure, and other variables. Output is change in resistance due to strain, which can be converted into voltage.

Thermistor - Also called a resistance thermometer; an instrument used to measure temperature. Operation is based on change in resistance as a function of temperature.

Thermocouple - widely used temperature transducer based on the Seebeck effect, in which a junction of two dissimilar metals emits emf related to temperature.

Turbine Flowmeter - transducer to measure flow rate. Output is pulse train.

Venturi Tube - device used to measure flow rates.

22.4 SUMMARY

- Selection of continuous sensors must include issues such as accuracy and resolution.
- Angular positions can be measured with potentiometers and encoders (more accurate).
- Tachometers are useful for measuring angular velocity.
- Linear positions can be measured with potentiometers (limited accuracy), LVDTs (limited range), moire fringes (high accuracy).
- Accelerometers measure acceleration of masses.
- Strain gauges and piezoelectric elements measure force.
- Pressure can be measured indirectly with bellows and Bourdon tubes.
- Flow rates can be measured with Venturi valves and pitot tubes.
- Temperatures can be measured with RTDs, thermocouples, and thermistors.
- Input signals can be single ended for more inputs or double ended for more accuracy.

22.4.1 References

Bryan, L.A. and Bryan, E.A., Programmable Controllers; Theory and Implementation, Industrial Text Co., 1988.

Swainston, F., A Systems Approach to Programmable Controllers, Delmar Publishers Inc., 1992.

22.5 PRACTICE PROBLEMS

(Note: Problem solutions are available at http://sites.google.com/site/automatedmanufacturingsystems/)

1. Name two types of inputs that would be analog input values (versus a digital value).

2. Search the web for common sensor manufacturers for 5 different types of continuous sensors. If possible identify prices for the units. Sensor manufacturers include (hyde park, banner, allen bradley, omron, etc.)

3. What is the resolution of an absolute optical encoder that has six binary tracks? nine tracks? twelve tracks?

4. Suggest a couple of methods for collecting data on the factory floor

5. If a thermocouple generates a voltage of 30mV at 800F and 40mV at 1000F, what voltage will be generated at 1200F?

6. A potentiometer is to be used to measure the position of a rotating robot link (as a voltage divider). The power supply connected across the potentiometer is 5.0 V, and the total wiper travel is 300 degrees. The wiper arm is directly connected to the rotational joint so that a given rotation of the joint corresponds to an equal rotation of the wiper arm.

      a) If the joint is at 42 degrees, what voltage will be output from the potentiometer?

      b) If the joint has been moved, and the potentiometer output is 2.765V, what is the position of the potentiometer?

7. A motor has an encoder mounted on it. The motor is driving a reducing gear box with a 50:1 ratio. If the position of the geared down shaft needs to be positioned to 0.1 degrees, what is the minimum resolution of the incremental encoder?

8. What is the difference between a strain gauge and an accelerometer? How do they work?

9. Use the equations for a permanent magnet DC motor to explain how it can be used as a tachometer.

10. What are the trade-offs between encoders and potentiometers?

11. A potentiometer is connected to a PLC analog input card. The potentiometer can rotate 300 degrees, and the voltage supply for the potentiometer is +/-10V. Write a ladder logic program to read the voltage from the potentiometer and convert it to an angle in radians stored in 'angle'.

22.6 ASSIGNMENT PROBLEMS

1. Write a simple C program to read incremental encoder inputs (A and B) to determine the current position of the encoder. Note: use the quadrature encoding to determine the position of the motor.

2. A high precision potentiometer has an accuracy of +/- 0.1% and can rotate 300degrees and is used as a voltage divider with a of 0V and 5V. The output voltage is being read by an A/D converter with a 0V to 10V input range. How many bits does the A/D converter need to accommodate the accuracy of the potentiometer?

3. The table of position and voltage values below were measured for an inexpensive potentiometer. Write a C subroutine that will accept a voltage value and interpolate the position value.

theta (deg)	V
0	0.1
67	0.6
145	1.6
195	2.4
213	3.4
296	4.2
315	5.0

22.1 PRACTICE PROBLEM SOLUTIONS

1. Temperature and displacement

2. Sensors can be found at www.ab.com, www.omron.com, etc

3. 360°/64steps, 360°/512steps, 360°/4096steps

4. data bucket, smart machines, PLCs with analog inputs and network connections

5.

$$V_{out} = \alpha(T - T_{ref}) \qquad 0.030 = \alpha(800 - T_{ref}) \qquad 0.040 = \alpha(1000 - T_{ref})$$

$$\frac{1}{\alpha} = \frac{800 - T_{ref}}{0.030} = \frac{1000 - T_{ref}}{0.040}$$

$$800 - T_{ref} = 750 - 0.75T_{ref}$$

$$50 = 0.25T_{ref} \qquad T_{ref} = 200F \qquad \alpha = \frac{0.040}{1000 - 200} = \frac{50\mu V}{F}$$

$$V_{out} = 0.00005(1200 - 200) = 0.050V$$

6.

a) $$V_{out} = (V_2 - V_1)\left(\frac{\theta_w}{\theta_{max}}\right) + V_1 = (5V - 0V)\left(\frac{42deg}{300deg}\right) + 0V = 0.7V$$

b) $$2.765V = (5V - 0V)\left(\frac{\theta_w}{300deg}\right) + 0V$$

$$2.765V = (5V - 0V)\left(\frac{\theta_w}{300deg}\right) + 0V$$

$$\theta_w = 165.9deg$$

7.

$$\theta_{output} = 0.1\frac{deg}{count} \qquad \frac{\theta_{input}}{\theta_{output}} = \frac{50}{1} \qquad \theta_{input} = 50\left(0.1\frac{deg}{count}\right) = 5\frac{deg}{count}$$

$$R = \frac{360\frac{deg}{rot}}{5\frac{deg}{count}} = 72\frac{count}{rot}$$

8.

        strain gauge measures strain in a material using a stretching wire that increases resistance - accelerometers measure acceleration with a cantilevered mass on a piezoelectric element.

9.

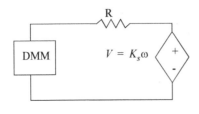

When the motor shaft is turned by another torque source a voltage is generated that is proportional to the angular velocity. This is the reverse emf. A DMM, or other high impedance instrument can be used to measure this, thus minimizing the loses in resistor R.

$$\dot{\omega} + \omega\left(\frac{K^2}{JR}\right) = V_s\left(\frac{K}{JR}\right)$$

$$V_s = \omega(K) + \dot{\omega}\left(\frac{JR}{K}\right)$$

10.

encoders cost more but can have higher resolutions. Potentiometers have limited ranges of motion

11.

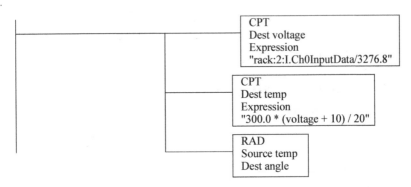

23CONTINUOUS ACTUATORS

Topics:

- Servo Motors; AC and DC
- Stepper motors
- Single axis motion control
- Hydraulic actuators

Objectives:

- To understand the main differences between continuous actuators
- Be able to select a continuous actuator
- To be able to plan a motion for a single servo actuator

Continuous actuators allow a system to position or adjust outputs over a wide range of values. Even in their simplest form, continuous actuators tend to be mechanically complex devices. For example, a linear slide system might be composed of a motor with an electronic controller driving a mechanical slide with a ball screw. The cost for such actuators can easily be higher than for the control system itself. These actuators also require sophisticated control techniques that will be discussed in later chapters. In general, when there is a choice, it is better to use discrete actuators to reduce costs and complexity.

23.1 ELECTRIC MOTORS

An electric motor is composed of a rotating center, called the rotor, and a stationary outside, called the stator. These motors use the attraction and repulsion of magnetic fields to induce forces, and hence motion. Typical electric motors use at least one electromagnetic coil, and sometimes permanent magnets to set up opposing fields. When a voltage is applied to these coils the result is a torque and rotation of an output shaft. There are a variety of motor configuration the yields motors suitable for different applications. Most notably, as the voltages supplied to the motors will vary the speeds and torques that they will provide.

- Motor Categories
  - AC motors - rotate with relatively constant speeds proportional to the frequency of the supply power
    - induction motors - squirrel cage, wound rotor - inexpensive, efficient.
    - synchronous - fixed speed, efficient
  - DC motors - have large torque and speed ranges
    - permanent magnet - variable speed
    - wound rotor and stator - series, shunt and compound (universal)
  - Hybrid
    - brushless permanent magnet -
    - stepper motors

- Contactors are used to switch motor power on/off

- Drives can be used to vary motor speeds electrically. This can also be done with mechanical or hydraulic machines.

- Popular drive categories
  - Variable Frequency Drives (VFD) - vary the frequency of the power delivered to the motor to vary speed.
  - DC motor controllers - variable voltage or current to vary the motor speed
  - Eddy Current Clutches for AC motors - low efficiency, uses a moving iron drum and windings
  - Wound rotor AC motor controllers - low efficiency, uses variable resistors to adjust the winding currents

A control system is required when a motor is used for an application that requires continuous position or velocity. A typical controller is shown in Figure 23.1. In any controlled system a command generator is required to specify a desired position. The controller will compare the feedback from the encoder to the desired position or velocity to determine the system error. The controller will then generate an output, based on the system error. The output is then passed through a power amplifier, which in turn drives the motor. The encoder is connected directly to the motor shaft to provide feedback of position.

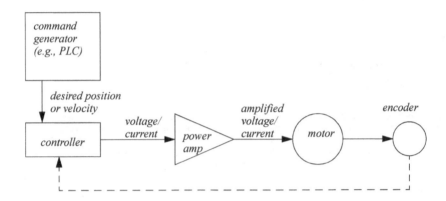

Figure 23.1   *A Typical Feedback Motor Controller*

23.1.1 Basic Brushed DC Motors

In a DC motor there is normally a set of coils on the rotor that turn inside a stator populated with permanent magnets. Figure 23.1 shows a simplified model of a motor. The magnets provide a permanent magnetic field for the rotor to push against. When current is run through the wire loop it creates a magnetic field.

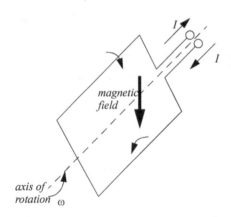

Figure 23.2   *A Simplified Rotor*

The power is delivered to the rotor using a commutator and brushes, as shown in Figure 23.1. In the figure the power is supplied to the rotor through graphite brushes rubbing against the commutator. The commutator is split so that every half revolution the polarity of the voltage on the rotor, and the induced magnetic field reverses to push against the permanent magnets.

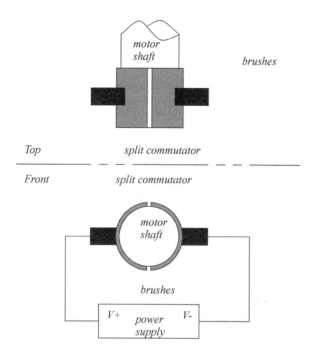

Top *split commutator*

Front *split commutator*

Figure 23.3 **A Split Ring Commutator**

The direction of rotation will be determined by the polarity of the applied voltage, and the speed is proportional to the voltage. A feedback controller is used with these motors to provide motor positioning and velocity control.

These motors are losing popularity to brushless motors. The brushes are subject to wear, which increases maintenance costs. In addition, the use of brushes increases resistance, and lowers the motors efficiency.

ASIDE: The controller to drive a servo motor normally uses a Pulse Width Modulated (PWM) signal. As shown below the signal produces an effective voltage that is relative to the time that the signal is on. The percentage of time that the signal is on is called the duty cycle. When the voltage is on all the time the effective voltage delivered is the maximum voltage. So, if the voltage is only on half the time, the effective voltage is half the maximum voltage. This method is popular because it can produce a variable effective voltage efficiently. The frequency of these waves is normally above 20KHz, above the range of human hearing.

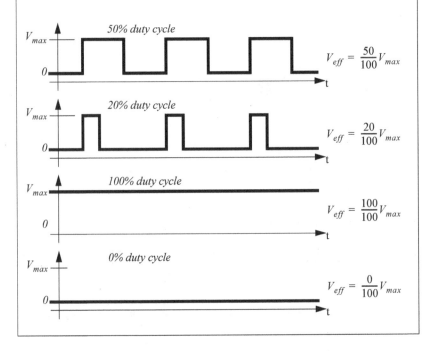

Figure 23.4   *Pulse Width Modulation (PWM) For Control*

ASIDE: A PWM signal can be used to drive a motor with the circuit shown below. The PWM signal switches the NPN transistor, thus switching power to the motor. In this case the voltage polarity on the motor will always be the same direction, so the motor may only turn in one direction.

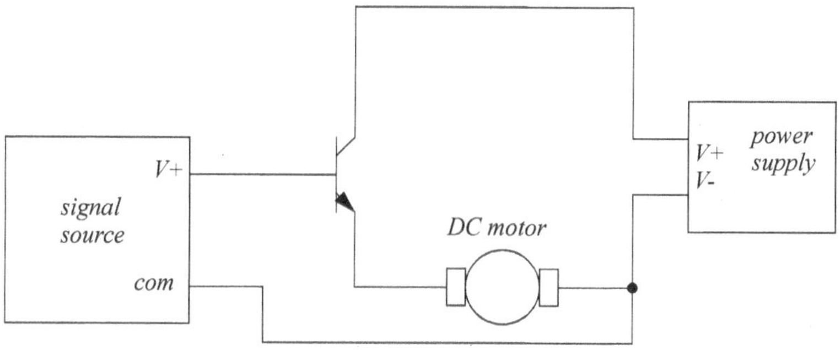

Figure 23.5 ***PWM Unidirectional Motor Control Circuit***

ASIDE: When a motor is to be controlled with PWM in two directions the H-bridge circuit (shown below) is a popular choice. These can be built with individual components, or purchased as integrated circuits for smaller motors. To turn the motor in one direction the PWM signal is applied to the Va inputs, while the Vb inputs are held low. In this arrangement the positive voltage is at the left side of the motor. To reverse the direction the PWM signal is applied to the Vb inputs, while the Va inputs are held low. This applies the positive voltage to the right side of the motor.

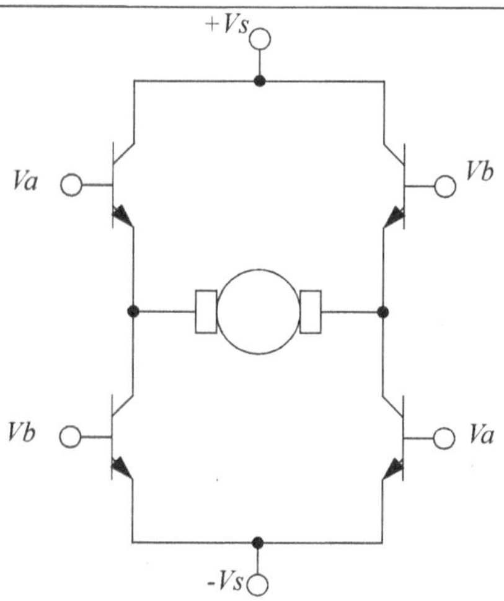

Figure 23.6 ***PWM Bidirectional Motor Control Circuit***

23.1.2 AC Motors

Power is normally generated as 3-phase AC, so using this increases the efficiency of electrical drives. In AC motors the AC current is used to create changing fields in the motor. Typically AC motors have windings on the stator with multiple poles. Each pole is a pair of windings. As the AC current reverses, the magnetic field in the rotor appears to rotate.

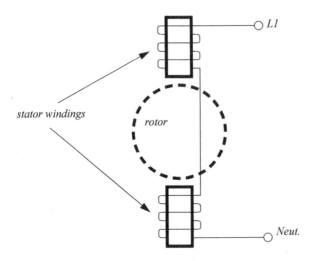

Figure 23.7   *A 2 Pole Single Phase AC Motor*

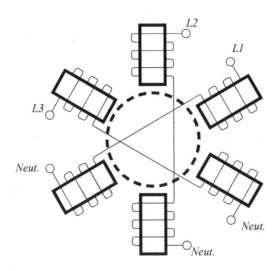

Figure 23.8   *A 6 Pole 3-Phase AC Motor*

The number of windings (poles) can be an integer multiple of the number of phases of power. More poles results in a lower rotational speed of the motor. Rotor types for induction motors are listed below. Their function is to intersect changing magnetic fields from the stator. The changing field induces currents in the rotor. These cur-

470

rents in turn set up magnetic fields that oppose fields from the stator, generating a torque.

Squirrel cage - has the shape of a wheel with end caps and bars
Wound Rotor - the rotor has coils wound. These may be connected to external contacts via commutator

Induction motors require slip. If the motor turns at the precise speed of the stator field, it will not see a changing magnetic field. The result would be a collapse of the rotor magnetic field. As a result an induction motor always turns slightly slower than the stator field. The difference is called the slip. This is typically a few percent. As the motor is loaded the slip will increase until the motor stalls.

An induction motor has the windings on the stator. The rotor is normally a squirrel cage design. The squirrel cage is a cast aluminum core that when exposed to a changing magnetic field will set up an opposing field. When an AC voltage is applied to the stator coils an AC magnetic field is created, the squirrel cage sets up an opposing magnetic field and the resulting torque causes the motor to turn.

The motor will turn at a frequency close to that of the applied voltage, but there is always some slip. It is possible to control the speed of the motor by controlling the frequency of the AC voltage. Synchronous motor drives control the speed of the motors by synthesizing a variable frequency AC waveform, as shown in Figure 23.1.

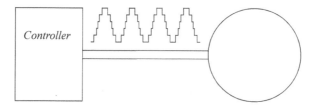

Figure 23.9   **AC Motor Speed Control**

These drives should be used for applications that only require a single rotational direction. The torque speed curve for a typical induction motor is shown in Figure 23.1. When the motor is used with a fixed frequency AC source the synchronous speed of the motor will be the frequency of AC voltage divided by the number of poles in the motor. The motor actually has the maximum torque below the synchronous speed. For example a 2 pole motor might have a synchronous speed of (2*60*60/2) 3600 RPM, but be rated for 3520 RPM. When a feedback controller is used the issue of slip becomes insignificant.

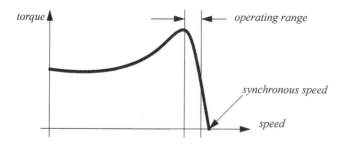

Figure 23.10   **Torque Speed Curve for an Induction Motor**

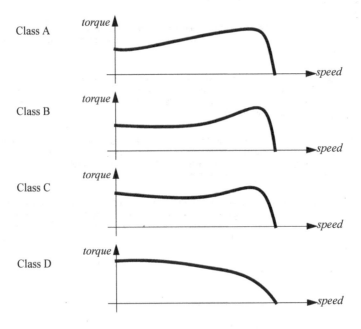

Class A

Class B

Class C

Class D

Figure 23.11   ***NEMA Squirrel Cage Torque Speed Curves***

Wound rotor induction motors use external resistors. varying the resistance allows the motors torque speed curve to vary. As the resistance value is increased the motor torque speed curve shifts from the Class A to Class D shapes. The figure below shows the relationship between the motor speed and applied power, slip, and number of poles. An ideal motor with no load would have a slip of 0%.

$$RPM = \frac{f120}{p}\left(1 - \frac{S}{100\%}\right)$$

where,

f = power frequency (60Hz typ.)

p = number of poles (2, 4, 6, etc...)

RPM = motor speed in rotations per minute

S = motor slip

Single phase AC motors can run in either direction. To compensate for this a shading pole is used on the stator windings. It basically acts as an inductor to one side of the field which slows the field buildup and collapse. The result is that the field strength seems to naturally rotate. Thermal protection is normally used in motors to prevent overheating.

Universal motors were presented earlier for DC applications, but they can also be used for AC power sources. This is because the field polarity in the rotor and stator both reverse as the AC current reverses. Synchronous motors are different from induction motors in that they are designed to rotate at the frequency of the fields, in other words there is no slip. Synchronous motors use generated fields in the rotor to oppose the stators field.

Starting AC motors can be hard because of the low torque at low speeds. To deal with this a switching arrangement is often used. At low speeds other coils or capacitors are connected into the circuits. At higher speeds centrifugal switches disconnect these and the motor behavior switches. Single phase induction motors are typically used for loads under 1HP. Various types (based upon their starting and running modes) are,

- split phase - there are two windings on the motor. A starting winding is used to provide torque at lower speeds.
- capacitor run -
- capacitor start
- capacitor start and run
- shaded pole - these motors use a small offset coil (such as a single copper winding) to encourage the field buildup to occur asymmetrically. These motors are for low torque applications much less than 1HP.
- universal motors (also used with DC) have a wound rotor and stator that are connected in series.

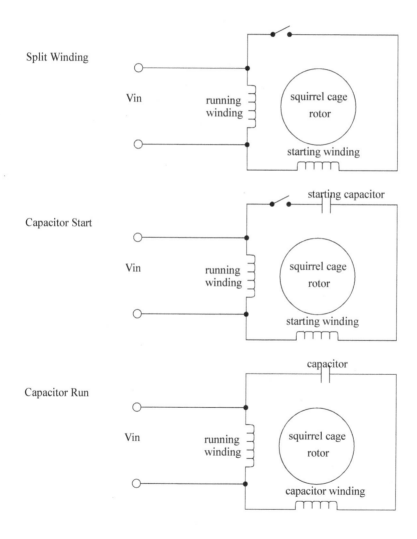

Figure 23.12   **Single Phase Motor Configurations**

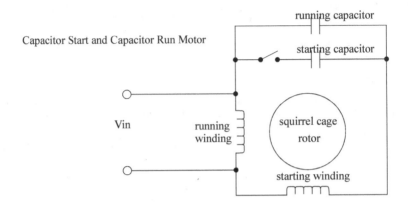

Capacitor Start and Capacitor Run Motor

running capacitor

starting capacitor

Vin

running winding

squirrel cage rotor

starting winding

Figure 23.13 **Single Phase Motor Configurations**

23.1.3 Brushless DC Motors

Brushless motors use a permanent magnet on the rotor, and use windings on the stator. Therefore there is no need to use brushes and a commutator to switch the polarity of the voltage on the coil. The lack of brushes means that these motors require less maintenance than the brushed DC motors.

A typical Brushless DC motor could have three poles, each corresponding to one power input, as shown in Figure 23.1. Each of coils is separately controlled. The coils are switched on to attract or repel the permanent magnet rotor.

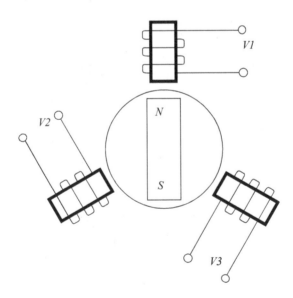

$V1$

$V2$

N

S

$V3$

Figure 23.14 **A Brushless DC Motor**

To continuously rotate these motors the current in the stator coils must alternate continuously. If the power supplied to the coils was a 3-phase AC sinusoidal waveform, the motor will rotate continuously. The applied voltage can also be trapezoidal, which will give a similar effect. The changing waveforms are controller using position feedback from the motor to select switching times. The speed of the motor is proportional to the frequency of the signal.

A typical torque speed curve for a brushless motor is shown in Figure 23.1.

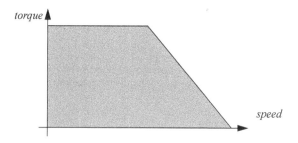

Figure 23.15　　　**Torque Speed Curve for a Brushless DC Motor**

23.1.4 Stepper Motors

Stepper motors are designed for positioning. They move one step at a time with a typical step size of 1.8 degrees giving 200 steps per revolution. Other motors are designed for step sizes of 1.8, 2.0, 2.5, 5, 15 and 30 degrees.

There are two basic types of stepper motors, unipolar and bipolar, as shown in Figure 23.1. The unipolar uses center tapped windings and can use a single power supply. The bipolar motor is simpler but requires a positive and negative supply and more complex switching circuitry.

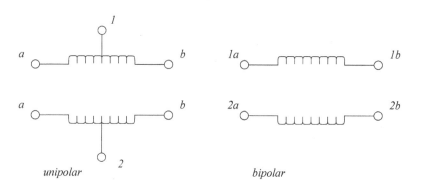

Figure 23.16　　　**Unipolar and Bipolar Stepper Motor Windings**

The motors are turned by applying different voltages at the motor terminals. The voltage change patterns for a unipolar motor are shown in Figure 23.1. For example, when the motor is turned on we might apply the voltages as shown in line 1. To rotate the motor we would then output the voltages on line 2, then 3, then 4, then 1, etc. Reversing the sequence causes the motor to turn in the opposite direction. The dynamics of the motor and load limit the maximum speed of switching, this is normally a few thousand steps per second. When not turning the output voltages are held to keep the motor in position.

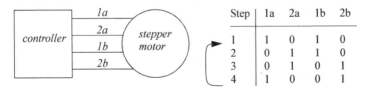

Step	1a	2a	1b	2b
1	1	0	1	0
2	0	1	1	0
3	0	1	0	1
4	1	0	0	1

To turn the motor the phases are stepped through 1, 2, 3, 4, and then back to 1. To reverse the direction of the motor the sequence of steps can be reversed, eg. 4, 3, 2, 1, 4, ... If a set of outputs is kept on constantly the motor will be held in position.

Figure 23.17   ***Stepper Motor Control Sequence for a Unipolar Motor***

Stepper motors do not require feedback except when used in high reliability applications and when the dynamic conditions could lead to slip. A stepper motor slips when the holding torque is overcome, or it is accelerated too fast. When the motor slips it will move a number of degrees from the current position. The slip cannot be detected without position feedback.

Stepper motors are relatively weak compared to other motor types. The torque speed curve for the motors is shown in Figure 23.1. In addition they have different static and dynamic holding torques. These motors are also prone to resonant conditions because of the stepped motion control.

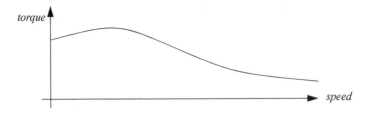

Figure 23.18   ***Stepper Motor Torque Speed Curve***

The motors are used with controllers that perform many of the basic control functions. At the minimum a *translator* controller will take care of switching the coil voltages. A more sophisticated *indexing* controller will accept motion parameters, such as distance, and convert them to individual steps. Other types of controllers also provide finer step resolutions with a process known as *microstepping*. This effectively divides the logical steps described in Figure 23.1 and converts them to sinusoidal steps.

translators - the user indicates maximum velocity and acceleration and a distance to move
indexer - the user indicates direction and number of steps to take
microstepping - each step is subdivided into smaller steps to give more resolution

23.1.5 Wound Field Motors

These use DC power on the rotor and stator to generate the magnetic field (i.e., no permanent magnets).

The shunt motor configuration

    - have the rotor and stator coils connected in parallel.
    - when the load on these motors is reduced the current flow increases slightly, increasing the field, and slowing the motor.
    - these motors have a relatively small variation in speed as they are varied, and are considered to have a relatively constant speed.
    - the speed of the motor can be controlled by changing the supply voltage, or by putting a rheostat/resistor in series with the stator windings.

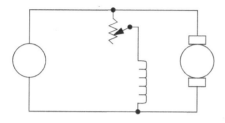

$$I_a = \frac{V_a}{R_a}$$

$$T = K_t I_a \phi$$

where,

I_a, V_a, R_a = *Armature current, voltage and resistance*

T = *Torque on motor shaft*

K_t = *Motor speed constant*

ϕ = *motor field flux*

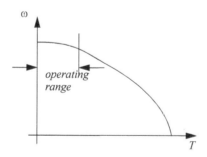

Series motors\

    - have the rotor and stator coils connected in series.
    - as the motor speed increases the current increases, the motor can theoretically accelerate to infinite speeds if unloaded. This makes the dangerous when used in applications where they are potentially unloaded.
    - these motors typically have greater starting torques that shunt motors

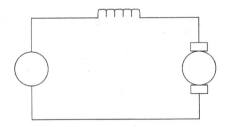

$$I_a = \frac{V_a}{R_a + R_f}$$

$$T = K_t I_a \phi = K_t I_a^2$$

where,

I_a, V_a = Armature current, voltage

R_a, R_f = Armature and field coil resistance

T = Torque on motor shaft

K_t = Motor speed constant

ϕ = motor field flux

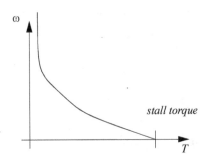

$$e_f = r_a i_a + D l_a i_a + e_m$$

$$e_m = K_e \theta D$$

$$T = K_T i_a$$

$$e_a = (r_a + l_a D) i_a + K_e D \theta$$

$$e_a = (r_a + l_a D)\left(\frac{T}{K_T}\right) + K_e D \theta$$

Figure 23.19   *Equations for an armature controlled DC motor*

Compound motors;

- have the rotor and stator coils connected in series.
- differential compound motors have the shunt and series winding field aligned so that they oppose each other.
- cumulative compound motors have the shunt and series winding fields aligned so that they add

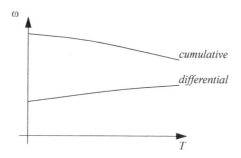

$$e_f = r_f i_f + l_f i_f D$$

$$T = K_T i_f$$

$$\frac{T}{\theta} = JD^2 + BD$$

$$\frac{\theta}{T} = \frac{1}{JD^2 + BD}$$

$$\frac{\theta}{i_f} = \frac{\theta\,T}{T i_f} = \frac{K_T}{JD^2 + BD}$$

$$\frac{\theta}{e_f} = \frac{\theta\,i_f}{i_f e_f} = \left(\frac{K_T}{JD^2 + BD}\right)\left(\frac{1}{r_f + l_f D}\right)$$

$$\frac{T}{e_f} = \frac{T i_f}{i_f e_f} = K_T\left(\frac{1}{r_f + l_f D}\right)$$

Figure 23.20   *Equations for a controlled field motor*

23.2 HYDRAULICS

Hydraulic systems are used in applications requiring a large amount of force and slow speeds. When used for continuous actuation they are mainly used with position feedback. An example system is shown in Figure 23.1. The controller examines the position of the hydraulic system, and drivers a servo valve. This controls the flow of fluid to the actuator. The remainder of the provides the hydraulic power to drive the system.

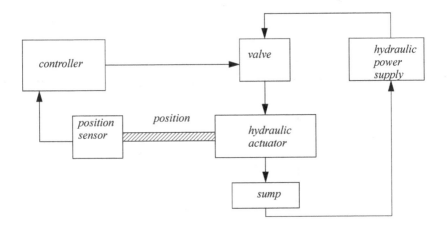

Figure 23.21   *Hydraulic Servo System*

The valve used in a hydraulic system is typically a solenoid controlled valve that is simply opened or closed. Newer, more expensive, valve designs use a scheme like pulse with modulation (PWM) which open/close the valve quickly to adjust the flow rate.

23.3 OTHER SYSTEMS

The continuous actuators discussed earlier in the chapter are the more common types. For the purposes of completeness additional actuators are listed and described briefly below.

> Heaters - to control a heater with a continuous temperature a PWM scheme can be used to limit a DC voltage, or an SCR can be used to supply part of an AC waveform.
> Pneumatics - air controlled systems can be used for positioning with suitable feedback. Velocities can also be controlled using fast acting valves.
> Linear Motors - a linear motor works on the same principles as a normal rotary motor. The primary difference is that they have a limited travel and their cost is typically much higher than other linear actuators.
> Ball Screws - rotation is converted to linear motion using balls screws. These are low friction screws that drive nuts filled with ball bearings. These are normally used with slides to bear mechanical loads.

23.4 SUMMARY

- AC motors are low cost and work in a relatively narrow speed range (e.g., around 1700RPM).
- DC motors work over a range of speeds and tend to be more controllable. Costs are higher for controls or maintenance.
- Motion control introduces velocity and acceleration objectives to servo control.

23.5 PRACTICE PROBLEMS

(Note: Problem solutions are available at http://sites.google.com/site/automatedmanufacturingsystems/)

1. A stepping motor is to be used to drive each of the three linear axes of a cartesian coordinate robot. The motor output shaft will be connected to a screw thread with a screw pitch of 0.125". It is desired that the control resolution of each of the axes be 0.025"
   a) to achieve this control resolution how many step angles are required on the stepper motor?
   b) What is the corresponding step angle?
   c) Determine the pulse rate that will be required to drive a given joint at a velocity of 3.0"/sec.

2. For the stepper motor in the previous question, a pulse train is to be generated by the robot controller.
   a) How many pulses are required to rotate the motor through three complete revolutions?
   b) If it is desired to rotate the motor at a speed of 25 rev/min, what pulse rate must be generated by the robot controller?

3. Explain the differences between stepper motors, variable frequency induction motors and DC motors using tables.

4. Short answer,
   a) Compare the various types of motors discussed in the class using a detailed table.
   b) When using a motor there are the static and kinetic friction limits. Will deadband correction allow the motor to move slower than both, one, or neither? Explain your answer.
   c) What is the purpose of a calibration curve?

23.6 ASSIGNMENT PROBLEMS

1. A stepper motor is to be used to actuate one joint of a robot arm in a light duty pick and place application. The step angle of the motor is 10 degrees. For each pulse received from the pulse train source the motor rotates through a distance of one step angle.
   a) What is the resolution of the stepper motor?
   b) Relate this value to the definitions of control resolution, spatial resolution, and accuracy, as discussed in class.
   c) For the stepper motor, a pulse train is to be generated by a motion controller. How many pulses are required to rotate the motor through three complete revolutions? If it is desired to rotate the motor at a speed of 25 rev/min, what pulse rate must be generated by the robot controller?

2. Describe the voltage ripple that would occur when using a permanent magnet DC motor as a tachometer. Hint: consider the use of the commutator to switch the polarity of the coil.

3. Compare the advantages/disadvantages of DC permanent magnet motors and AC induction motors.

23.1 PRACTICE PROBLEM SOLUTIONS

1.

a) $\quad P = 0.125\left(\dfrac{in}{rot}\right) \qquad R = 0.025\dfrac{in}{step}$

$$\theta = \frac{R}{P} = \frac{0.025\dfrac{in}{step}}{0.125\left(\dfrac{in}{rot}\right)} = 0.2\frac{rot}{step} \qquad \text{Thus} \qquad \frac{1}{0.2\dfrac{rot}{step}} = 5\frac{step}{rot}$$

b) $\quad \theta = 0.2\dfrac{rot}{step} = 72\dfrac{deg}{step}$

c)
$$PPS = \frac{3\dfrac{in}{s}}{0.025\dfrac{in}{step}} = 120\frac{steps}{s}$$

2.

a)
$$pulses = (3rot)\left(5\frac{step}{rot}\right) = 15 steps$$

b) $\quad \dfrac{pulses}{s} = \left(25\dfrac{rot}{min}\right)\left(5\dfrac{step}{rot}\right) = 125\dfrac{steps}{min} = 125\left(\dfrac{1min}{60s}\right)\dfrac{steps}{min} = 2.08\dfrac{step}{s}$

3.

	speed	torque
stepper motor	very low speeds	low torque
vfd	limited speed range	good at rated speed
dc motor	wide range	decreases at higher speeds

4.

a)

(ans.

Motor Type	Cost	Torque	Speed	Applications
AC/Induction	low	med	limited	consumer applications/large power
DC Brushed	low/med	med	variable	short life
DC Brushless	high	med	variable	high precision
Stepper	low/med	low	low	positioning
Shunt	med	med	varies	
Series	med	high	varies	large break away torques

b) Deadband correction allows the motor to break free of the static friction. Once moving freely the torque required to 'stick' the motor is determined by the lower kinetic friction. Generally this means that the motor can move slightly slower than the static friction minimum speed, but not the kinetic friction minimum speed.

c) Calibration is a process where instrumentation outputs are related to inputs. These results are then used later to relate measurement equipment outputs with actual phenomenon. For example, in the laboratory, tachometers are calibrated by turning them at a steady speed. The speed is measured with a strobe tachometer and the voltage output is also recorded. These are then used to make a graph relating voltage and speed. Later the strobe tachometer is not used and the voltage output of the tach. is used to calculate the speed.

24CONTINUOUS CONTROL

Topics:

- Feedback control of continuous systems
- Control of systems with logical actuators
- PID control with continuous actuators
- Analysis of PID controlled systems
- PID control with a PLC
- Design examples

Objectives:

- To understand the concepts behind continuous control
- Be able to control a system with logical actuators
- Be able to analyze and control system with a PID controller

Continuous processes require continuous sensors and/or actuators. For example, an oven temperature can be measured with a thermocouple. Simple decision-based control schemes can use continuous sensor values to control logical outputs, such as a heating element. Linear control equations can be used to examine continuous sensor values and set outputs for continuous actuators, such as a variable position gas valve.

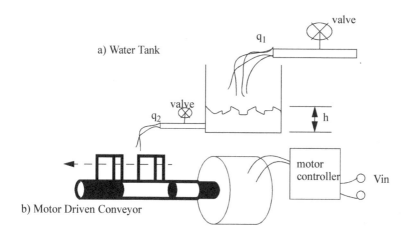

Figure 24.1   **Continuous Systems**

Two continuous control systems are shown in Figure 24.1. The water tank can be controlled valves. In a simple control scheme, one of the valves is set by the process, but we control the other to maximize some control object. If the water tank was actually a city water tank, the outlet valve would be the domestic and industrial water users. The inlet valve would be set to keep the tank level at maximum. If the level drops there will be a reduced water pressure at the outlet, and if the tank becomes too full it could overflow. The conveyor will move boxes between stations. Two common choices are to have it move continuously, or to move the boxes between positions, and then stop. When starting and stopping the boxes should be accelerated quickly, but not so quickly that they slip. And, the conveyor should stop at precise positions. In both of these systems, a good control system design will result in better performance.

A mechanical control system is pictured in Figure 24.1 that could be used for the water tank in Figure 24.1. This controller will adjust the valve position, therefore controlling the flow rate into the tank. The height of the fluid in the tank will change the hydrostatic pressure at the bottom of the tank. A pressure line is connected to a pressure cell. As the pressure inside the cell changes, the cell will expand and contract, opening and closing the valve. As the tank fills the pressure becomes higher, the cell expands, and the valve closes, reducing the flow in. The desired height of the tank can be adjusted by sliding the pressure cell up/down a distance *x*. In this example the height *x* is called the setpoint. The *control variable* is the position of the valve, and, the *feedback* variable is the water pressure from the tank. The *controller* is the pressure cell.

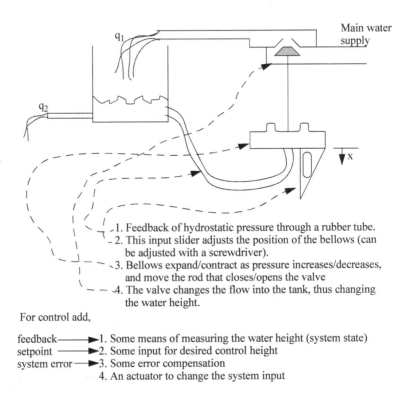

1. Feedback of hydrostatic pressure through a rubber tube.
2. This input slider adjusts the position of the bellows (can be adjusted with a screwdriver).
3. Bellows expand/contract as pressure increases/decreases, and move the rod that closes/opens the valve
4. The valve changes the flow into the tank, thus changing the water height.

For control add,

feedback———►1. Some means of measuring the water height (system state)
setpoint ———►2. Some input for desired control height
system error——►3. Some error compensation
4. An actuator to change the system input

Figure 24.2   *A Feedback Controller*

Continuous control systems typically need a target value, this is called a *setpoint*. The controller should be designed with some objective in mind. Typical objectives are listed below.

fastest response - reach the setpoint as fast as possible (e.g., hard drive speed)
smooth response - reduce acceleration and jerks (e.g., elevators)
energy efficient - minimize energy usage (e.g., industrial oven)
noise immunity - ignores disturbances in the system (e.g., variable wind gusts)

An engineer can design a controller mathematically when performance and stability are important issues. A common industrial practice is to purchase a *PID* unit, connect it to a process, and tune it through trial and error. This is suitable for simpler systems, but these systems are less efficient and prone to instability. In other words it is quick and easy, but these systems can go *out-of-control*.

24.1 CONTROL OF LOGICAL ACTUATOR SYSTEMS

Many continuous systems will be controlled with logical actuators. Common examples include building HVAC (Heating, Ventilation and Air Conditioning) systems. The system setpoint is entered on a *thermostat*. The controller will then attempt to keep the temperature within a few degrees as shown in Figure 24.1. If the temperature is below the bottom limit the heater is turned on. When it passes the upper limit it is turned off, and it will stay off until if passes the lower limit. If the gap between the upper and lower the boundaries is larger, the heater will

486

turn on less often, but be on for longer, and the temperature will vary more. This technique is not exact, and time lags will often lead to overshoot above and below the temperature limits.

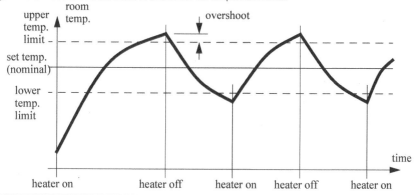

Note: This system turns on/off continuously. This behavior is known as hunting. If the limits are set too close to the nominal value, the system will hunt at a faster rate. Therefore, to prevent wear and improve efficiency we normally try to set the limits as far away from nominal as possible.

Figure 24.3   *Continuous Control with a Logical Actuator*

Figure 24.1 shows a controller that will keep the temperature between 72 and 74 (degrees presumably). The temperature will be read and stored in *temp*, and the output to turn the heater on is connected to *heater*.

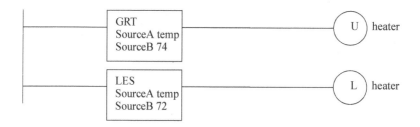

Figure 24.4   *A Ladder Logic Controller for a Logical Actuator*

24.2 CONTROL OF CONTINUOUS ACTUATOR SYSTEMS

24.2.1 Block Diagrams

Figure 24.1 shows a simple block diagram for controlling arm position. The system setpoint, or input, is the desired position for the arm. The arm position is expressed with the joint angles. The input enters a summation block, shown as a circle, where the actual joint angles are subtracted from the desired joint angles. The resulting difference is called the *error*. The *error* is transformed to joint torques by the first block labeled *neural system and muscles*. The next block, *arm structure and dynamics*, converts the torques to new arm positions. The new arm positions are converted back to joint angles by the *eyes*.

487

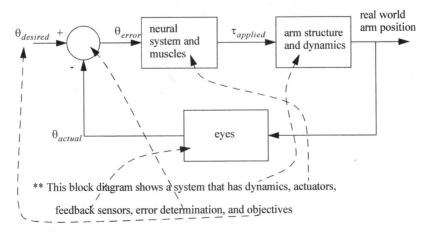

$\theta_{desired}$ + θ_{error} neural system and muscles $\tau_{applied}$ arm structure and dynamics real world arm position

θ_{actual} eyes

** This block diagram shows a system that has dynamics, actuators, feedback sensors, error determination, and objectives

Figure 24.5   **A Block Diagram**

The blocks in block diagrams represent real systems that have inputs and outputs. The inputs and outputs can be real quantities, such as fluid flow rates, voltages, or pressures. The inputs and outputs can also be calculated as values in computer programs. In continuous systems the blocks can be described using differential equations. Laplace transforms and transfer functions are often used for linear systems.

24.2.2 Feedback Control Systems

As introduced in the previous section, feedback control systems compare the desired and actual outputs to find a system error. A controller can use the error to drive an actuator to minimize the error. When a system uses the output value for control, it is called a feedback control system. When the feedback is subtracted from the input, the system has negative feedback. A negative feedback system is desirable because it is generally more stable, and will reduce system errors. Systems without feedback are less accurate and may become unstable.

A car is shown in Figure 24.1, without and with a velocity control system. First, consider the car by itself, the control variable is the gas pedal angle. The output is the velocity of the car. The negative feedback controller is shown inside the dashed line. Normally the driver will act as the control system, adjusting the speed to get a desired velocity. But, most automobile manufacturers offer *cruise control* systems that will automatically control the speed of the system. The driver will activate the system and set the desired velocity for the cruise controller with buttons. When running, the cruise control system will observe the velocity, determine the speed error, and then adjust the gas pedal angle to increase or decrease the velocity.

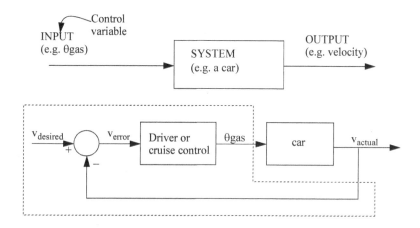

Figure 24.6   *Addition of a Control System to a Car*

The control system must perform some type of calculation with *Verror*, to select a new θgas. This can be implemented with mechanical mechanisms, electronics, or software. Figure 24.1 lists a number of rules that a person would use when acting as the controller. The driver will have some target velocity (that will occasionally be based on speed limits). The driver will then compare the target velocity to the actual velocity, and determine the difference between the target and actual. This difference is then used to adjust the gas pedal angle.

1. If v_{error} is a little positive/negative, increase/decrease θ_{gas} a little.

2. If v_{error} is very big/small, increase/decrease θ_{gas} a lot.

3. If v_{error} is near zero, keep θ_{gas} the same.

4. If v_{error} suddenly becomes bigger/smaller, then increase/decrease θ_{gas} quickly.

Figure 24.7   *Human Control Rules for Car Speed*

Mathematical rules are required when developing an automatic controller. The next two sections describe different approaches to controller design.

24.2.3 Proportional Controllers

Figure 24.1 shows a block diagram for a common servo motor controlled positioning system. The input is a numerical position for the motor, designated as *C*. (Note: The relationship between the motor shaft angle and *C* is determined by the encoder.) The difference between the desired and actual *C* values is the system error. The controller then converts the error to a control voltage *V*. The current amplifier keeps the voltage *V* the same, but increases the current (and power) to drive the servomotor. The servomotor will turn in response to a voltage, and drive an encoder and a ball screw. The encoder is part of the negative feedback loop. The ball screw converts the rotation into a linear displacement *x*. In this system, the position *x* is not measured directly, but it is estimated using the motor shaft angle.

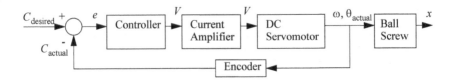

Figure 24.8 *A Servomotor Feedback Controller*

The blocks for the system in Figure 24.1 could be described with the equations in Figure 24.1. The summation block becomes a simple subtraction. The control equation is the simplest type, called a proportional controller. It will simply multiply the error by a constant Kp. A larger value for Kp will give a faster response. The current amplifier keeps the voltage the same. The motor is assumed to be a permanent magnet DC servo motor, and the ideal equation for such a motor is given. In the equation J is the polar mass moment of inertia, R is the resistance of the motor coils, and Km is a constant for the motor. The velocity of the motor shaft must be integrated to get position. The ball screw will convert the rotation into a linear position if the angle is divided by the Threads Per Inch (TPI) on the screw. The encoder will count a fixed number of Pulses Per Revolution (PPR).

$$\text{Summation Block:} \quad e = C_{desired} - C_{actual} \quad (1)$$

$$\text{Controller:} \quad V_c = K_p e \quad (2)$$

$$\text{Current Amplifier:} \quad V_m = V_c \quad (3)$$

$$\text{Servomotor:} \quad \left(\frac{d}{dt}\right)\omega + \left(\frac{K_m}{JR}\right)^2 \omega = \left(\frac{K_m}{JR}\right)V_m \quad (4)$$

$$\omega = \frac{d}{dt}\theta_{actual} \quad (5)$$

$$\text{Ball Screw:} \quad x = \frac{\theta_{actual}}{TPI} \quad (6)$$

$$\text{Encoder:} \quad C_{actual} = PPR(\theta_{actual}) \quad (7)$$

Figure 24.9 *A Servomotor Feedback Controller*

The system equations can be combined algebraically to give a single equation for the entire system as shown in Figure 24.1. The resulting equation (12) is a second order non-homogeneous differential equation that can be solved to model the performance of the system.

$(21.4), (21.5) \quad \left(\frac{d}{dt}\right)^2 \theta_{actual} + \left(\frac{K_m}{JR}\right)^2 \left(\frac{d}{dt}\right) \theta_{actual} = \left(\frac{K_m}{JR}\right) V_m \hfill (21.8)$

$(21.2), (21.3) \quad V_m = K_p e \hfill (21.9)$

$(21.1), (21.9) \quad V_m = K_p (C_{desired} - C_{actual}) \hfill (21.10)$

$(21.8), (21.10) \quad \left(\frac{d}{dt}\right)^2 \theta_{actual} + \left(\frac{K_m}{JR}\right)^2 \left(\frac{d}{dt}\right) \theta_{actual} = \left(\frac{K_m}{JR}\right) K_p (C_{desired} - C_{actual}) \hfill (21.11)$

$(21.7), (21.11) \quad \left(\frac{d}{dt}\right)^2 \theta_{actual} + \left(\frac{K_m}{JR}\right)^2 \left(\frac{d}{dt}\right) \theta_{actual} = \left(\frac{K_m}{JR}\right) K_p (C_{desired} - PPR\theta_{actual})$

$$\left(\frac{d}{dt}\right)^2 \theta_{actual} + \left(\frac{K_m}{JR}\right)^2 \left(\frac{d}{dt}\right) \theta_{actual} + \left(\frac{K_m(PPR)K_p}{JR}\right) \theta_{actual} = \left(\frac{K_p K_m}{JR}\right) C_{desired} \quad (21.12)$$

Figure 24.10 **A Combined System Model**

A proportional control system can be implemented with the ladder logic shown in Figure 24.1. The control system has a start/stop button. When the system is active *Run* will be on, and the proportional controller calculation will be performed with the *SUB* and *MUL* functions. When the system is inactive the *MOV* function will set the output to zero.

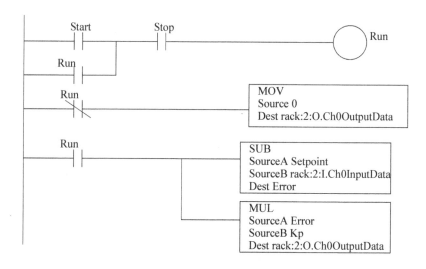

Figure 24.11 *Implementing a Proportional Controller with Ladder Logic*

This controller may be able to update a few times per second. This is an important design consideration - recall that the Nyquist Criterion requires that the control system response be much faster than the system being controlled. Typically this controller will only be suitable for systems that don't change more than 10 times per second. (Note: The speed limitation is a practical limitation for a SoftLogix processor with reasonable update times for analog inputs and outputs.) This must also be considered if you choose to do a numerical analysis of the control system.

24.2.4 PID Control Systems

Proportional-Integral-Derivative (PID) controllers are the most common controller choice. The basic controller equation is shown in Figure 24.1. The equation uses the system error e, to calculate a control variable u. The equation uses three terms. The proportional term, Kp, will push the system in the right direction. The derivative term, Kd will respond quickly to changes. The integral term, Ki will respond to long-term errors. The values of Kc, Ki and Kp can be selected, or tuned, to get a desired system response.

$$u = K_c e + K_i \int e\,dt + K_d \left(\frac{de}{dt}\right)$$

$$\left.\begin{matrix} Kc \\ Ki \\ Kd \end{matrix}\right\rangle \text{Relative weights of components}$$

Figure 24.12   **PID Equation**

Figure 24.1 shows a (partial) block diagram for a system that includes a PID controller. The desired set-point for the system is a potentiometer set up as a voltage divider. A summer block will subtract the input and feedback voltages. The error then passes through terms for the proportional, integral and derivative terms; the results are summed together. An amplifier increases the power of the control variable u, to drive a motor. The motor then turns the shaft of another potentiometer, which will produce a feedback voltage proportional to shaft position.

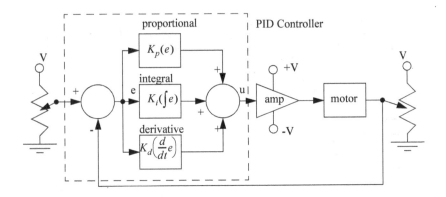

Figure 24.13   **A PID Control System**

Recall the cruise control system for a car. Figure 24.1 shows various equations that could be used as the controller.

492

PID Controller
$$\theta_{gas} = K_p v_{error} + K_i \int v_{error} dt + K_d \left(\frac{dv_{error}}{dt} \right)$$

PI Controller
$$\theta_{gas} = K_p v_{error} + K_i \int v_{error} dt$$

PD Controller
$$\theta_{gas} = K_p v_{error} + K_d \left(\frac{dv_{error}}{dt} \right)$$

P Controller
$$\theta_{gas} = K_p v_{error}$$

Figure 24.14   **Different Controllers**

When implementing these equations in a computer program the equations can be rewritten as shown in Figure 24.1. To do this calculation, previous error and control values must be stored. The calculation also require the scan time T between updates.

$$u_n = u_{n-1} + e_n \left(K_p + K_i T + \frac{K_d}{T} \right) + e_{n-1} \left(-K_p - 2\frac{K_d}{T} \right) + e_{n-2} \left(\frac{K_d}{T} \right)$$

Figure 24.15   **A PID Calculation**

The PID calculation is available as a ladder logic function, as shown in Figure 24.1. This can be used in place of the *SUB* and *MUL* functions in Figure 24.1. In this example the calculation uses the feedback variable stored in *Proc Variable* (as read from the analog input rack:2:I.Ch0InputData). The result is stored in the analog output rack:2:O.Ch0OutputData. The control block uses the parameters stored in *pid_control* to perform the calculations. Most PLC programming software will provide dialogues to set these value.

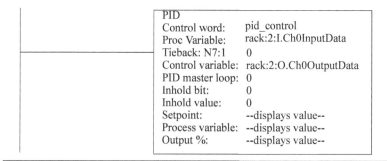

PID	
Control word:	pid_control
Proc Variable:	rack:2:I.Ch0InputData
Tieback: N7:1	0
Control variable:	rack:2:O.Ch0OutputData
PID master loop:	0
Inhold bit:	0
Inhold value:	0
Setpoint:	--displays value--
Process variable:	--displays value--
Output %:	--displays value--

Note: When entering the ladder logic program into the computer you will be able to enter the PID parameters on a popup screen.

Figure 24.16   **PLC-5 PID Control Block**

A description of important PID parameters is given in the following list assuming that we have defined 'pid:PID'. At the upper end the parameters can be set to generate alarms and verify system operation. For example, many of the limit values are a function of the integers used for analog IO values, and will be limited to -4096 to 4095.

pid.CTL:DINT
    pid.EN:BOOL - the PID function is enabled and running
    pid.PVT:BOOL -
    pid.DOE:BOOL - 0=d/dtPV; 1=d/dtError
    pid.SWM:BOOL - 0 = automatic, 1 = manual
    pid.MO:BOOL
    pid.PE:BOOL - 0=independent PID eqn; 1=dependent
    pid.NDF:BOOL - 0=no derivative smoothing; 1=derivative smoothing
    pid.NOBC:BOOL - 0=no bias calculation, 1=yes
    pid.NOZC:BOOL - 0=no zero crossing calculation; 1=yes
    pid.INI:BOOL - 0=not initialized; 1=initialized
    pid.SPOR:BOOL - 0=setpoint not out of range, 1=within
    pid.OLL:BOOL - 0=above minimum CV limit; 1=outside
    pid.OLH:BOOL - 0=below maximum CV limit; 1=inside
    pid.EWD:BOOL - 0=error outside deadband; 1=error inside
    pid.DVNA:BOOL - 0=ok; 1=Error is below lower limit
    pid.DVPA:BOOL - 0=ok; 1=Error is above upper limit
    pid.PVLA:BOOL - 0=ok; 1=PV is below lower limit
    pid.PVHA:BOOL - 0=ok; 1=PV is above upper limit
pid.SP:REAL - setpoint
pid.KP:REAL - proportional gain
pid.KI:REAL - integral gain
pid.KD:REAL - derivative gain
pid.BIAS:REAL - feed forward bias
pid.MAXS:REAL - maximum scaling
pid.MINS:REAL - minimum scaling
pid.DB:REAL - deadband
pid.SO:REAL - set output percentage
pid.MAXO:REAL - maximum output limit percentage
pid.MINO:REAL - minimum output limit percentage
pid.UPD:REAL - loop update time in seconds
pid.PV:REAL - scaled PV value
pid.ERR:REAL - scaled Error value
pid.OUT:REAL - scaled output value
pid.PVH:REAL - process variable high alarm
pid.PVL:REAL - process variable low alarm
pid.DVP:REAL - positive deviation alarm
pid.DVN:REAL - negative deviation alarm
pid.PVDB:REAL - process variable deadband alarm
pid.DVDB:REAL - error alarm deadband
pid.MAXI:REAL - maximum PV value
pid.MINI:REAL - minimum PV value
pid.TIE:REAL - tieback value for manual control
pid.MAXCV:REAL - maximum CV value
pid.MINCV:REAL - minimum CV value
pid.MINTIE:REAL - maximum tieback value
pid.MAXTIE:REAL - minimum tieback value
pid.DATA:REAL[17] - temporary and workspace (e.g. integration sums)

When a controller is off it can drift far from the setpoint and have a large. If the controller is reengaged this error will be integrated, potentially resulting in a very large integral value. As the PID equation approaches the setpoint it may not be able to handle the large error and shoot past the setpoint. This phenomenon is known as windup. The tieback value is used to overcome this problem by allowing a smooth transfer from manual to automatic mode.

PID controllers can also be purchased as cards or stand-alone modules that will perform the PID calculations in hardware. These are useful when the response time must be faster than is possible with a PLC and ladder logic.

24.3 DESIGN CASES

24.3.1 Oven Temperature Control

Problem: Design an analog controller that will read an oven temperature between 1200F and 1500F. When it passes 1500 degrees the oven will be turned off, when it falls below 1200F it will be turned on again. The voltage from the thermocouple is passed through a signal conditioner that gives 1V at 500F and 3V at 1500F. The controller should have a start button and E-stop.

Solution:

Select a 12 bit 1794 module and use the 0V to 10V range on channel 1 with double ended inputs.

$$R = 2^N = 4096$$

$$V_{1V} = INT\left[\left(\frac{V_{in} - V_{min}}{V_{max} - V_{min}}\right)R\right] = 410$$

$$V_{3V} = INT\left[\left(\frac{V_{in} - V_{min}}{V_{max} - V_{min}}\right)R\right] = 1229$$

Cards: rack:0 - Analog Input
                 rack:1 - DC Inputs
                 rack:2 - DC Outputs

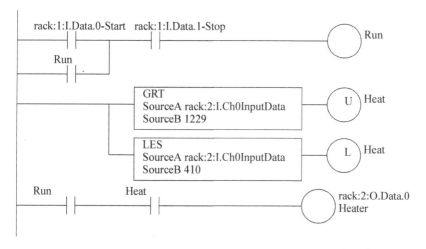

Figure 24.17   *Oven Control Program with Analog Inputs and Logical Outputs*

24.3.2 Water Tank Level Control

Problem: The system in Figure 24.1 will control the height of the water in a tank. The input from the pressure transducer, Vp, will vary between 0V (empty tank) and 5V (full tank). A voltage output, Vo, will position a valve to change the tank fill rate. Vo varies between 0V (no water flow) and 5V (maximum flow). The system will always be on: the emergency stop is connected electrically. The desired height of a tank is specified by another voltage, Vd. The output voltage is calculated using Vo = 0.5 (Vd - Vp). If the output voltage is greater than 5V is will be made 5V, and below 0V is will be made 0V.

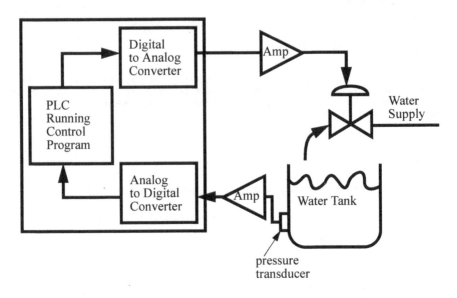

Figure 24.18   *Water Tank Level Controller*

SOLUTION   Analog Input: Select a 12 bit 1794 output module and use the 0V to 10V
                range on channel 1 with double ended inputs.

$$R = 2^N = 4096$$

Analog Output: Select a 12 bit 1794 output module and use the 0V to 10V
                range on channel 1.

$$R = 2^N = 4096$$

Card: rack:0 - Analog Input / Output

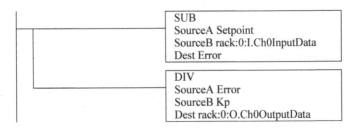

Figure 24.19   *A Water Tank Level Control Program*

24.4 SUMMARY

- Negative feedback controllers make a continuous system stable.
- When controlling a continuous system with a logical actuator set points can be used.
- Block diagrams can be used to describe controlled systems.
- Block diagrams can be converted to equations for analysis.
- Continuous actuator systems can use P, PI, PD, PID controllers.

24.5 PRACTICE PROBLEMS

(Note: Problem solutions are available at http://sites.google.com/site/automatedmanufacturingsystems/)

1. What is the advantage of feedback in a control system?

2. Can PID control solve problems of inaccuracy in a machine?

3. If a control system should respond to long term errors, but not respond to sudden changes, what type of control equation should be used?

4. Develop a ladder logic program that implements a PID controller using the discrete equation.

5. Why is logical control so popular when continuous control allows more precision?

6. Design the complete ladder logic for a control system that implements the control equation below for motor speed control. Assume that the motor speed is read from a tachometer, into an analog input card in rack 0, slot 0, input 1. The tachometer voltage will be between 0 and 8Vdc, for speeds between 0 and 1000rpm. The voltage output to drive the motor controller is output from an analog output card in rack 0, slot 1, output 1. Assume the desired RPM is stored in 'rpm'.

$$V_{motor} = (rpm_{moter} - rpm_{desired})0.02154$$

where,

$$V_{motor} = \text{The voltage output to the motor}$$
$$rpm_{moter} = \text{The RPM of the motor}$$
$$rpm_{desired} = \text{The desired RPM of the motor}$$

7. Write a ladder logic control program to keep a water tank at a given height. The control system will be active after the Start button is pushed, but it can be stopped by a Stop button. The water height in the tank is measured with an ultrasonic sensor that will output 10V at 1m depth, and 1V at 10cm depth. A solenoid controlled valve will open and close to allow water to enter. The water height setpoint is put in *height*, in centimeters, and the actual height should be +/-5cm.

8. Implement a program that will input an analog voltage Vi and output half that voltage, Vi/2. If the input voltage is between 3V and 5V the output 'warning' will be turned on. Include start and stop buttons that will force the output voltage to zero when not running. Do not show the bits that would be set in memory, but list the settings that should be made for the cards (e.g. voltage range).

9. List and describe the most important control memory parameters required to enable a PID function.

10. Implement the system in the block diagram below. Indicate all of the settings required for the analog IO cards. The calculations are to be done with voltage values, therefore input values must be converted from their integer

values.

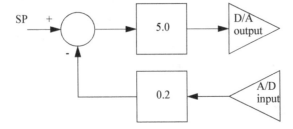

24.6 ASSIGNMENT PROBLEMS

1. Design a basic feedback control system for temperature control of an oven. Indicate major components, and where they are used.

2. Develop ladder logic for a system that adjusts the height of a box of plastic pellets. An ultrasonic sensor detects the top surface of the plastic pellets. The ultrasonic sensor has been calibrated so that when the output is above 5V the box is in the right height range. When it is less than 5V, a motor should be turned on until the box height results in an input of 6V.

3. Write a program that implements a simple proportional controller. The analog input card is in slot 0 of the PLC rack, and the analog output card is in slot 1. The setpoint for the controller is stored in 'Setpoint'. The gain constant is stored in 'Kgain'.

4. A conveyor line is to be controlled with either a variable frequency drive, or a brushless servo motor. Workers will place boxes on the inlet side of the conveyor, these will be detected with a 'box present' sensor. The box position is also detected with an ultrasonic sensor with a range from 10cm to 1m . When present, boxes on the conveyor will be moved until they are 55cm from the sensor. Once in place, the system will stop until the box is removed. After this, the process can begin again when a new box is detected. Design all of the required ladder logic for the process.

5. A temperature control system is being developed to control the water flow rate for cooling a mold set. Unfortunately the sensor in the dies doesn't allow us to measure the temperature. But it does provide a set of bimetallic contacts that close when the die is above 110C. Luckily a Variable Frequency Drive (VFD) is available for controlling the flow rate of the water. The control scheme will increase the water flow rate when the die temperature input, HOT, is active. When the HOT input if off the flow rate will be decreased, until the flow rate is zero. In other words, when the HOT input is on, a timer will start. The time accumulated, DELAY, will be proportional to a voltage output to control the VFD. If the HOT sensors turns off the DELAY value will be decreased until it has a value of zero. Write the ladder logic for this controller.

24.1 PRACTICE PROBLEM SOLUTIONS

1. Feedback control, more specifically negative feedback, can improve the stability and accuracy of a control system.

2. A PID controller will compare a setpoint and output variable. If there is a persistent error, the integral part of the controller will adjust the output to reduce long term errors.

3. A PI controller

4.

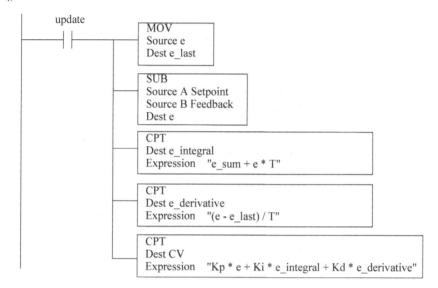

5. Logical control is more popular because the system is more controllable. This means either happen, or they don't happen. If a system requires a continuous control system then it will tend to be unstable, and even when controlled a precise values can be hard to obtain. The need for control also implies that the system requires some accuracy, thus the process will tend to vary, and be a source of quality control problems.

6.

```
         update                          CPT
  |       | |                            Dest analog_output
  |-------| |--------------------------- Expression
  |       | |                            "0.02154*(analog_input - SP)"
```

499

7.

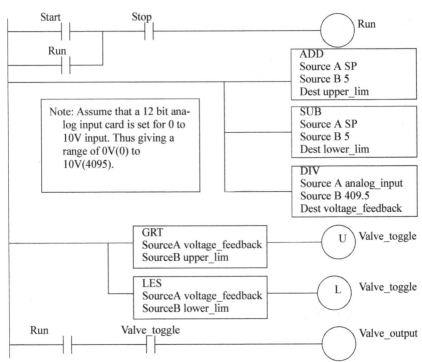

8.

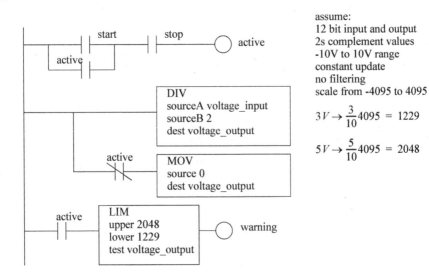

assume:
12 bit input and output
2s complement values
-10V to 10V range
constant update
no filtering
scale from -4095 to 4095

$$3V \rightarrow \frac{3}{10} 4095 = 1229$$

$$5V \rightarrow \frac{5}{10} 4095 = 2048$$

9.

```
pid.CTL:DINT
        pid.EN:BOOL = TRUE
        pid.DOE:BOOL - TRUE=d/dtError
        pid.SWM:BOOL - 0 = automatic
        pid.INI:BOOL - 1=initialized
pid.SP:REAL - setpoint
pid.KP:REAL - proportional gain
pid.KI:REAL - integral gain
pid.KD:REAL - derivative gain
pid.MAXS:REAL - maximum scaling
pid.MINS:REAL - minimum scaling
pid.SO:REAL - set output percentage
pid.MAXO:REAL - maximum output limit percentage
pid.MINO:REAL - minimum output limit percentage
pid.UPD:REAL - loop update time in seconds
pid.PV:REAL - scaled PV value
pid.ERR:REAL - scaled Error value
pid.OUT:REAL - scaled output value
pid.MAXI:REAL - maximum PV value
pid.MINI:REAL - minimum PV value
pid.MAXCV:REAL - maximum CV value
pid.MINCV:REAL - minimum CV value
```

10.

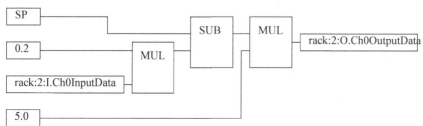

25FUZZY LOGIC

Topics:

- Fuzzy logic theory; sets, rules and solving

Objectives:

- To understand fuzzy logic control.
- Be able to implement a fuzzy logic controller.

Fuzzy logic is well suited to implementing control rules that can only be expressed verbally, or systems that cannot be modelled with linear differential equations. Rules and membership sets are used to make a decision. A simple verbal rule set is shown in Figure 25.1. These rules concern how fast to fill a bucket, based upon how full it is.

1. If (bucket is full) then (stop filling)
2. If (bucket is half full) then (fill slowly)
3. If (bucket is empty) then (fill quickly)

Figure 25.1   *A Fuzzy Logic Rule Set*

The outstanding question is "What does it mean when the bucket is empty, half full, or full?" And, what is meant by filling the bucket slowly or quickly. We can define sets that indicate when something is true (1), false (0), or a bit of both (0-1), as shown in Figure 25.1. Consider the *bucket is full* set. When the height is 0, the set membership is 0, so nobody would think the bucket is full. As the height increases more people think the bucket is full until they all think it is full. There is no definite line stating that the bucket is full. The other bucket states have similar functions. Notice that the *angle* function relates the valve angle to the fill rate. The sets are shifted to the right. In reality this would probably mean that the valve would have to be turned a large angle before flow begins, but after that it increases quickly.

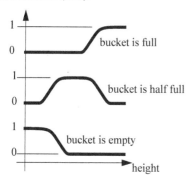
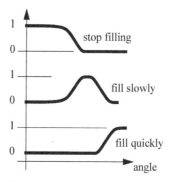

Figure 25.2   *Fuzzy Sets*

Now, if we are given a height we can examine the rules, and find output values, as shown in Figure 25.1. This begins be comparing the bucket height to find the membership for *bucket is full* at 0.75, *bucket is half full* at 1.0 and *bucket is empty* at 0. Rule 3 is ignored because the membership was 0. The result for rule 1 is 0.75, so the 0.75 membership value is found on the *stop filling* and a value of *a1* is found for the valve angle. For rule 2 the result was 1.0, so the *fill slowly* set is examined to find a value. In this case there is a range where *fill slowly* is 1.0, so the center point is chosen to get angle *a2*. These two results can then be combined with a weighted average to get angle $= \dfrac{0.75(a1) + 1.0(a2)}{0.75 + 1.0}$.

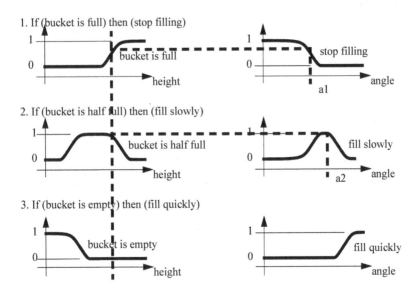

1. If (bucket is full) then (stop filling)

2. If (bucket is half full) then (fill slowly)

3. If (bucket is empty) then (fill quickly)

Figure 25.3   **Fuzzy Rule Solving**

An example of a fuzzy logic controller for controlling a servomotor is shown in Figure 25.1 [Lee and Lau, 1988]. This controller rules examines the system error, and the rate of error change to select a motor voltage. In this example the set memberships are defined with straight lines, but this will have a minimal effect on the controller performance.

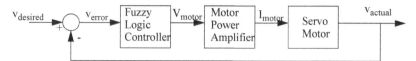

The rules for the fuzzy logic controller are;

1. If v_{error} is LP and $^d/_{dt}v_{error}$ is any then V_{motor} is LP.
2. If v_{error} is SP and $^d/_{dt}v_{error}$ is SP or ZE then V_{motor} is SP.
3. If v_{error} is ZE and $^d/_{dt}v_{error}$ is SP then V_{motor} is ZE.
4. If v_{error} is ZE and $^d/_{dt}v_{error}$ is SN then V_{motor} is SN.
5. If v_{error} is SN and $^d/_{dt}v_{error}$ is SN then V_{motor} is SN.
6. If v_{error} is LN and $^d/_{dt}v_{error}$ is any then V_{motor} is LN.

The sets for v_{error}, $^d/_{dt}v_{error}$, and V_{motor} are;

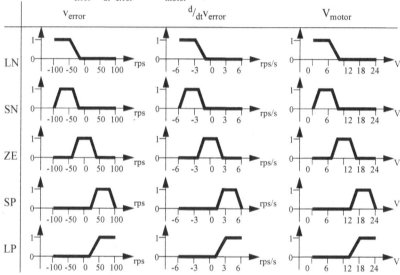

Figure 25.4   **A Fuzzy Logic Servo Motor Controller**

Consider the case where v_{error} = 30 rps and $^d/_{dt}\,v_{error}$ = 1 $^{rps}/_s$. Rule 1to 6 are calculated in Figure 25.1.

1. If v_{error} is LP and $^d/_{dt}v_{error}$ is any then V_{motor} is LP.

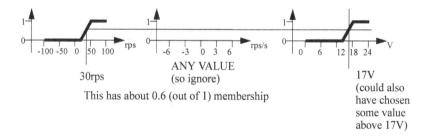

2. If v_{error} is SP and $^d/_{dt}v_{error}$ is SP or ZE then V_{motor} is SP.

the OR means take the highest of the two memberships

the AND means take the lowest of the two memberships

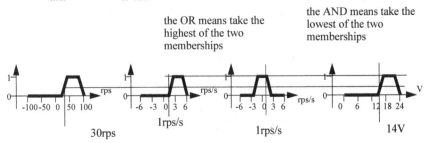

30rps   1rps/s   1rps/s   14V

This has about 0.4 (out of 1) membership

3. If v_{error} is ZE and $^d/_{dt}v_{error}$ is SP then V_{motor} is ZE.

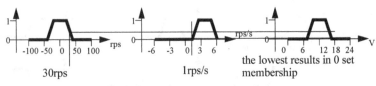

30rps   1rps/s   the lowest results in 0 set membership

This has about 0.0 (out of 1) membership

4. If v_{error} is ZE and $^d/_{dt}v_{error}$ is SN then V_{motor} is SN.

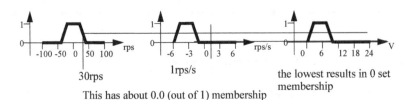

30rps   1rps/s   the lowest results in 0 set membership

This has about 0.0 (out of 1) membership

5. If v_{error} is SN and $^d/_{dt}v_{error}$ is SN then V_{motor} is SN.

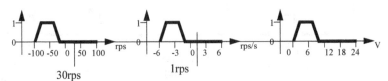

30rps   1rps

This has about 0.0 (out of 1) membership

6. If v_{error} is LN and $\frac{d}{dt}v_{error}$ is any then V_{motor} is LN.

30rps   ANY VALUE

This has about 0 (out of 1) membership

Figure 25.5   *Rule Calculation*

The results from the individual rules can be combined using the calculation in Figure 25.1. In this case only two of the rules matched, so only two terms are used, to give a final motor control voltage of 15.8V.

$$V_{motor} = \frac{\sum_{i=1}^{n}(V_{motor_i})(membership_i)}{\sum_{i=1}^{n}(membership_i)}$$

$$V_{motor} = \frac{0.6(17V) + 0.4(14V)}{0.6 + 0.4} = 15.8V$$

Figure 25.6   *Rule Results Calculation*

25.1 COMMERCIAL CONTROLLERS

At the time of writing Allen Bradley did not offer any Fuzzy Logic systems for their PLCs. But, other vendors such as Omron offer commercial controllers. Their controller has 8 inputs and 2 outputs. It will accept up to 128 rules that operate on sets defined with polygons with up to 7 points.

It is also possible to implement a fuzzy logic controller manually, possible in structured text.

25.2 SUMMARY

- Fuzzy rules can be developed verbally to describe a controller.
- Fuzzy sets can be developed statistically or by opinion.
- Solving fuzzy logic involves finding fuzzy set values and then calculating a value for each rule. These values for each rule are combined with a weighted average.

25.2.1 References

Li, Y.F., and Lau, C.C., "Application of Fuzzy Control for Servo Systems", IEEE International Confer-

25.3 ASSIGNMENT PROBLEMS

1. Find products that include fuzzy logic controllers in their designs.

2. Suggest 5 control problems that might be suitable for fuzzy logic control.

3. Two fuzzy rules, and the sets they use are given below. If v_{error} = 30rps, and $\frac{d}{dt}v_{error}$ = 3$^{rps}/_s$, find V_{motor}.

1. If (v_{error} is ZE) and ($\frac{d}{dt}v_{error}$ is ZE) then (V_{motor} is ZE).
2. If (v_{error} is SP) or ($\frac{d}{dt}v_{error}$ is SP) then (V_{motor} is SP).

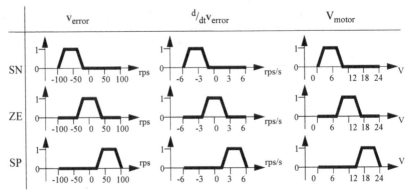

4. Develop a set of fuzzy control rules adjusting the water temperature in a sink.

5. Develop a fuzzy logic control algorithm and implement it in structured text. The fuzzy rule set below is to be used to control the speed of a motor. When the error (difference between desired and actual speeds) is large the system will respond faster. When the difference is smaller the response will be smaller. Calculate the outputs for

the system given errors of 5, 20 and 40.

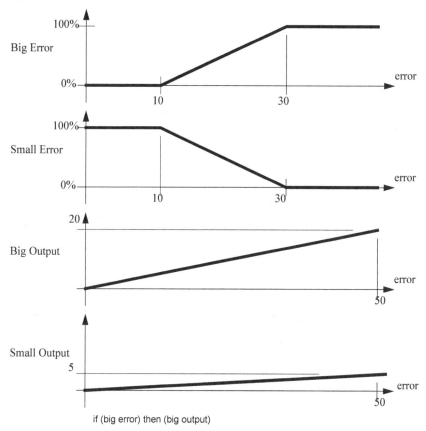

if (big error) then (big output)

26SERIAL COMMUNICATION

Topics:

- Serial communication and RS-232c
- ASCII ladder logic functions
- Design case

Overview:

- To understand serial communications with RS-232
- Be able to use serial communications with a PLC

Multiple control systems will be used for complex processes. These control systems may be PLCs, but other controllers include robots, data terminals and computers. For these controllers to work together, they must communicate. This chapter will discuss communication techniques between computers, and how these apply to PLCs.

The simplest form of communication is a direct connection between two computers. A network will simultaneously connect a large number of computers on a network. Data can be transmitted one bit at a time in series, this is called serial communication. Data bits can also be sent in parallel. The transmission rate will often be limited to some maximum value, from a few bits per second, to billions of bits per second. The communications often have limited distances, from a few feet to thousands of miles/kilometers.

Data communications have evolved from the 1800's when telegraph machines were used to transmit simple messages using Morse code. This process was automated with teletype machines that allowed a user to type a message at one terminal, and the results would be printed on a remote terminal. Meanwhile, the telephone system began to emerge as a large network for interconnecting users. In the late 1950s Bell Telephone introduced data communication networks, and Texaco began to use remote monitoring and control to automate a polymerization plant. By the 1960s data communications and the phone system were being used together. In the late 1960s and 1970s modern data communications techniques were developed. This included the early version of the Internet, called ARPAnet. Before the 1980s the most common computer configuration was a centralized mainframe computer with remote data terminals, connected with serial data line. In the 1980s the personal computer began to displace the central computer. As a result, high speed networks are now displacing the dedicated serial connections. Serial communications and networks are both very important in modern control applications.

An example of a networked control system is shown in Figure 26.1. The computer and PLC are connected with an RS-232 (serial data) connection. This connection can only connect two devices. Devicenet is used by the Computer to communicate with various actuators and sensors. Devicenet can support up to 63 actuators and sensors. The PLC inputs and outputs are connected as normal to the process.

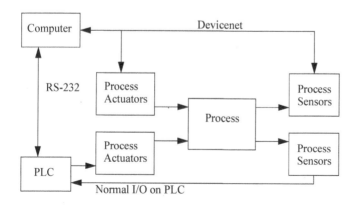

Figure 26.1   *A Communication Example*

26.1 SERIAL COMMUNICATIONS

Serial communications send a single bit at a time between computers. This only requires a single communication channel, as opposed to 8 channels to send a byte. With only one channel the costs are lower, but the communication rates are slower. The communication channels are often wire based, but they may also be can be optical and radio. Figure 26.1 shows some of the standard electrical connections. RS-232c is the most common standard that is based on a voltage change levels. At the sending computer an input will either be true or false. The *line driver* will convert a false value *in* to a *Txd* voltage between +3V to +15V, true will be between -3V to -15V. A cable connects the *Txd* and *com* on the sending computer to the *Rxd* and *com* inputs on the receiving computer. The receiver converts the positive and negative voltages back to logic voltage levels in the receiving computer. The cable length is limited to 50 feet to reduce the effects of electrical noise. When RS-232 is used on the factory floor, care is required to reduce the effects of electrical noise - careful grounding and shielded cables are often used.

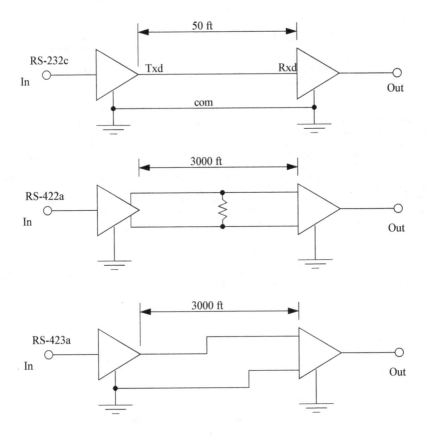

Figure 26.2   *Serial Data Standards*

The RS-422a cable uses a 20 mA current loop instead of voltage levels. This makes the systems more immune to electrical noise, so the cable can be up to 3000 feet long. The RS-423a standard uses a differential voltage level across two lines, also making the system more immune to electrical noise, thus allowing longer cables. To provide serial communication in two directions these circuits must be connected in both directions.

To transmit data, the sequence of bits follows a pattern, like that shown in Figure 26.1. The transmission starts at the left hand side. Each bit will be true or false for a fixed period of time, determined by the transmission speed.

A typical data byte looks like the one below. The voltage/current on the line is made true or false. The width of the bits determines the possible bits per second (bps). The value shown before is used to transmit a single byte. Between bytes, and when the line is idle, the *Txd* is kept true, this helps the receiver detect when a sender is present. A single start bit is sent by making the *Txd* false. In this example the next eight bits are the transmitted data, a byte with the value 17. The data is followed by a parity bit that can be used to check the byte. In this example there are two data bits set, and even parity is being used, so the parity bit is set. The parity bit is followed by two stop bits to help separate this byte from the next one.

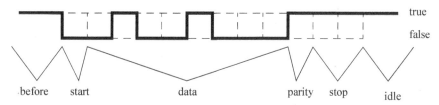

Descriptions:

before - this is a period where no bit is being sent and the line is true.

start - a single bit to help get the systems synchronized.

data - this could be 7 or 8 bits, but is almost always 8 now. The value shown here is a byte with the binary value 00010010 (the least significant bit is sent first).

parity - this lets us check to see if the byte was sent properly. The most common choices here are no parity bit, an even parity bit, or an odd parity bit. In this case there are two bits set in the data byte. If we are using even parity the bit would be true. If we are using odd parity the bit would be false.

stop - the stop bits allow a pause at the end of the data. One or two stop bits can be used.

idle - a period of time where the line is true before the next byte.

Figure 26.3   ***A Serial Data Byte***

Some of the byte settings are optional, such as the number of data bits (7 or 8), the parity bit (none, even or odd) and the number of stop bits (1 or 2). The sending and receiving computers must know what these settings are to properly receive and decode the data. Most computers send the data asynchronously, meaning that the data could be sent at any time, without warning. This makes the bit settings more important.

Another method used to detect data errors is half-duplex and full-duplex transmission. In half-duplex transmission the data is only sent in one direction. But, in full-duplex transmission a copy of any byte received is sent back to the sender to verify that it was sent and received correctly. (Noto: if you type and nothing shows up on a screen, or characters show up twice you may have to change the half/full duplex setting.)

The transmission speed is the maximum number of bits that can be sent per second. The units for this is *baud*. The baud rate includes the start, parity and stop bits. For example a 9600 baud transmission of the data in Figure 26.1 would transfer up to $\dfrac{9600}{(1 + 8 + 1 + 2)} = 800$ bytes each second. Lower baud rates are 120, 300, 1.2K, 2.4K and 9.6K. Higher speeds are 19.2K, 28.8K and 33.3K. (Note: When this is set improperly you will get many transmission errors, or *garbage* on your screen.)

Serial lines have become one of the most common methods for transmitting data to instruments: most personal computers have two serial ports. The previous discussion of serial communications techniques also applies to devices such as modems.

26.1.1 RS-232

The RS-232c standard is based on a low/false voltage between +3 to +15V, and an high/true voltage between -3 to -15V (+/-12V is commonly used). Figure 26.1 shows some of the common connection schemes. In all methods the *txd* and *rxd* lines are crossed so that the sending *txd* outputs are into the listening *rxd* inputs when communicating between computers. When communicating with a communication device (modem), these lines are not crossed. In the *modem* connection the *dsr* and *dtr* lines are used to control the flow of data. In the *computer* the *cts* and *rts* lines are connected. These lines are all used for handshaking, to control the flow of data from sender to receiver. The *null-modem* configuration simplifies the handshaking between computers. The three wire configuration is a crude way to connect to devices, and data can be lost.

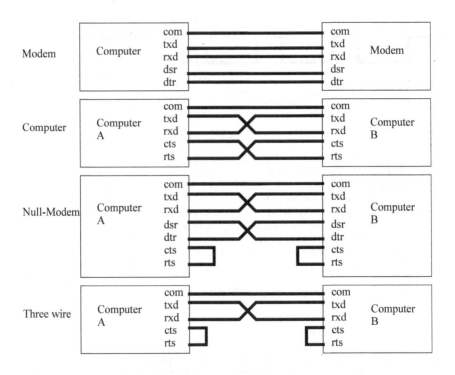

Figure 26.4   **Common RS-232 Connection Schemes**

Common connectors for serial communications are shown in Figure 26.1. These connectors are either male (with pins) or female (with holes), and often use the assigned pins shown. The DB-9 connector is more common now, but the DB-25 connector is still in use. In any connection the *RXD* and *TXD* pins must be used to transmit and receive data. The *COM* must be connected to give a common voltage reference. All of the remaining pins are used for *handshaking*.

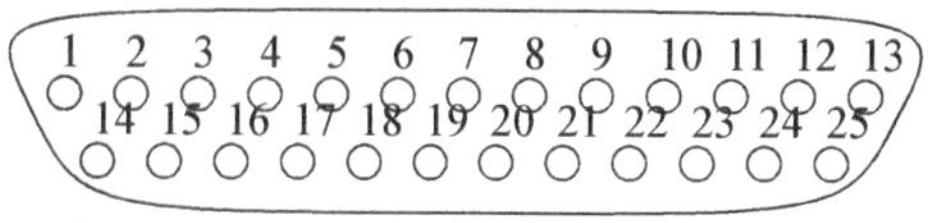

Commonly used pins

```
1 - GND (chassis ground)
2 - TXD (transmit data)
3 - RXD (receive data)
4 - RTS (request to send)
5 - CTS (clear to send)
6 - DSR (data set ready)
7 - COM (common)
8 - DCD (Data Carrier Detect)
20 - DTR (data terminal ready)
```

Other pins

```
9 - Positive Voltage
10 - Negative Voltage
11 - not used
12 - Secondary Received Line Signal Detector
13 - Secondary Clear to Send
14 - Secondary Transmitted Data
15 - Transmission Signal Element Timing (DCE)
16 - Secondary Received Data
17 - Receiver Signal Element Timing (DCE)
18 - not used
19 - Secondary Request to Send
21 - Signal Quality Detector
22 - Ring Indicator (RI)
23 - Data Signal Rate Selector (DTE/DCE)
24 - Transmit Signal Element Timing (DTE)
25 - Busy
```

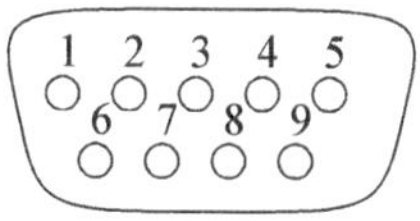

```
1 - DCD
2 - RXD
3 - TXD
4 - DTR
5 - COM
6 - DSR
7 - RTS
8 - CTS
9 - RI
```

Note: these connec-
tors often have
very small num-
bers printed on
them to help you
identify the pins.

Figure 26.5   *Typical RS-232 Pin Assignments and Names*

The *handshaking* lines are to be used to detect the status of the sender and receiver, and to regulate the flow of data. It would be unusual for most of these pins to be connected in any one application. The most common pins are provided on the DB-9 connector, and are also described below.

> TXD/RXD - (transmit data, receive data) - data lines
> DCD - (data carrier detect) - this indicates when a remote device is present
> RI - (ring indicator) - this is used by modems to indicate when a connection is about to be made.
> CTS/RTS - (clear to send, ready to send)
> DSR/DTR - (data set ready, data terminal ready) these handshaking lines indicate when the
>   remote machine is ready to receive data.
> COM - a common ground to provide a common reference voltage for the TXD and RXD.

When a computer is ready to receive data it will set the *CTS* bit, the remote machine will notice this on the *RTS* pin. The *DSR* pin is similar in that it indicates the modem is ready to transmit data. *XON* and *XOFF* characters are used for a software only flow control scheme.

Many PLC processors have an RS-232 port that is normally used for programming the PLC. Figure 26.1 shows a PLC connected to a personal computer with a Null-Modem line. It is connected to the *channel 0* serial connector on the PLC processor, and to the *com 1* port on the computer. In this example the *terminal* could be a personal computer running a terminal emulation program. The ladder logic below will send a string to the serial port channel 0 when *A* goes true. In this case the string is stored is string memory *'example'* and has a length of 4 characters. If the string stored in *example* is "*HALFLIFE*", the terminal program will display the string "*HALF*".

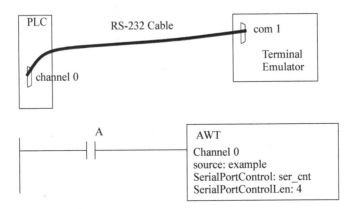

Figure 26.6   *Serial Output Using Ladder Logic*

The AWT (Ascii WriTe) function below will write to serial ports on the CPU only.

26.1.2 ASCII Functions

ASCII functions allow programs to manipulate strings in the memory of the PLC. The basic functions are listed in Figure 26.1.

AWA(channel, string, control, length) - append characters to the output buffer
ABL(channel, control)- reports the number of ASCII characters including line endings
ACB(channel, control) - reports the numbers of ASCII characters in buffer
AHL(channel, mask, mask, control) - does data handshaking
ARD(channel, dest, control, length) - will get characters from the ASCII buffer
ARL(channel, dest, control, length) - will get characters from an ASCII buffer
ASR(string, string) - compares two strings
AWT(channel, string, control, length) - will write characters to an ASCII output
CONCAT(string, string, dest) - concatenate strings
DELETE(string, len, start, dest) - deletes characters from a larger string
DTOS(integer, string) - convert an integer to a string
FIND(string, string, start) - find one string inside another
INSERT(string, string, start, dest) - puts characters inside a string
LOWER(integer, string) - convert a string to lower case
MID(string, start, length, dest) - this will copy a segment of a string out of a larger string
RTOS(integer, string) - convert a real to a string
STOD(string, dest) - convert ASCII string to integer
STOR(string, dest) - convert ASCII string to real
UPPER(string, dest) - convert a string to upper case

Figure 26.7   *PLC ASCII Functions*

In the example in Figure 26.1, the characters "Hi " are placed into string memory *str_in*. The ACB function checks to see how many characters have been received, and are waiting in channel 0. When the number of characters equals 2, the ARD (Ascii ReaD) function will then copy those characters into memory *str_0*, and bit *real_ctl.DN* will be set. This done bit will cause the two characters to be concatenated to the "Hi ", and the result written back to the serial port. So, if I typed in my initial "HJ", I would get the response "HI HJ".

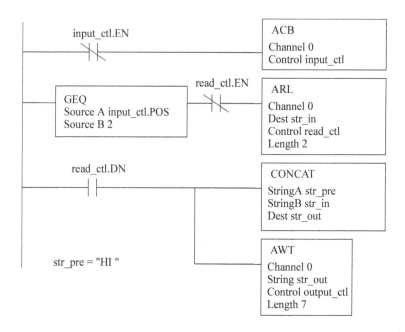

Figure 26.8   An ASCII String Example

The ASCII functions can also be used to support simple number conversions. The example in Figure 26.1 will convert the strings in *str_a* and *str_b* to integers, add the numbers, and store the result as a string in *str_c*.

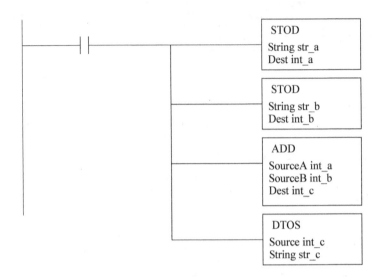

Figure 26.9 *A String to Integer Conversion Example*

Many of the remaining string functions are illustrated in Figure 26.1. When *A* is true the *ABL* and *ACB* functions will check for characters that have arrived on channel 1, but have not been retrieved with an *ARD* function. If the characters "ABC<*CR*>" have arrived (<*CR*> is an ASCII carriage return) the *ACB* would count the three characters, and store the value in *cnt_1.POS*. The *ABL* function would also count the <*CR*> and store a value of four in *cnt_2.POS*. If *B* is true, and the string in *str_a* is "ABCDEFGHIJKL", then "EF" will be stored in *str_b*. The last function will compare the strings in *str_c* and *str_d*, and if they are equal, output *string_match* will be turned on.

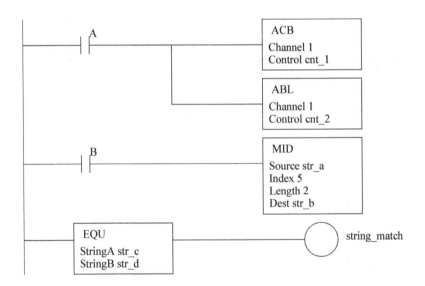

Figure 26.10 *String Manipulation Functions*

The *AHL* function can be used to do handshaking with a remote serial device.

26.2 PARALLEL COMMUNICATIONS

Parallel data transmission will transmit multiple bits at the same time over multiple wires. This does allow faster data transmission rates, but the connectors and cables become much larger, more expensive and less flexible. These interfaces still use handshaking to control data flow.

These interfaces are common for computer printer cables and short interface cables, but they are uncommon on PLCs. A list of common interfaces follows.

Centronics printer interface - These are the common printer interface used on most personal computers. It was made popular by the now defunct Centronics printer company.
GPIB/IEEE-488 - (General Purpose Instruments Bus) This bus was developed by Hewlett Packard Inc. for connecting instruments. It is still available as an option on many new instruments.

26.3 DESIGN CASES

26.3.1 PLC Interface To a Robot

Problem: A robot will be loading parts into a box until the box reaches a prescribed weight. A PLC will feed parts into a pickup fixture when it is empty. The PLC will tell the robot when to pick up a part and load it into the box by passing it an ASCII string, "pickup".

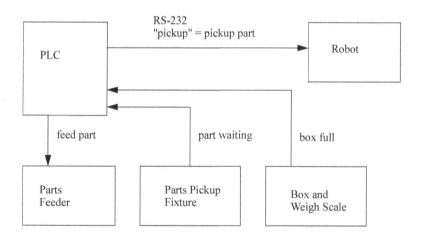

Figure 26.11   **Box Loading System**

Solution: The following ladder logic will implement part of the control system for the system in Figure 26.1.

519

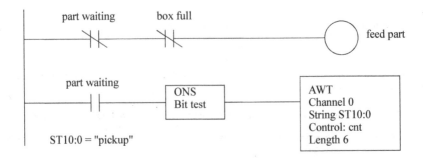

Figure 26.12   *A Box Loading System*

26.4 SUMMARY

- Serial communications pass data one bit at a time.
- RS-232 communications use voltage levels for short distances. A variety of communications cables and settings were discussed.
- ASCII functions are available of PLCs making serial communications possible.

26.5 PRACTICE PROBLEMS

(Note: Problem solutions are available at http://sites.google.com/site/automatedmanufacturingsystems/)

1. Describe what the bits would be when an *A* (ASCII 65) is transmitted in an RS-232 interface with 8 data bits, even parity and 1 stop bit.

2. Divide the string in 'str_a' by the string in 'str_b' and store the results in 'str_c'. Check for a divide by zero error.

   str_a "100"
   str_b "10"
   str_c

3. How long would it take to transmit an ASCII file over a serial line with 8 data bits, no parity, 1 stop bit? What if the data were 7 bits long?

4. Write a number guessing program that will allow a user to enter a number on a terminal that transmits it to a PLC where it is compared to a value in *target*. If the guess is above "Hi" will be returned. If below "Lo" will be returned. When it matches "ON" will be returned.

5. Write a structured text program that reads inputs from 'channel 0'. An input string of 'CLEAR' will clear a storage array. Up to 100 real values with the format 'XXX.XX' will arrive on 'channel 0' and are to be stored in the array. If the string 'AVG' is received, the average of the array contents will be calculated and written out 'Channel 0'.

26.6 ASSIGNMENT PROBLEMS

1. Describe an application of ASCII communications.

2. Write a ladder logic program to output an ASCII warning message on channel 1 when the value in 'temp' is less than 10, or greater than 20. The message should be "out of temp range".

3. Write a program that will send an ASCII message every minute. The message should begin with the word 'count', followed by a number. The number will be 1 on the first scan of the PLC, and increment once each minute.

4. A PLC will be controlled by ASCII commands received through the RS-232C communications port. The commands will cause the PLC to move between the states shown in the state diagram. Implement the ladder logic.

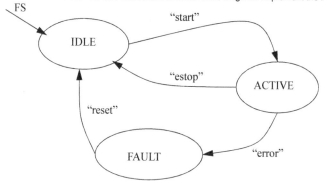

5. A program is to be written to control a robot through an RS-232c interface. The robot has already been programmed to respond to two ASCII strings. When the robot receives the string 'start' it will move a part from a feeder to a screw machine. When the robot receives an 'idle' command it will become inactive (safe). The PLC has 'start' and 'end' inputs to control the process. The PLC also has two other inputs that indicate when the parts feeder has parts available ('part present') and when the screw machine is done ('machine idle'). The 'start' button will start a cycle where the robot repeatedly loads parts into the screw machine whenever the 'machine idle' input is true. If the 'part present' sensor is off (i.e., no parts), or the 'end' input is off (a stop requested), the screw machine will be allowed to finish, but then the process will stop and the robot will be sent the idle command. Use a structured design method (e.g., state diagrams) to develop a complete ladder logic program to perform the task.

6. A PLC is connected to a scale that measures weights and then sends an ASCII string. The string format is 'XXXX.XX'. So a weight of 29.9 grams would result in a string of '0029.90'. The PLC is to read the string and then check to see if the weight is between 18.23 and 18.95 grams. If it is not then an error output light should be set until a reset button is pushed.

7. Write a program that will convert a numerical value stored in the 'REAL' value *float* and write it out the RS-232 output with 3 decimal places.

8. A system for testing hydraulic reservoirs is to be designed and built using a PLC. Part of the test will be conducted using a computer based Data AQuisition (DAQ) system for high speed analog inputs. When the test begins a command of 'S' is sent to the DAQ system, and an output 'pump' will be turned on. The test is started with a 'start' input and stopped with a 'stop' input. The test will be shut down and an error light turned on if the flow sensor does not turn on within 0.1s, or if the pressure input rises above 4V. When the test is done the DAQ system will send a 'D' to the PLC. The PLC will retrieve the data by sending and 'R' to the DAQ system. The data is returned in the format 'xxxx.x<cr><lf>'. The last data line will be 'END'. The array of data should be analyzed and the results stored in the real variables 'maximum', 'average', 'standard_deviation', and 'median'. These variables will displayed on an HMI. Write a structured text program for the control system.

26.1 PRACTICE PROBLEM SOLUTIONS

1.

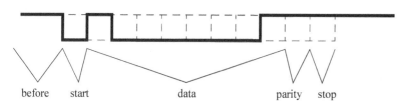

before   start   data   parity   stop

2.

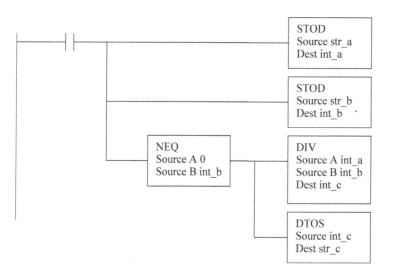

3. If we assume 9600 baud, for (1start+8data+0parity+1stop)=10 bits/byte we get 960 bytes per second. If there are only 7 data bits per byte this becomes 9600/9 = 1067 bytes per second.

4.

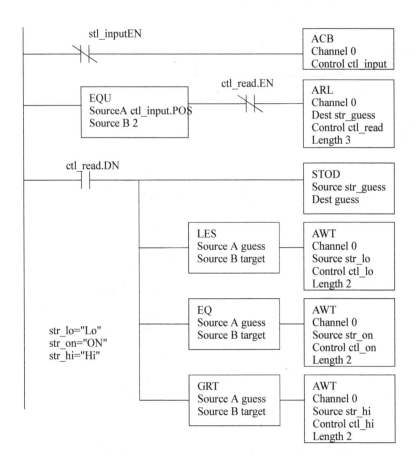

5.

```
SBR();
        IF S:FS THEN
        i = 0;
        END_IF;
        ACB(0, c);
        IF c.POS = 6 THEN
        ARL(0, str_in, s);
        IF i < 100 THEN
        r[i] = STOR(str_in);
        i := i + 1;
        END_IF;
        ELSE
        ARL(0, str_in, s);
        IF str_in = str_clear THEN
        i := 0;
        END_IF
        IF str_in = str_avg THEN
        sum := 0;
        FOR j = 0 to length-1 DO
        sum := sum + r[j];
        END_FOR;
        str_out := RTOS(sum / i);
        AWT(0, str_out, s);
        END_IF;
        END_IF;
RET();
```

Tags:

```
r:REAL[100]
i:INT
j:INT
sum:REAL
c:SerialPortControl
s:SerialPortControl
str_in:STRING
str_out:STRING
str_clear:STRING = "CLEAR"
str_avg:STRING = "AVG"
```

27NETWORKING

A computer with a single network interface can communicate with many other computers. This economy and flexibility has made networks the interface of choice, eclipsing point-to-point methods such as RS-232. Typical advantages of networks include resource sharing and ease of communication. But, networks do require more knowledge and understanding.

Small networks are often called Local Area Networks (LANs). These may connect a few hundred computers within a distance of hundreds of meters. These networks are inexpensive, often costing $100 or less per network node. Data can be transmitted at rates of millions of bits per second. Many controls system are using networks to communicate with other controllers and computers. Typical applications include;

    • taking quality readings with a PLC and sending the data to a database computer.
    • distributing recipes or special orders to batch processing equipment.
    • remote monitoring of equipment.

Larger Wide Area Networks (WANs) are used for communicating over long distances between LANs. These are not common in controls applications, but might be needed for a very large scale process. An example might be an oil pipeline control system that is spread over thousands of miles.

27.1 TOPOLOGY

The structure of a network is called the topology. Figure 27.1 shows the basic network topologies. The *Bus* and *Ring* topologies both share the same network wire. In the *Star* configuration each computer has a single wire that connects it to a central hub.

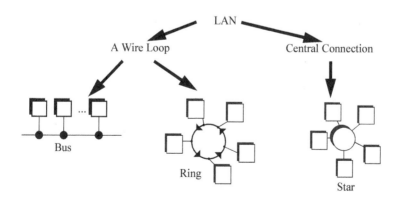

Figure 27.1   *Network Topologies*

In the *Ring* and *Bus* topologies the network control is distributed between all of the computers on the network. The wiring only uses a single loop or run of wire. But, because there is only one wire, the network will slow down significantly as traffic increases. This also requires more sophisticated network interfaces that can determine when a computer is allowed to transmit messages. It is also possible for a problem on the network wires to halt the entire network.

The *Star* topology requires more wire overall to connect each computer to an intelligent hub. But, the network interfaces in the computer become simpler, and the network becomes more reliable. Another term commonly used is that it is deterministic, this means that performance can be predicted. This can be important in critical applications.

For a factory environment the bus topology is popular. The large number of wires required for a star configuration can be expensive and confusing. The loop of wire required for a ring topology is also difficult to connect, and it can lead to ground loop problems. Figure 27.1 shows a tree topology that is constructed out of smaller bus networks. Repeaters are used to boost the signal strength and allow the network to be larger.

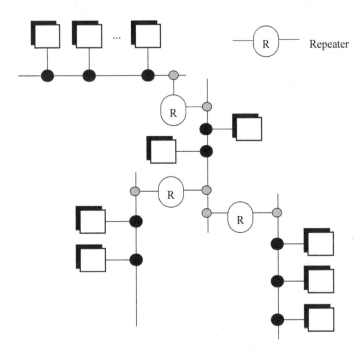

Figure 27.2   **The Tree Topology**

27.2 OSI NETWORK MODEL

The Open System Interconnection (OSI) model in Figure 27.1 was developed as a tool to describe the various hardware and software parts found in a network system. It is most useful for educational purposes, and explaining the things that should happen for a successful network application. The model contains seven layers, with the hardware at the bottom, and the software at the top. The darkened arrow shows that a message originating in an application program in computer #1 must travel through all of the layers in both computers to arrive at the application in computer #2. This could be part of the process of reading email.

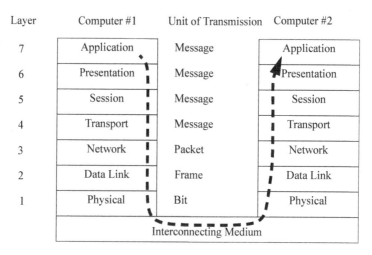

Layer	Computer #1	Unit of Transmission	Computer #2
7	Application	Message	Application
6	Presentation	Message	Presentation
5	Session	Message	Session
4	Transport	Message	Transport
3	Network	Packet	Network
2	Data Link	Frame	Data Link
1	Physical	Bit	Physical

Interconnecting Medium

Application - This is high level software on the computer.
Presentation - Translates application requests into network operations.
Session - This deals with multiple interactions between computers.
Transport - Breaks up and recombines data to small packets.
Network - Network addresses and routing added to make frame.
Data Link - The encryption for many bits, including error correction added to a frame.
Physical - The voltage and timing for a single bit in a frame.
Interconnecting Medium - (not part of the standard) The wires or transmission medium of the network.

Figure 27.3   **The OSI Network Model**

The *Physical* layer describes items such as voltage levels and timing for the transmission of single bits. The *Data Link* layer deals with sending a small amount of data, such as a byte, and error correction. Together, these two layers would describe the serial byte shown in the previous chapter. The *Network* layer determines how to move the message through the network. If this were for an internet connection this layer would be responsible for adding the correct network address. The *Transport* layer will divide small amounts of data into smaller packets, or recombine them into one larger piece. This layer also checks for data integrity, often with a checksum. The *Session* layer will deal with issues that go beyond a single block of data. In particular it will deal with resuming transmission if it is interrupted or corrupted. The *Session* layer will often make long term connections to the remote machine. The *Presentation* layer acts as an application interface so that syntax, formats and codes are consistent between the two networked machines. For example this might convert '\' to '/' in HTML files. This layer also provides subroutines that the user may call to access network functions, and perform functions such as encryption and compression. The *Application* layer is where the user program resides. On a computer this might be a web browser, or a ladder logic program on a PLC.

Most products can be described with only a couple of layers. Some networking products may omit layers in the model.

27.3 NETWORKING HARDWARE

The following is a description of most of the hardware that will be needed in the design of networks.

• Computer (or network enabled equipment)

529

- Network Interface Hardware - The network interface may already be built into the computer/PLC/ sensor/etc. These may cost $15 to over $1000.
- The Media - The physical network connection between network nodes.
  - 10baseT (twisted pair) is the most popular. It is a pair of twisted copper wires terminated with an RJ-45 connector.
  - 10base2 (thin wire) is thin shielded coaxial cable with BNC connectors
  - 10baseF (fiber optic) is costly, but signal transmission and noise properties are very good.
- Repeaters (Physical Layer) - These accept signals and retransmit them so that longer networks can be built.
- Hub/Concentrator - A central connection point that network wires will be connected to. It will pass network packets to local computers, or to remote networks if they are available.
- Router (Network Layer) - Will isolate different networks, but redirect traffic to other LANs.
- Bridges (Data link layer) - These are intelligent devices that can convert data on one type of network, to data on another type of network. These can also be used to isolate two networks.
- Gateway (Application Layer) - A Gateway is a full computer that will direct traffic to different networks, and possibly screen packets. These are often used to create firewalls for security.

Figure 27.1 shows the basic OSI model equivalents for some of the networking hardware described before.

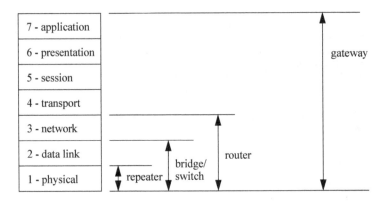

Figure 27.4   **Network Devices and the OSI Model**

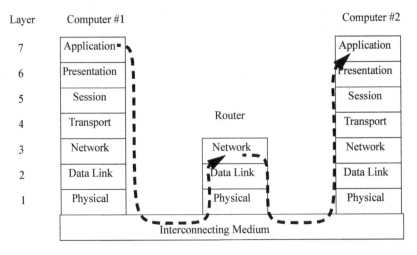

Figure 27.5   *The OSI Network Model with a Router*

27.4 NETWORK CONTROLS

A wide variety of networks are commercially available, and each has particular strengths and weaknesses. The differences arise from their basic designs. One simple issue is the use of the network to deliver power to the nodes. Some control networks will also supply enough power to drive some sensors and simple devices. This can eliminate separate power supplies, but it can reduce the data transmission rates on the network. The use of network taps or tees to connect to the network cable is also important. Some taps or tees are simple *passive* electrical connections, but others involve sophisticated *active* tees that are more costly, but allow longer networks.

The transmission type determines the communication speed and noise immunity. The simplest transmission method is baseband, where voltages are switched off and on to signal bit states. This method is subject to noise, and must operate at lower speeds. RS-232 is an example of baseband transmission. Carrierband transmission uses FSK (Frequency Shift Keying) that will switch a signal between two frequencies to indicate a true or false bit. This technique is very similar to FM (Frequency Modulation) radio where the frequency of the audio wave is transmitted by changing the frequency of a carrier frequency about 100MHz. This method allows higher transmission speeds, with reduced noise effects. Broadband networks transmit data over more than one channel by using multiple carrier frequencies on the same wire. This is similar to sending many cable television channels over the same wire. These networks can achieve very large transmission speeds, and can also be used to guarantee real time network access.

The bus network topology only uses a single transmission wire for all nodes. If all of the nodes decide to send messages simultaneously, the messages would be corrupted (a collision occurs). There are a variety of methods for dealing with network collisions, and arbitration.

CSMA/CD (Collision Sense Multiple Access/Collision Detection) - if two nodes start talking and detect a collision then they will stop, wait a random time, and then start again.
CSMA/BA (Collision Sense Multiple Access/Bitwise Arbitration) - if two nodes start talking at the same time the will stop and use their node addresses to determine which one goes first.
Master-Slave - one device one the network is the master and is the only one that may start communication. slave devices will only respond to requests from the master.
Token Passing - A token, or permission to talk, is passed sequentially around a network so that only one station may talk at a time.

The token passing method is deterministic, but it may require that a node with an urgent message wait to receive the token. The master-slave method will put a single machine in charge of sending and receiving. This can be restrictive if multiple controllers are to exist on the same network. The CSMA/CD and CSMA/BA methods will both allow nodes to talk when needed. But, as the number of collisions increase the network performance degrades quickly.

27.5 NETWORK STANDARDS

Bus types are listed below.

Low level busses - these are low level protocols that other networks are built upon.
    RS-485, Bitbus, CAN bus, Lonworks, Arcnet
General open buses - these are complete network types with fully published standards.
    ASI, Devicenet, Interbus-S, Profibus, Smart Distributed System (SDS), Seriplex
Specialty buses - these are buses that are proprietary.
    Genius I/O, Sensoplex

27.5.1 Devicenet

Devicenet has become one of the most widely supported control networks. It is an open standard, so components from a variety of manufacturers can be used together in the same control system. It is supported and promoted by the Open Devicenet Vendors Association (ODVA) (see http://www.odva.org). This group includes members from all of the major controls manufacturers.

This network has been designed to be noise resistant and robust. One major change for the control engineer is that the PLC chassis can be eliminated and the network can be connected directly to the sensors and actuators. This will reduce the total amount of wiring by moving I/O points closer to the application point. This can also simplify the connection of complex devices, such as HMIs. Two way communications inputs and outputs allow diagnosis of network problems from the main controller.

Devicenet covers all seven layers of the OSI standard. The protocol has a limited number of network address, with very small data packets. But this also helps limit network traffic and ensure responsiveness. The length of the network cables will limit the maximum speed of the network. The basic features of are listed below.

- A single bus cable that delivers data and power.
- Up to 64 nodes on the network.
- Data packet size of 0-8 bytes.
- Lengths of 500m/250m/100m for speeds of 125kbps/250kbps/500kbps respectively.
- Devices can be added/removed while power is on.
- Based on the CANbus (Controller Area Network) protocol for OSI levels 1 and 2.
- Addressing includes peer-to-peer, multicast, master/slave, polling or change of state.

An example of a Devicenet network is shown in Figure 27.1. The dark black lines are the network cable. Terminators are required at the ends of the network cable to reduce electrical noise. In this case the PC would probably be running some sort of software based PLC program. The computer would have a card that can communicate with Devicenet devices. The *FlexIO rack* is a miniature rack that can hold various types of input and output modules. Power taps (or tees) split the signal to small side branches. In this case one of the taps connects a power supply, to provide the 24Vdc supply to the network. Another two taps are used to connect a *smart sensor* and another *FlexIO rack*. The *Smart sensor* uses power from the network, and contains enough logic so that it is one node on the network. The network uses *thin trunk line* and *thick trunk line* which may limit network performance.

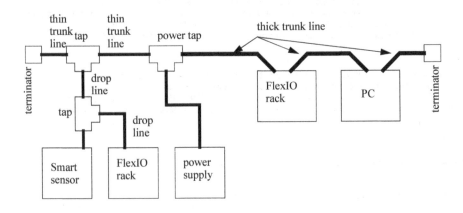

Figure 27.6   **A Devicenet Network**

The network cable is important for delivering power and data. Figure 27.1 shows a basic cable with two wires for data and two wires for the power. The cable is also shielded to reduce the effects of electrical noise. The two basic types are thick and thin trunk line. The cables may come with a variety of connections to devices.

- bare wires
- unsealed screw connector
- sealed mini connector
- sealed micro connector
- vampire taps

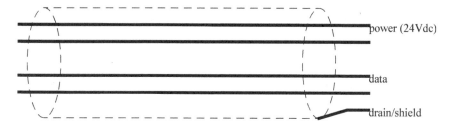

power (24Vdc)

data

drain/shield

Thick trunk - carries up to 8A for power up to 500m
Thin trunk - up to 3A for power up to 100m

Figure 27.7   *Shielded Network Cable*

Some of the design issues for this network include;

- Power supplies are directly connected to the network power lines.
- Length to speed is 156m/78m/39m to 125Kbps/250Kbps/500Kbps respectively.
- A single drop is limited to 6m.
- Each node on the network will have its own address between 0 and 63.

If a PLC-5 was to be connected to Devicenet a scanner card would need to be placed in the rack. The ladder logic in Figure 27.1 would communicate with the sensors through a scanner card in slot 3. The read and write blocks would read and write the Devicenet input values to integer memory from *N7:40* to *N7:59*. The outputs would be copied from the integer memory between *N7:20* to *N7:39*. The ladder logic to process inputs and outputs would need to examine and set bits in integer memory.

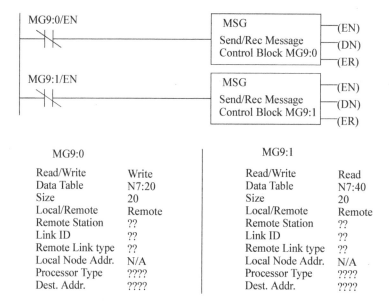

Figure 27.8   *Communicating with Devicenet Inputs and Outputs*

On an Allen Bradley Softlogix PLC the I/O will be copied into blocks of integer memory. These blocks are selected by the user in setup software. The ladder logic would then using integer memory for inputs and outputs, as shown in Figure 27.1. Here the inputs are copied into N9 integer memory, and the outputs are set by copying the N10 block of memory back to the outputs.

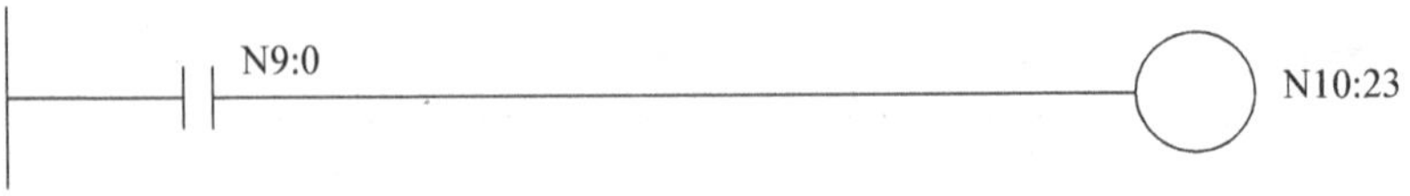

Figure 27.9   *Devicenet Inputs and Outputs in Software Based PLCs*

27.5.2 CANbus

The CANbus (Controller Area Network bus) standard is part of the Devicenet standard. Integrated circuits are now sold by many of the major vendors (Motorola, Intel, etc.) that support some, or all, of the standard on a single chip. This section will discuss many of the technical details of the standard.

CANbus covers the first two layers of the OSI model. The network has a bus topology and uses bit wise resolution for collisions on the network (i.e., the lower the network identifier, the higher the priority for sending). A data frame is shown in Figure 27.1. The frame is like a long serial byte, like that seen in the previous chapter. The frame begins with a start bit. This is then followed with a message identifier. For Devicenet this is a 5 bit address code (for up to 64 nodes) and a 6 bit command code. The *ready to receive it* bit will be set by the receiving machine. (Note: both the sender and listener share the same wire.) If the receiving machine does not set this bit the remainder of the message is aborted, and the message is resent later. While sending the first few bits, the sender monitors the bits to ensure that the bits send are heard the same way. If the bits do not agree, then another node on the network has tried to write a message at the same time - there was a collision. The two devices then wait a period of time, based on their identifier and then start to resend. The second node will then detect the message, and wait until it is done. The next 6 bits indicate the number of bytes to be sent, from 0 to 8. This is followed by two sets of bits for CRC (Cyclic Redundancy Check) error checking, this is a checksum of earlier bits. The next bit *ACK slot* is set by the receiving node if the data was received correctly. If there was a CRC error this bit would not be set, and the message would be resent. The remaining bits end the transmission. The *end of frame* bits are equivalent to stop bits. There must be a delay of at least 3 bits before the next message begins.

1 bit	start of frame
11 bits	identifier
1 bit	ready to receive it
6 bits	control field - contains number of data bytes
0-8 bytes	data - the information to be passed
15 bits	CRC sequence
1 bit	CRC delimiter
1 bit	ACK slot - other listeners turn this on to indicate frame received
1 bit	ACK delimiter
7 bits	end of frame
>= 3 bits	delay before next frame

(arbitration field — pointing to identifier and ready to receive it)

Figure 27.10   A CANbus Data Frame

Because of the bitwise arbitration, the address with the lowest identifier will get the highest priority, and be able to send messages faster when there is a conflict. As a result the controller is normally put at address *0*. And, lower priority devices are put near the end of the address range.

27.5.3 Controlnet

Controlnet is complimentary to Devicenet. It is also supported by a consortium of companies, (http://www.controlnet.org) and it conducts some projects in cooperation with the Devicenet group. The standard is designed for communication between controllers, and permits more complex messages than Devicenet. It is not suitable for communication with individual sensors and actuators, or with devices off the factory floor.

Controlnet is more complicated method than Devicenet. Some of the key features of this network include,

- Multiple controllers and I/O on one network
- Deterministic
- Data rates up to 5Mbps
- Multiple topologies (bus, star, tree)
- Multiple media (coax, fiber, etc.)
- Up to 99 nodes with addresses, up to 48 without a repeater
- Data packets up to 510 bytes
- Unlimited I/O points
- Maximum length examples
     1000m with coax at 5Mbps - 2 nodes
     250m with coax at 5Mbps - 48 nodes
     5000m with coax at 5Mbps with repeaters
     3000m with fiber at 5Mbps
     30Km with fiber at 5Mbps and repeaters
- 5 repeaters in series, 48 segments in parallel
- Devices powered individually (no network power)

• Devices can be removed while network is active

This control network is unique because it supports a real-time messaging scheme called Concurrent Time Domain Multiple Access (CTDMA). The network has a scheduled (high priority) and unscheduled (low priority) update. When collisions are detected, the system will wait a time of at least 2ms, for unscheduled messages. But, scheduled messages will be passed sooner, during a special time window.

27.5.4 Ethernet

Ethernet has become the predominate networking format. Version I was released in 1980 by a consortium of companies. In the 1980s various versions of ethernet frames were released. These include Version II and Novell Networking (IEEE 802.3). Most modern ethernet cards will support different types of frames.

The ethernet frame is shown in Figure 27.1. The first six bytes are the destination address for the message. If all of the bits in the bytes are set then any computer that receives the message will read it. The first three bytes of the address are specific to the card manufacturer, and the remaining bytes specify the remote address. The address is common for all versions of ethernet. The source address specifies the message sender. The first three bytes are specific to the card manufacturer. The remaining bytes include the source address. This is also identical in all versions of ethernet. The *ethernet type* identifies the frame as a Version II ethernet packet if the value is greater than 05DChex. The other ethernet types use these to bytes to indicate the datalength. The *data* can be between 46 to 1500 bytes in length. The frame concludes with a *checksum* that will be used to verify that the data has been transmitted correctly. When the end of the transmission is detected, the last four bytes are then used to verify that the frame was received correctly.

6 bytes	destination address
6 bytes	source address
2 bytes	ethernet type
46-1500 bytes	data
4 bytes	checksum

Figure 27.11   *Ethernet Version II Frame*

Ethernet protocols and hardware are the primary influences in forming the Internet. On the Internet each computer is given an address. Currently this address is a four byte address under the IPV4 standard, for example '192.168.1.4'. In the near future these addresses will be extended to six bytes under the IPV6 standard. However, users normally refer to machines using names such as 'www.gvsu.edu' which is translated to an IPV4 address '148.61.1.10' by a Directory Name Server (DNS).

When any computer (or PLC) sends a message on Ethernet, the destination address is part of that message. The message will then be routed through the network to the destination address. Within companies (and control systems) there are often local networks hidden behind firewalls that cannot be accessed directly from the Internet. When ethernet is used for control systems (Ethernet/IP) a sub-network is normally used. In this case a router is used for a group of network addresses with the same three first bytes, such as '192.168.1.__'. This also calls for a netmask of '255.255.255.0' that indicates what addresses are on the sub-network. The network will also have a broadcast or gateway assigned for the router (192.168.1.1 or 192.168.1.154 would be common choices). In a case where the network address is outside the sub-network, the router will send it out to the greater network, and return the responses.

When setting up a control network using ethernet you will need to assign a unique IPV4 address to each device. This can be done by setting a permanent address in the device configuration, this is called a Static IP address. Another alternative is to automatically assign the addresses using DHCP or BOOTP protocols. Each device on a network is assigned a unique Media Access Control (MAC) number during manufacturing. Most routers have the ability to accept DHCP requests with MAC numbers and assign IP addresses. Names can also be assigned by the BOOTP and DHCP servers.

27.5.5 Profibus

Another control network that is popular in europe, but also available world wide. It is also promoted by a consortium of companies (http://www.profibus.com). General features include;

- A token passing between up to three masters
- Maximum of 126 nodes
- Straight bus topology
- Length from 9600m/9.6Kbps with 7 repeaters to 500m/12Mbps with 4 repeaters
- With fiber optic cable lengths can be over 80Km
- 2 data lines and shield
- Power needed at each station
- Uses RS-485, ethernet, fiber optics, etc.
- 2048 bits of I/O per network frame

27.5.6 Sercos

The SErial Real-time COmmunication System (SERCOS) is an open standard designed for multi-axis motion control systems. The motion controller and axes can be implemented separately and then connected using the SERCOS network. Many vendors offer cards that allow PLCs to act as clients and/or motion controllers.

- Deterministic with response times as small as a few nanoseconds
- Data rates of 2, 4, 8 and 16 Mbaud
- Documented with IEC 61491 in 1995 and 2002
- Uses a fiber optic rings, RS-485 and buses

27.6 PROPRIETARY NETWORKS

27.6.1 Data Highway

Allen-Bradley has developed the Data Highway II (DH+) network for passing data and programs between PLCs and to computers. This bus network allows up to 64 PLCs to be connected with a single twisted pair in a shielded cable. Token passing is used to control traffic on the network. Computers can also be connected to the DH+ network, with a network card to download programs and monitor the PLC. The network will support data rates of 57.6Kbps and 230 Kbps

The DH+ basic data frame is shown in Figure 27.1. The frame is byte oriented. The first byte is the *DLE* or delimiter byte, which is always $10. When this byte is received the PLC will interpret the next byte as a command. The *SOH* identifies the message as a DH+ message. The next byte indicates the destination station - each node one the network must have a unique number. This is followed by the *DLE* and *STX* bytes that identify the start of the data. The data follows, and its' length is determined by the command type - this will be discussed later. This is then followed by a *DLE* and *ETX* pair that mark the end of the message. The last byte transmitted is a checksum to determine the correctness of the message.

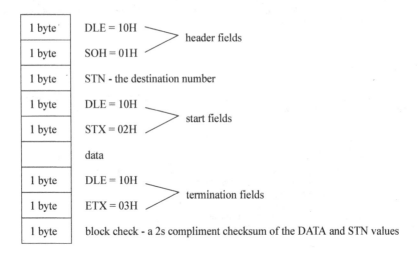

1 byte	DLE = 10H	⎫ header fields
1 byte	SOH = 01H	⎭
1 byte	STN - the destination number	
1 byte	DLE = 10H	⎫ start fields
1 byte	STX = 02H	⎭
	data	
1 byte	DLE = 10H	⎫ termination fields
1 byte	ETX = 03H	⎭
1 byte	block check - a 2s compliment checksum of the DATA and STN values	

Figure 27.12   *The Basic DH+ Data Frame*

The general structure for the data is shown in Figure 27.1. This packet will change for different commands. The first two bytes indicate the destination, *DST*, and source, *SRC*, for the message. The next byte is the command, *CMD*, which will determine the action to be taken. Sometimes, the function, *FNC*, will be needed to modify the command. The transaction, *TNS*, field is a unique message identifier. The two address, *ADDR*, bytes identify a target memory location. The *DATA* fields contain the information to be passed. Finally, the *SIZE* of the data field is transmitted.

	1 byte	DST - destination node for the message
	1 byte	SRC - the node that sent the message
	1 byte	CMD - network command - sometime FNC is required
	1 byte	STS - message send/receive status
	2 byte	TNS - transaction field (a unique message ID)
optional	1 byte	FNC may be required with some CMD values
optional	2 byte	ADDR - a memory location
optional	variable	DATA - a variable length set of data
optional	1 byte	SIZE - size of a data field

Figure 27.13   *Data Filed Values*

Examples of commands are shown in Figure 27.1. These focus on moving memory and status information between the PLC, and remote programming software, and other PLCs. More details can be found in the Allen-Bradley DH+ manuals.

CMD	FNC	Description
00		Protected write
01		Unprotected read
02		Protected bit write
05		Unprotected bit write
06	00	Echo
06	01	Read diagnostic counters
06	02	Set variables
06	03	Diagnostic status
06	04	Set timeout
06	05	Set NAKs
06	06	Set ENQs
06	07	Read diagnostic counters
08		Unprotected write
0F	00	Word range write
0F	01	Word range read
0F	02	Bit write
0F	11	Get edit resource
0F	17	Read bytes physical
0F	18	Write bits physical
0F	26	Read-modify-write
0F	29	Read section size
0F	3A	Set CPU mode
0F	41	Disable forces
0F	50	Download all request
0F	52	Download completed
0F	53	Upload all request
0F	55	Upload completed
0F	57	Initialize memory
0F	5E	Modify PLC-2 compatibility file
0F	67	typed write
0F	68	typed read
0F	A2	Protected logical read - 3 address fields
0F	AA	Protected logical write - 3 addr. fields

Figure 27.14 **DH+ Commands for a PLC-5 (all numbers are hexadecimal)**

The ladder logic in Figure 27.1 can be used to copy data from the memory of one PLC to another. Unlike other networking schemes, there are no *login* procedures. In this example the first MSG instruction will write the message from the local memory *N7:20 - N7:39* to the remote PLC-5 (node 2) into its memory from *N7:40* to *N7:59*. The second MSG instruction will copy the memory from the remote PLC-5 memory *N7:40* to *N7:59* to the remote PLC-5 memory *N7:20* to *N7:39*. This transfer will require many scans of ladder logic, so the *EN* bits will prevent a read or write instruction from restarting until the previous *MSG* instruction is complete.

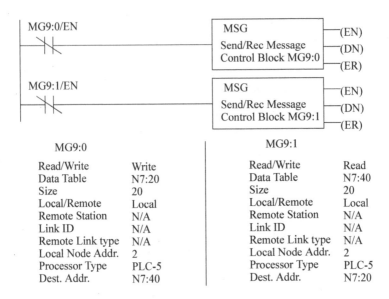

	MG9:0		MG9:1	
Read/Write	Write		Read/Write	Read
Data Table	N7:20		Data Table	N7:40
Size	20		Size	20
Local/Remote	Local		Local/Remote	Local
Remote Station	N/A		Remote Station	N/A
Link ID	N/A		Link ID	N/A
Remote Link type	N/A		Remote Link type	N/A
Local Node Addr.	2		Local Node Addr.	2
Processor Type	PLC-5		Processor Type	PLC-5
Dest. Addr.	N7:40		Dest. Addr.	N7:20

Figure 27.15   **Ladder Logic for Reading and Writing to PLC Memory**

The DH+ data packets can be transmitted over other data links, including ethernet and RS-232.

27.7 NETWORK COMPARISONS

No one network is ideal for solving all controls problems. Table 1 shows a variety of network types and criteria. Generally there is a tradeoff between length, speed, cost, and reliability. For example, slower networks such as Controlnet/Devicenet, Lonworks, Modbus, and Profibus are designed to be highly predictable (deterministic) and work well between a central processor and distributed IO modules. Other network types such as Ethernet provide a low cost alternative for connecting IO components but it can be susceptible to electrical noise. Other network types such as Sercos are designed for motion control systems and provide outstanding interoperability between manufacturers.

Table 1: Network Comparison

Network	topology	addresses	length	speed	packet size
Bluetooth	wireless	8	10	64Kbps	continuous
CANopen	bus	127	25m-1000m	1Mbps-10Kbps	8 bytes
ControlNet	bus or star	99	250m-1000m wire, 3-30km fiber	5Mbps	0-510 bytes
Devicenet	bus	64	500m	125-500Kbps	8 bytes

Network	topology	addresses	length	speed	packet size
Ethernet	bus, star	1024	85m coax, 100m twisted pair, 400m-50km fiber	10-1000Gbps	46-1500bytes
Foundation Fieldbus	star	unlimited	100m twisted pair, 2km fiber	100Mbps	<=1500 bytes
Interbus	bus	512	12.8km with 400m segments	500-2000 Kbps	0-246 bytes
Lonworks	bus, ring, star	32,000	<=2km	78Kbps-1.25Mbps	228 bytes
Modbus	bus, star	250	350m	300bps-38.4Kbps	0-254 bytes
Profibus	bus, star, ring	126	100-1900m	9.6Kbps-12Mbps	0-244bytes
Sercos	rings	254	800m	2-16Mbps	32bits
USB	star	127	5m	>100Mbps	1-1000bytes

27.8 DESIGN CASES

27.8.1 Devicenet

Problem: A robot will be loading parts into a box until the box reaches a prescribed weight. A PLC will feed parts into a pickup fixture when it is empty. The PLC will tell the robot when to pick up a part and load it using Devicenet.

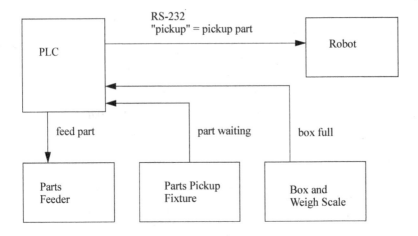

Figure 27.16   **Box Loading System**

27.9 SUMMARY

- Networks come in a variety of topologies, but buses are most common on factory floors.
- The OSI model can help when describing network related hardware and software.
- Networks can be connected with a variety of routers, bridges, gateways, etc.
- Devicenet is designed for interfacing to a few inputs and outputs.
- Controlnet is designed for interfacing between controllers.
- Controlnet and devicenet are based on CANbus.
- Ethernet is common, and can be used for high speed communication.
- Profibus is another control network.

27.10 PRACTICE PROBLEMS

(Note: Problem solutions are available at http://sites.google.com/site/automatedmanufacturingsystems/)

1. Explain why networks are important in manufacturing controls.

2. We will use a PLC to control a cereal box filling machine. For single runs the quantities of cereal types are controlled using timers. There are 6 different timers that control flow, and these result in different ratios of product. The values for the timer presets will be downloaded from another PLC using the DH+ network. Write the ladder logic for the PLC.

3.

    a) We are developing ladder logic for an oven to be used in a baking facility. A PLC is controlling the temperature of an oven using an analog voltage output. The oven must be started with a push button and can be stopped at any time with a stop push button. A recipe is used to control the times at each temperature (this is written into the PLC memory by another PLC). When idle, the output voltage should be 0V, and during heating the output voltages, in sequence, are 5V, 7.5V, 9V. The timer preset values, in sequence, are in N7:0, N7:1, N7:2. When the oven is on, a value of 1 should be stored in N7:3, and when the oven is off, a value of 0 should be stored in N7:3. Draw a state diagram and write the ladder logic for this station.

    b) We are using a PLC as a master controller in a baking facility. It will update recipes in remote PLCs using DH+. The master station is #1, the remote stations are #2 and #3. When an operator pushes one of three buttons, it will change the recipes in two remote PLCs if both of the remote PLCs are idle. While the remote PLCs are running they will change words in their internal memories (N7:3=0 means idle and N7:3=1 means active). The new recipe values will be written to the remote PLCs using DH+. The table below shows the values for each PLC. Write the ladder logic for the master controller.

	button A	button B	button C
	13	17	14
PLC #2	690	235	745
	45	75	34
	76	72	56
	345	234	645
PLC #3	987	12	23
	345	34	456
	764	456	568
	87	67	8

4. A controls network is to be 1500m long. Suggest three different types of networks that would meet the specifications.

5 How many data bytes (maximum) could be transferred in one second with DH+?

6. Is the OSI model able to describe all networked systems?

7. What are the different methods for resolving collisions on a bus network?

27.11 ASSIGNMENT PROBLEMS

1. Describe an application for DH networking.

2. The response times of hydraulic switches is being tested in a PLC controlled station. When the units arrive a 'part present' sensor turns on. The part is then clamped in place by turning on a 'clamp' output. 1 second after clamping, a 'flow' output is turned on to start the test. The response time is the delay between when 'flow' is turned on, and the 'engaged' input turns on. When the unit has responded, up to 10 seconds later, the 'flow' output is turned off, and the system is allowed to sit for 5 seconds to discharge before unclamping. The result of the test is written to one of the memory locations from F8:0 to F8:39, for a total of 40 separate tests. When 40 tests have been done, the memory block from F8:0 to F8:39 is sent to another PLC using DH+, and the process starts again. Write the ladder logic to control the station.

3. a) Controls are to be developed for a machine that packages golf tees. Each container will normally hold 1000 tees filled from three different hoppers, each containing a different color. For marketing purposes the ratio of colors is changed frequently. To make the controller easy to reconfigure, the number of tees from each hopper are stored in the memory locations N7:0, N7:1 and N7:2. The process is activated when an empty package arrives, activating a PRESENT input. When filling the package, the machine opens a single hopper with a solenoid, and counts the tees with an optical sensor, until the specified count has been surpassed. It then repeats the operation with the two other hoppers. When done, it activates a SEAL for 2 seconds to advance a heated ram that seals the package. After that, the DONE output is turned on until the PRESENT sensor turns off. Write the ladder logic for this process.

b) Write a ladder logic program that will read and parse values from an RS-232 input. The format of the input will be an eleven character line with three integer numbers separated by commas. The integers will be padded to three characters by padding with zeros. The line will be terminated with a CR and a LF. The three integers are to be parsed and stored in the memory locations N7:0, N7:1 and N7:2 to be used in a golf tee packaging machine.

4. A master PLC is located at the top of a mine shaft and controls an elevator system. A second PLC is located half a mile below to monitor the bottom of the elevator shaft. At the top of the mine shaft the PLC has inputs for the door (D), a top limit switch (T), and start (G) and stop (S) pushbuttons. The PLC has two outputs to apply power (P) to the motor, or reverse (R) the motor direction. The PLC at the bottom of the elevator shaft checks a bottom limit switch (B) and a door closed (C) sensor. The two PLCs are connected using DH+. Write ladder logic for both PLCs and indicate the communication settings. Use structured design techniques.

27.1 PRACTICE PROBLEM SOLUTIONS

1. These networks allow us to pass data between devices so that individually controlled systems can be integrated into a more complex manufacturing facility. An example might be a serial connection to a PLC so that SPC data can be collected as product is made, or recipes downloaded as they are needed.

2.

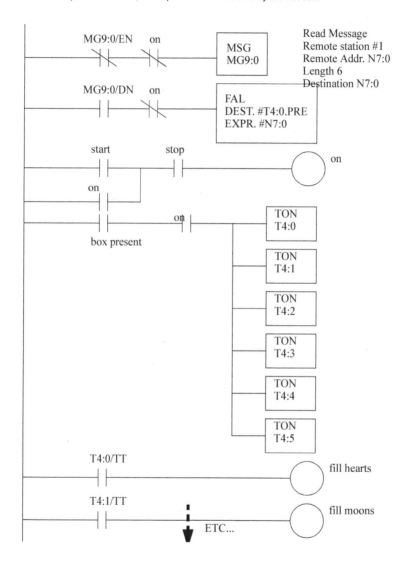

3.

a)

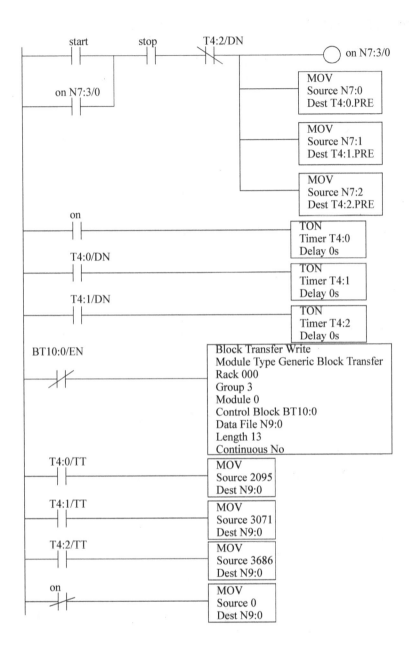

b)

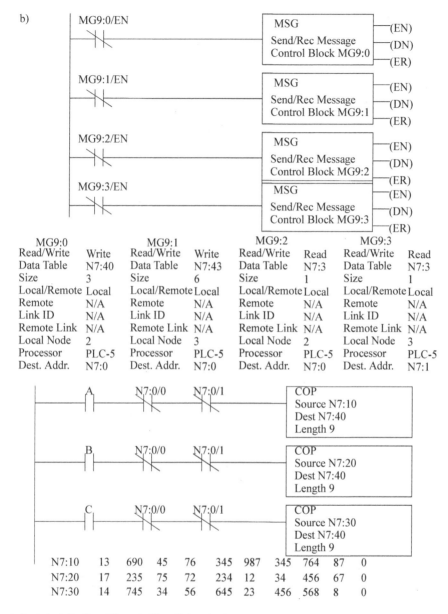

MG9:0		MG9:1		MG9:2		MG9:3	
Read/Write	Write	Read/Write	Write	Read/Write	Read	Read/Write	Read
Data Table	N7:40	Data Table	N7:43	Data Table	N7:3	Data Table	N7:3
Size	3	Size	6	Size	1	Size	1
Local/Remote	Local	Local/Remote	Local	Local/Remote	Local	Local/Remote	Local
Remote	N/A	Remote	N/A	Remote	N/A	Remote	N/A
Link ID	N/A	Link ID	N/A	Link ID	N/A	Link ID	N/A
Remote Link	N/A	Remote Link	N/A	Remote Link	N/A	Remote Link	N/A
Local Node	2	Local Node	3	Local Node	2	Local Node	3
Processor	PLC-5	Processor	PLC-5	Processor	PLC-5	Processor	PLC-5
Dest. Addr.	N7:0	Dest. Addr.	N7:0	Dest. Addr.	N7:0	Dest. Addr.	N7:1

N7:10	13	690	45	76	345	987	345	764	87	0
N7:20	17	235	75	72	234	12	34	456	67	0
N7:30	14	745	34	56	645	23	456	568	8	0

4. Controlnet, Profibus, Ethernet with multiple subnets

5 the maximum transfer rate is 230 Kbps, with 11 bits per byte (1start+8data+2+stop) for 20909 bytes per second. Each memory write packet contains 17 overhead bytes, and as many as 2000 data bytes. Therefore as many as 20909*2000/(2000+17) = 20732 bytes could be transmitted per second. Note that this is ideal, the actual maximum rates would be actually be a fraction of this value.

6. The OSI model is just a model, so it can be used to describe parts of systems, and what their functions are. When used to describe actual networking hardware and software, the parts may only apply to one or two layers. Some parts may implement all of the layers in the model.

7. When more than one client tries to start talking simultaneously on a bus network they interfere, this is called a collision. When this occurs they both stop, and will wait a period of time before starting again. If they both wait different amounts of time the next one to start talking will get priority, and the other will have to wait. With CSMA/CD the clients wait a random amount of time. With CSMA/BA the clients wait based upon their network address, so their priority is related to their network address. Other networking methods prevent collisions by limiting communications. Master-slave networks require that client do not less talk, unless they are responding to a request from a master machine. Token passing only permits the holder of the token to talk.

28HUMAN MACHINE INTERFACES (HMI)

Topics:

- General features of an HMI
- Structured programming of an HMI
- A formal approach to HMI design

Objectives:

- To be able to methodically design an HMI for an application

For simpler control systems, buttons and switches are quite suitable for operator interfaces. However as the number of operator options increases, or the interface becomes more complicated it may be preferable to replace many of the buttons, dials, and indicators with a Human Machine Interface (HMI). These units can be as simple as a single line of text and a couple of push buttons. More complicated units use large color monitors with touch screen capabilities. Ultimately these units are very powerful because the display contents can be changed to match the mode of operation.

An HMI is a simple to program graphical interface, very much like modern computer software. The simplest control pair are a button and indicator. Consider the example in Figure 28.1. The button can be used as a simple input to the PLC, while the output status can be shown with an indicator. The programmer will set up the Ethernet connection to pass tag/variable data between the HMI and PLC so that a change in one appears in the other. So, if the button is touched on the screen of the HMI, the value is changed in the memory of the HMI. On the next data update cycle it is sent to the PLC. The program in the PLC reads the value change and then sets a new indicator value. The updated indicator is then sent to the HMI on a subsequent communication update. The newly changed value in the HMI is then used to update the indicator on the screen.

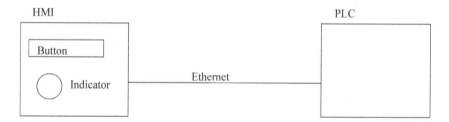

Figure 28.1   **A Simple HMI Application**

Obviously the first example is quite simple and only replaces and output or input, but much more capabilities are possible. Typical elements found on HMIs include,

- Multiple screens - each screen is given a function such as idle, maintenance, auto, and manual
- Logical inputs and indicators - simple on off inputs and outputs. These will be the most common component, often used by operators to start and stop operations.
- Numeric inputs and indicators - analog inputs and outputs. These can be used for applications such as monitoring temperatures, or setting part counts.
- Graphs - often to see trends over time for counts or analog values
- Text and Images - pictures and text are often used for operator instructions and help.
- Colors - some units have colors that can be used to highlight or differentiate items. For example a bright red object on the screen can indicate a fault.
- Sounds - some more advanced HMIs offer sound outputs

The general implementation steps for implementing and HMI are listed below. To control the HMI from a PLC the user input will set bits in the PLC memory, and other bits in the PLC memory can be set to turn on/off items on the HMI screen.

1. Design work
2. Enter or load the PLC variable/tag names that the HMI will use.
3. Layout screens, buttons, etc. on the programing software.
4. Download the program to the HMI unit.
5. Connect the unit to a PLC.
6. Run the system - read and write to the HMI using PLC memory locations to get input and update screens.

28.1 HMI/MMI DESIGN

There are a few basic approaches that will help when designing any Graphical User Interface (GUI), such as those running on HMIs. As normal the design process begins with gathering information, providing a structure, and then implementing the structure. A good set of introductory questions are given below.

1. Who needs what information?
2. How do they expect to see it presented?
3. When does information need to be presented?
4. Do the operators have any special needs?
5. Is sound important?
6. What choices should the operator have?

After this information gathering stage the HMI Functionality can be designed using a state diagram and a list of input/output requirements for each screen. A simple example is given in Figure 28.1. Each of the four states in the diagram will become one of four HMI screens. The transitions between the screens are normally touch-screen buttons, but could also be values set by the PLC. For each of the screens a requirements list must be developed that shows the important information and actions available to the user.

State Diagram:

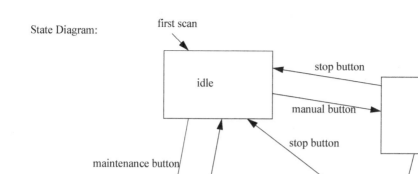

Screen Requirements:

    Idle:
        temperature
    Manual:
        advance and retract inputs
        temperature
        jam condition indicator
    Auto:
        temperature
        cycle count
    Maintenance:
        jog advance
        jog retract
        temperature
        advance limit switch indicator
        retract limit switch indicator

Figure 28.2   **State Diagram for a Simple HMI**

HMI design also requires a limited amount of artistic skill to increase usability. A few simple rules-of-thumb a good HMI design are given.

    Common Look and Feel - each screen should look very similar.
        Components with similar function should appear in the same place on each screen
        Colors and logos should be consistent between screens
        Use colors conservatively - especially green, red, and yellow
        Avoid crowding

    Provide helpful notes (or popup help)

    Indicate the current state of operation clearly on each screen

    Provide feedback to the user
        buttons - change the shading or similar when pressed

Adjust the interface to the user
> maintenance screen have more technical content
> operator screens simpler language, task oriented
> other languages for non-english speakers
> icons instead of words for reduced literacy

A clear interface design can be presented to customers and non-technical people to get feedback before the implementation. These should include i) the State Diagram, ii) a list of Screen Requirements, and iii) a Look-and-Feel in general, and for each screen. Once these have been set the process of programming the HMI becomes a trivial matter.

28.2 SUMMARY

- HMIs allow customized controls that can change based upon the operation mode of a machine.
- State Diagrams and Screen Requirements can be used with a Look-and-Feel before implementation.

29ELECTRICAL DESIGN AND CONSTRUCTION

Topics:

- Electrical wiring issues; cabinet wiring and layout, grounding, shielding and inductive loads
- Enclosures for industrial environments

Objectives:

- To learn the major issues in designing controllers including; electrical schematics, panel layout, grounding, shielding, enclosures.

It is uncommon for engineers to build their own controller designs. For example, once the electrical designs are complete, they must be built by an electrician. Therefore, it is your responsibility to effectively communicate your design intentions to the electricians through drawings. In some factories, the electricians also enter the ladder logic and do debugging. This chapter discusses the design issues in implementation that must be considered by the designer.

29.1 ELECTRICAL WIRING DIAGRAMS

In an industrial setting a PLC is not simply "plugged into a wall socket". The electrical design for each machine must include at least the following components.

transformers - to step down AC supply voltages to lower levels
power contacts - to manually enable/disable power to the machine with e-stop buttons
terminals - to connect devices
fuses or breakers - will cause power to fail if too much current is drawn
grounding - to provide a path for current to flow when there is an electrical fault
enclosure - to protect the equipment, and users from accidental contact

A control system will normally use AC and DC power at different voltage levels. Control cabinets are often supplied with single phase AC at 220/440/550V, or two phase AC at 220/440Vac, or three phase AC at 330/550V. This power must be dropped down to a lower voltage level for the controls and DC power supplies. 110Vac is common in North America, and 220Vac is common in Europe and the Commonwealth countries. It is also common for a controls cabinet to supply a higher voltage to other equipment, such as motors.

An example of a wiring diagram for a motor controller is shown in Figure 29.1 (note: the symbols are discussed in detail later). Dashed lines indicate a single purchased component. This system uses 3 phase AC power (L1, L2 and L3) connected to the terminals. The three phases are then connected to a power interrupter. Next, all three phases are supplied to a motor starter that contains three contacts, *M*, and three thermal overload relays (breakers). The contacts, *M*, will be controlled by the coil, *M*. The output of the motor starter goes to a three phase AC motor. Power is supplied by connecting a step down transformer to the control electronics by connecting to phases *L2* and *L3*. The lower voltage is then used to supply power to the left and right rails of the ladder below. The *neutral* rail is also grounded. The logic consists of two push buttons. The *start* push button is normally open, so that if something fails the motor cannot be started. The *stop* push button is normally closed, so that if a wire or connection fails the system halts safely. The system controls the *motor starter* coil *M*, and uses a spare contact on the starter, *M*, to seal in the motor stater.

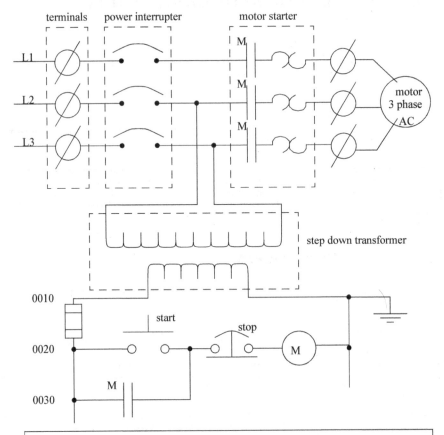

Figure 29.1   *A Motor Controller Schematic*

The diagram also shows numbering for the wires in the device. This is essential for industrial control systems that may contain hundreds or thousands of wires. These numbering schemes are often particular to each
facility, but there are tools to help make wire labels that will appear in the final controls cabinet.

Once the electrical design is complete, a layout for the controls cabinet is developed, as shown in Figure
29.1. The physical dimensions of the devices must be considered, and adequate space is needed to *run* wires
between components. In the cabinet the AC power would enter at the *terminal block*, and be connected to the *main
breaker*. It would then be connected to the *contactors* and *fuses*. Two of the phases are also connected to the
transformer to power the logic. The start and stop buttons are at the left of the box (note: normally these are

mounted elsewhere, and a separate layout drawing would be needed).

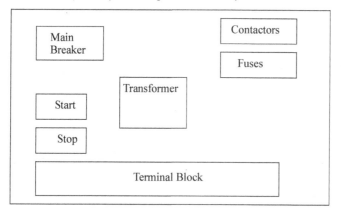

Figure 29.2   ***A Physical Layout for the Control Cabinet***

The final layout in the cabinet might look like the one shown in Figure 29.1.

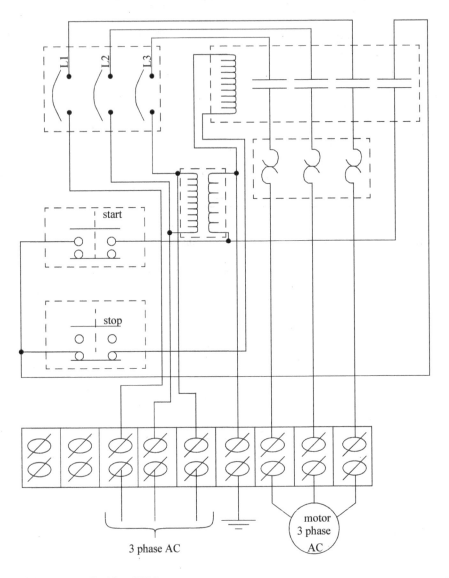

Figure 29.3   *Final Panel Wiring*

When being built the system will follow certain standards that may be company policy, or legal require-ments. This often includes items such as;

       hold downs - the will secure the wire so they don't move
       labels - wire labels help troubleshooting
       strain reliefs - these will hold the wire so that it will not be pulled out of screw terminals
       grounding - grounding wires may be needed on each metal piece for safety

A photograph of an industrial controls cabinet is shown in Figure 29.1.

Figure 29.4   *An Industrial Controls Cabinet*

29.1.1 Selecting Voltages

When selecting voltage ranges and types for inputs and outputs of a PLC some care can save time, money and effort. Figure 29.1 that shows three different voltage levels being used, therefore requiring three different input cards. If the initial design had selected a *standard* supply voltage for the system, then only one power supply, and PLC input card would have been required.

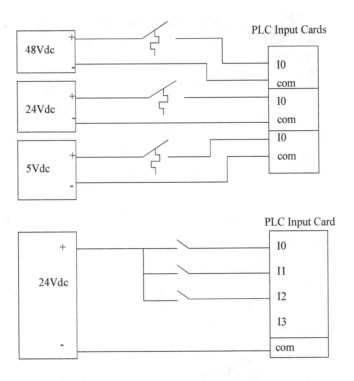

Figure 29.5   *Standardized Voltages*

29.1.2 Grounding

The terms *ground* and *common* are often interchanged (I do this often), but they do mean different things. The term, ground, comes from the fact that most electrical systems find a local voltage level by placing some metal in the earth (ground). This is then connected to all of the electrical outlets in the building. If there is an electrical fault, the current will be drawn off to the ground. The term, common, refers to a reference voltage that components of a system will use as common zero voltage. Therefore the function of the ground is for safety, and the common is for voltage reference. Sometimes the common and ground are connected.

The most important reason for grounding is human safety. Electrical current running through the human body can have devastating effects, especially near the heart. Figure 29.1 shows some of the different current levels, and the probable physiological effects. The current is dependant upon the resistance of the body, and the contacts. A typical scenario is, a hand touches a high voltage source, and current travels through the body and out a foot to ground. If the person is wearing rubber gloves and boots, the resistance is high and very little current will flow. But, if the person has a sweaty hand (salty water is a good conductor), and is standing barefoot in a pool of water their resistance will be much lower. The voltages in the table are suggested as reasonable for a healthy adult in normal circumstances. But, during design, you should assume that no voltage is safe.

current in body (mA)	effect
0-1	negligible (normal circumstances, 5VDC)
1-5	uncomfortable (normal circumstances, 24VDC)
10-20	possibility for harm (normal circumstances, 120VAC)
20-50	muscles contract (normal circumstances, 220VAC)
50-100	pain, fainting, physical injuries
100-300	heart fibrillates
300+	burns, breathing stops, etc.

Figure 29.6   **Current Levels**

Aside: Step potential is another problem. Electron waves from a fault travel out in a radial direction through the ground. If a worker has two feet on the ground at different radial distances, there will be a potential difference between the feet that will cause a current to flow through the legs. The gist of this is - if there is a fault, don't run/walk away/ towards.

Figure 29.1 shows a grounded system with a metal enclosures. The left-hand enclosure contains a transformer, and the enclosure is connected directly to ground. The wires enter and exit the enclosure through insulated strain reliefs so that they don't contact the enclosure. The second enclosure contains a load, and is connected in a similar manner to the first enclosure. In the event of a major fault, one of the "live" electrical conductors may come loose and touch the metal enclosure. If the enclosure were not grounded, anybody touching the enclosure would receive an electrical shock. When the enclosure is grounded, the path of resistance between the case and the ground would be very small (about 1 ohm). But, the resistance of the path through the body would be much higher (thousands of ohms or more). So if there were a fault, the current flow through the ground might "blow" a fuse. If a worker were touching the case their resistance would be so low that they might not even notice the fault.

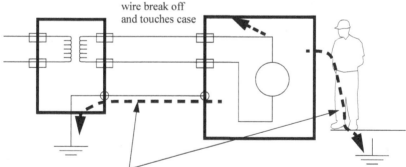

wire break off
and touches case

Current can flow two ways, but most will follow the path of least resistance, good grounding will keep the worker relatively safe in the case of faults.

Figure 29.7   **Grounding for Safety**

Note: Always ground systems first before applying power. The first time a system is activated it will have a higher chance of failure.

When improperly grounded a system can behave erratically or be destroyed. Ground loops are caused when too many separate connections to ground are made creating loops of wire. Figure 29.1 shows ground wires as darker lines. A ground loop caused because an extra ground was connected between *device A* and ground. The last connection creates a loop. If a current is induced, the loop may have different voltages at different points. The connection on the right is preferred, using a *tree* configuration. The grounds for devices *A* and *B* are connected back to the power supply, and then to the ground.

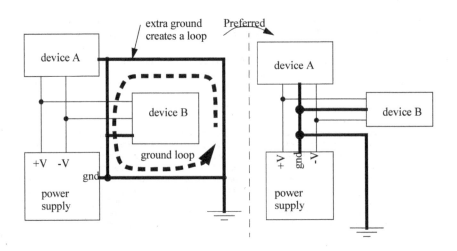

Figure 29.8   **Eliminating Ground Loops**

Problems often occur in large facilities because they may have multiple ground points at different end of large buildings, or in different buildings. This can cause current to flow through the ground wires. As the current flows it will create different voltages at different points along the wire. This problem can be eliminated by using electrical isolation systems, such as optocouplers.

When designing and building electrical control systems, the following points should prove useful.

- Avoid ground loops
  - Connect the enclosure to the ground bus.
  - Each PLC component should be grounded back to the main PLC chassis. The PLC chassis should be grounded to the backplate.
  - The ground wire should be separated from power wiring inside enclosures.
  - Connect the machine ground to the enclosure ground.
- Ensure good electrical connection
  - Use star washers to ensure good electrical connection.
  - Mount ground wires on bare metal, remove paint if needed.
  - Use 12AWG stranded copper for PLC equipment grounds and 8AWG stranded copper for enclosure backplate grounds.
  - The ground connection should have little resistance (<0.1 ohms is good).

29.1.3 Wiring

As the amount of current carried by a wire increases, it is important to use a wire with a larger cross section. A larger cross section results in a lower resistance, and less heating of the wire. The standard wire gages are listed in Figure 29.1.

AWG #	Dia. (mil)	Res. 25C (ohm/1000 ft)	Rated Current (A)
4	204	0.25	
6	162	0.40	
8	128	0.64	
10	102	1.0	
12	81	1.6	
14	64	2.6	
16	51	4.1	
18	40	6.5	
20	32	10	
22	25	17	
24	20	26	

Figure 29.9   *American Wire Gage (AWG) Copper Wire Sizes*

29.1.4 Suppressors

Most of us have seen a Vandegraff generator, or some other inductive device that can generate large sparks using inductive coils. On the factory floor there are some massive inductive loads that make this a significant design problem. This includes devices such as large motors and inductive furnaces. The root of the problem is that coils of wire act as inductors and when current is applied they build up magnetic fields, requiring energy. When the applied voltage is removed and the fields collapse the energy is dumped back out into the electrical system. As a result, when an inductive load is turned on it draws an excess amount of current (and lights dim), and when it is turn it off there is a power surge. In practical terms this means that large inductive loads will create voltage spikes that will damage our equipment.

Surge suppressors can be used to protect equipment from voltage spikes caused by inductive loads. Figure 29.1 shows the schematic equivalent of an uncompensated inductive load. For this to work reliably we would need to over design the system above the rated loads. The second schematic shows a technique for compensating for an AC inductive load using a resistor capacitor pair. It effectively acts as a high pass filter that allows a high frequency voltage spike to be short circuited. The final surge suppressor is common for DC loads. The diode allows current to flow from the negative to the positive. If a negative voltage spike is encountered it will short circuit through the diode.

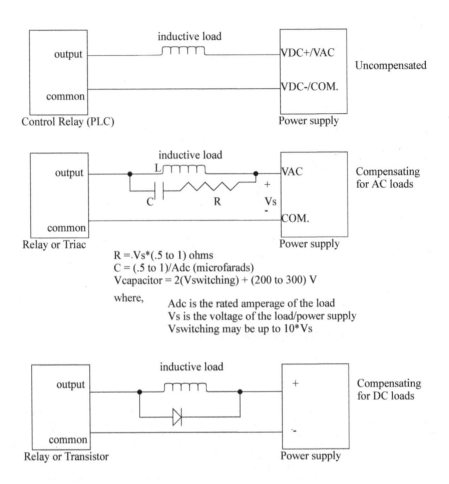

R = .Vs*(.5 to 1) ohms
C = (.5 to 1)/Adc (microfarads)
Vcapacitor = 2(Vswitching) + (200 to 300) V

where,   Adc is the rated amperage of the load
          Vs is the voltage of the load/power supply
          Vswitching may be up to 10*Vs

Figure 29.10   *Surge Suppressors*

29.1.5 PLC Enclosures

PLCs are well built and rugged, but they are still relatively easy to damage on the factory floor. As a result, enclosures are often used to protect them from the local environment. Some of the most important factors are listed below with short explanations.

Dirt - Dust and grime can enter the PLC through air ventilation ducts. As dirt clogs internal circuitry, and external circuitry, it can effect operation. A storage cabinet such as Nema 4 or 12 can help protect the PLC.

Humidity - Humidity is not a problem with many modern materials. But, if the humidity condenses, the water can cause corrosion, conduct current, etc. Condensation should be avoided at all costs.

Temperature - The semiconductor chips in the PLC have operating ranges where they are operational. As the temperature is moved out of this range, they will not operate properly, and the PLC will shut down. Ambient heat generated in the PLC will help keep the PLC operational at lower temperatures (generally to 0°C). The upper range for the devices is about 60°C, which is generally sufficient for sealed cabinets, but warm temperatures, or other heat sources (e.g. direct irradiation from the sun) can raise the temperature above acceptable limits. In extreme

562

conditions heating, or cooling units may be required. (This includes "cold-starts" for PLCs before their semiconductors heat up).

Shock and Vibration - The nature of most industrial equipment is to apply energy to change work-pieces. As this energy is applied, shocks and vibrations are often produced. Both will travel through solid materials with ease. While PLCs are designed to withstand a great deal of shock and vibration, special elastomer/spring or other mounting equipment may be required. Also note that careful consideration of vibration is also required when wiring.

Interference - Electromagnetic fields from other sources can induce currents.

Power - Power will fluctuate in the factory as large equipment is turned on and off. To avoid this, various options are available. Use an isolation transformer. A UPS (Uninterruptable Power Supply) is also becoming an inexpensive option, and are widely available for personal computers.

A standard set of enclosures was developed by NEMA (National Electric Manufacturers Association). These enclosures are intended for voltage ratings below 1000Vac. Figure 29.1 shows some of the rated cabinets. Type 12 enclosures are a common choice for factory floor applications.

Type 1 - General purpose - indoors
Type 2 - Dirt and water resistant - indoors
Type 3 - Dust-tight, rain-tight and sleet (ice) resistant - outdoors
Type 3R- Rainproof and sleet (ice) resistant - outdoors
Type 3S- Rainproof and sleet (ice) resistant - outdoors
Type 4 - Water-tight and dust-tight - indoors and outdoors
Type 4X - Water-tight and Dust-tight - indoors and outdoors
Type 5 - Dust-tight and dirt resistant - indoors
Type 6 - Waterproof - indoors and outdoors
Type 6P - Waterproof submersible - indoors and outdoors
Type 7 - Hazardous locations - class I
Type 8 - Hazardous locations - class I
Type 9 - Hazardous locations - class II
Type 10 - Hazardous locations - class II
Type 11 - Gas-tight, water-tight, oiltight - indoors
Type 12 - Dust-tight and drip-tight - indoors
Type 13 - Oil-tight and dust-tight - indoors

Factor	1	2	3	3R	3S	4	4X	5	6	6P	11	12	12K	13
Prevent human contact	x	x	x	x	x	x	x	x	x	x	x	x	x	x
falling dirt	x	x	x	x	x	x	x	x	x	x	x	x	x	x
liquid drop/light splash		x				x	x		x	x	x	x	x	x
airborne dust/particles						x	x	x	x	x		x	x	x
wind blown dust			x		x	x	x		x	x				
liquid heavy stream/splash						x	x		x	x				
oil/coolant seepage												x	x	x
oil/coolant spray/splash														x
corrosive environment							x			x	x			
temporarily submerged									x	x				
prolonged submersion										x				

Figure 29.11   **NEMA Enclosures**

29.1.6 Wire and Cable Grouping

In a controls cabinet the conductors are passed through channels or bundled. When dissimilar conductors are run side-by-side problems can arise. The basic categories of conductors are shown in Figure 29.1. In general category 1 conductors should not be grouped with other conductor categories. Care should be used when running category 2 and 3 conductors together.

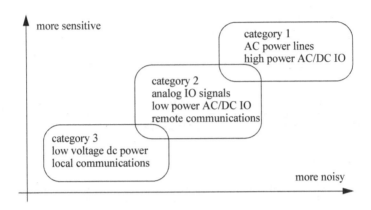

Figure 29.12   ***Wire and Cable Categories***

29.2 FAIL-SAFE DESIGN

All systems will fail eventually. A fail-safe design will minimize the damage to people and equipment. Consider the selection electrical connections. If wires are cut or connections fail, the equipment should still be safe. For example, if a normally closed stop button is used, and the connector is broken, it will cause the machine to stop as if the stop button has been pressed.

NO (Normally open) - When wiring switches or sensors that start actions, use normally open switches so that if there is a problem the process will not start.

NC (Normally Closed) - When wiring switches that stop processes use normally closed so that if they fail the process will stop. E-Stops must always be NC, and they must cut off the master power, not just be another input to the PLC.

Hardware
- Use redundancy in hardware.
- Directly connect emergency stops to the PLC, or the main power supply.
- Use well controlled startup procedures that check for problems.
- Shutdown buttons must be easily accessible from all points around the machine.

29.3 SAFETY RULES SUMMARY

A set of safety rules was developed by Jim Rowell (http://www.mrplc.com, "Industrial Control Safety; or How to Scare the Bejesus Out of Me"). These are summarized below.

Grounding and Fuses
- Always ground power supplies and transformers.
- Ground all metal enclosures, casings, etc.
- All ground connections should be made with dedicated wires that are exposed so that their presence is obvious.
- Use fuses for all AC power lines, but not on the neutrals or grounds.
- If ground fault interrupts are used they should respond faster than the control system.

Hot vs. Neutral Wiring
- Use PNP wiring schemes for systems, especially for inputs that can initiate actions.
- Loads should be wired so that the ground/neutral is always connected, and the power is

switched.
- Sourcing and sinking are often confused, so check the diagrams or look for PNP/NPN markings.

AC / DC
- Use lower voltages when possible, preferably below 50V.
- For distant switches and sensors use DC.

Devices
- Use properly rated isolation transformers and power supplies for control systems. Beware autotransformers.
- Use Positive or Force-Guided Relays and contacts can fail safely and prevent operation in the event of a failure.
- Some 'relay replacement' devices do not adequately isolate the inputs and output and should not be used in safety critical applications.

Starts
- Use NO buttons and wiring for inputs that start processes.
- Select palm-buttons, and other startup hardware carefully to ensure that they are safety rated and will ensure that an operator is clear of the machine.
- When two-hand start buttons are used, use both the NO and NC outputs for each button. The ladder logic can then watch both for a completed actuation.

Stops
- E-stop buttons should completely halt all parts of a machine that are not needed for safety.
- E-stops should be hard-wired to kill power to electrically actuated systems.
- Use many red mushroom head E-stop buttons that are easy to reach.
- Use red non-mushroom head buttons for regular stops.
- A restart sequence should be required after a stop button is released.
- E-stop buttons should release pressure in machines to allow easy 'escape'.
- An 'extraction procedure' should be developed so that trapped workers can be freed.
- If there are any power storage devices (such as a capacitor bank) make sure they are disabled by the E-stops.
- Use NC buttons and wiring for inputs that stop processes.
- Use guards that prevent operation when unsafe, such as door open detection.
- If the failure of a stop input could cause a catastrophic failure, add a backup.

Construction
- Wire so that the power enters at the top of a device.
- Take special care to review regulations when working with machines that are like presses or brakes.
- Check breaker ratings for overload cases and supplemental protection.
- A power disconnect should be located on or in a control cabinet.
- Wires should be grouped by the power/voltage ratings. Run separate conduits or race-ways for different voltages.
- Wire insulation should be rated for the highest voltage in the cabinet.
- Use colored lights to indicate operational states. Green indicates in operation safely, red indicates problems.
- Construct cabinets to avoid contamination from materials such as oils.
- Conduits should be sealed with removable compounds if they lead to spaces at different temperatures and humidity levels.
- Position terminal strips and other components above 18" for ergonomic reasons.
- Cabinets should be protected with suitably rated fuses.
- Finger sized objects should not be able to reach any live voltages in a finished cabinet, however DMM probes should be able to measure voltages.

29.4 SUMMARY

- Electrical schematics used to layout and wire controls cabinets.
- JIC wiring symbols can be used to describe electrical components.
- Grounding and shielding can keep a system safe and running reliably.
- Failsafe designs ensure that a controller will cause minimal damage in the event of a failure.
- PLC enclosure are selected to protect a PLC from its environment.

29.5 PRACTICE PROBLEMS

(Note: Problem solutions are available at http://sites.google.com/site/automatedmanufacturingsystems/)

1. What steps are required to replace a defective PLC?

2. What are the trade-offs between 3-phase and single-phase AC power.

29.6 ASSIGNMENT PROBLEMS

1. Where is the best location for a PLC enclosure?

2. What is a typical temperature and humidity range for a PLC?

3. Draw the electrical schematic and panel layout for the relay logic below. The system will be connected to 3 phase power. Be sure to include a master power disconnect.

4. Why are nodes and wires labelled on a schematic, and in the controls cabinet?

5. Locate at least 10 JIC symbols for the sensors and actuators in earlier chapters.

6. How are shielding and grounding alike? Are shields and grounds connected?

7. What are significant grounding problems?

8. Why should grounds be connected in a tree configuration?

29.1 PRACTICE PROBLEM SOLUTIONS

1. in a rack the defective card is removed and replaced. If the card has wiring terminals these are removed first, and connected to the replacement card.

2. 3-phase power is ideal for large loads such as motors. Single phase power is suited to small loads, and the power usage on each phase must be balanced someplace on the electrical grid.

30 SOFTWARE ENGINEERING

Topics:

* Electrical wiring issues; cabinet wiring and layout, grounding, shielding and inductive loads
* Controller design; failsafe, debugging, troubleshooting, forcing
* Process modelling with the ANSI/ISA-S5.1-1984 standard
* Programming large systems
* Documentation

Objectives:

* To learn the major issues in program design.
* Be able to document a process with a process diagram.
* Be able to document a design project.
* Be able to develop a project strategy for large programs.

A careful, structured approach to designing software will cut the total development time, and result in a more reliable system.

30.1 FAIL SAFE DESIGN

It is necessary to predict how systems will fail. Some of the common problems that will occur are listed below.

Component jams - An actuator or part becomes jammed. This can be detected by adding sensors for actuator positions and part presence.

Operator detected failure - Some unexpected failures will be detected by the operator. In those cases the operator must be able to shut down the machine easily.

Erroneous input - An input could be triggered unintentionally. This could include something falling against a start button.

Unsafe modes - Some systems need to be entered by the operators or maintenance crew. People detectors can be used to prevent operation while people are present.

Programming errors - A large program that is poorly written can behave erratically when an unanticipated input is encountered. This is also a problem with assumed startup conditions.

Sabotage - For various reasons, some individuals may try to damage a system. These problems can be minimized preventing access.

Random failure - Each component is prone to random failure. It is worth considering what would happen if any of these components were to fail.

Some design rules that will help improve the safety of a system are listed below.

Programs
* A fail-safe design - Programs should be designed so that they check for problems, and shut down in safe ways. Most PLC's also have imminent power failure sensors, use these whenever danger is present to shut down the system safely.
* Proper programming techniques and modular programming will help detect possible problems on paper instead of in operation.
* Modular well designed programs.
* Use predictable, non-configured programs.
* Make the program inaccessible to unauthorized persons.
* Check for system OK at start-up.
* Use PLC built in functions for error and failure detection.

People
* Provide clear and current documentation for maintenance and operators.
* Provide training for new users and engineers to reduce careless and uninformed mistakes.

30.2 DEBUGGING

Most engineers have taken a programming course where they learned to write a program and then debug it. Debugging involves running the program, testing it for errors, and then fixing them. Even for an experienced programmer it is common to spend more time debugging than writing software. For PLCs this is not acceptable! If you are running the program and it is operating irrationally it will often damage hardware. Also, if the error is not obvious, you should go back and reexamine the program design. When a program is debugged by trial and error, there are probably errors remaining in the logic, and the program is very hard to trust. Remember, a bug in a PLC program might kill somebody.

Note: when running a program for the first time it can be a good idea to keep one hand on the E-stop button.

30.2.1 Troubleshooting

After a system is in operation it will eventually fail. When a failure occurs it is important to be able to identify and solve problems quickly. The following list of steps will help track down errors in a PLC system.

1. Look at the process and see if it is in a normal state. i.e. no jammed actuators, broken parts, etc. If there are visible problems, fix them and restart the process.
2. Look at the PLC to see which error lights are on. Each PLC vendor will provide documents that indicate which problems correspond to the error lights. Common error lights are given below. If any off the warning lights are on, look for electrical supply problems to the PLC.
HALT - something has stopped the CPU
RUN - the PLC thinks it is OK (and probably is)
ERROR - a physical problem has occurred with the PLC
3. Check indicator lights on I/O cards, see if they match the system. i.e., look at sensors that are on/off, and actuators on/off, check to see that the lights on the PLC I/O cards agree. If any of the light disagree with the physical reality, then interface electronics/mechanics need inspection.
4. Consult the manuals, or use software if available. If no obvious problems exist the problem is not simple, and requires a technically skilled approach.
5. If all else fails call the vendor (or the contractor) for help.

30.2.2 Forcing

Most PLCs will allow a user to *force* inputs and outputs. This means that they can be turned on, regardless of the physical inputs and program results. This can be convenient for debugging programs, and, it makes it easy to break and destroy things! When forces are used they can make the program perform erratically. They can also make outputs occur out of sequence. If there is a logic problem, then these don't help a programmer identify these problems.

Many companies will require extensive paperwork and permissions before forces can be used. I don't recommend forcing inputs or outputs, except in the most extreme circumstances.

30.3 PROCESS MODELING

There are many process modeling techniques, but only a few are suited to process control. The ANSI/ISA-S5.1-1984 Piping and Instrumentation Diagram (P&ID) standard provides good tools for documenting processes. The symbols used on the diagrams are shown in Figure 30.1. Note that the modifier used for the instruments can be applied to other discrete devices.

Discrete Device Symbols

Instruments

    field mounted

    panel mounted

    unaccessible or embedded

    auxilliary location, operator accessible

Controls

    Computer Function

    PLC

    Shared Display/Control

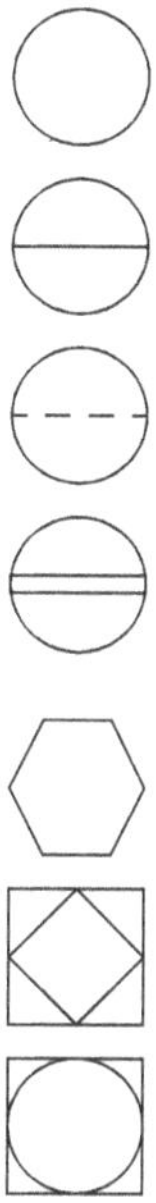

Figure 30.1   *Symbols for Functions and Instruments*

The process model is carefully labeled to indicate the function of each of the function on the diagram. Table 2 shows a list of the different instrumentation letter codes.

Table 1: ANSI/ISA-S5.1-1984 Instrumentation Symbols and Identification

LETTER	FIRST LETTER	SECOND LETTER
A	Analysis	Alarm
B	Burner, Combustion	User's Choice
C	User's Choice	Control
D	User's Choice	
E	Voltage	Sensor (Primary Element)
F	Flow Rate	
G	User's Choice	Glass (Sight Tube)
H	Hand (Manually Initiated)	
I	Current (Electric)	Indicate
J	Power	
K	Time or Time Schedule	Control Station
L	Level	Light (pilot)

Table 1: ANSI/ISA-S5.1-1984 Instrumentation Symbols and Identification

LETTER	FIRST LETTER	SECOND LETTER
M	User's Choice	
N	User's Choice	User's Choice
O	User's Choice	Orifice, Restriction
P	Pressure, Vacuum	Point (Test Connection)
Q	Quantity	
R	Radiation	Record or Print
S	Speed or Frequency	Switch
T	Temperature	Transmit
U	Multivariable	Multifunction
V	Vibration, Mechanical Analysis	Valve, Damper, Louver
W	Weight, Force	Well
X	Unclassified	Unclassified
Y	Event, State or Presence	Relay, Compute
Z	Position, Dimension	Driver, Actuator, Unclassified

The line symbols also describe the type of flow. Figure 30.1 shows a few of the popular flow lines.

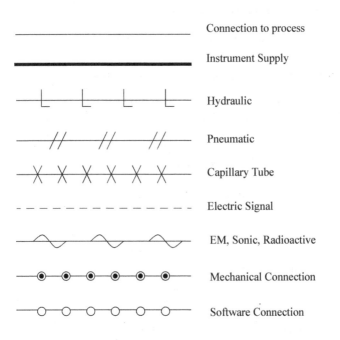

	Connection to process
	Instrument Supply
	Hydraulic
	Pneumatic
	Capillary Tube
	Electric Signal
	EM, Sonic, Radioactive
	Mechanical Connection
	Software Connection

Figure 30.2   **Flow Line Symbols and Types**

Figure 30.1 shows some of the more popular sensor and actuator symbols.

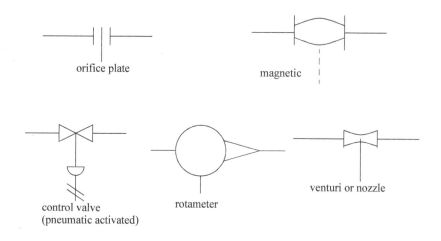

Figure 30.3   *Sensor and Actuator Symbols and Types*

30.4 PROGRAMMING FOR LARGE SYSTEMS

Previous chapters have explored design techniques to solve large problems using techniques such as state diagrams and SFCs. Large systems may contain hundreds of those types of problems. This section will attempt to lay a philosophical approach that will help you approach these designs. The most important concepts are clarity and simplicity.

30.4.1 Developing a Program Structure

Understanding the process will simplify the controller design. When the system is only partially understood, or vaguely defined the development process becomes iterative. Programs will be developed, and modified until they are acceptable. When information and events are clearly understood the program design will become obvious. Questions that can help clarify the system include;

> "What are the inputs?"
> "What are the outputs?"
> "What are the sequences of inputs and outputs?"
> "Can a diagram of the system operation be drawn?"
> "What information does the system need?"
> "What information does the system produce?"

When possible a large controls problems should be broken down into smaller problems. This often happens when parts of the system operate independent of each other. This may also happen when operations occur in a fixed sequence. If this is the case the controls problem can be divided into the two smaller (and simpler) portions. The questions to ask are;

> "Will these operations ever occur at the same time?"
> "Will this operation happen regardless of other operations?"
> "Is there a clear sequence of operations?"
> "Is there a physical division in the process or machine?"

After examining the system the controller should be broken into operations. This can be done with a tree structure as shown in Figure 30.1. This breaks control into smaller tasks that need to be executed. This technique is only used to divide the programming tasks into smaller sections that are distinct.

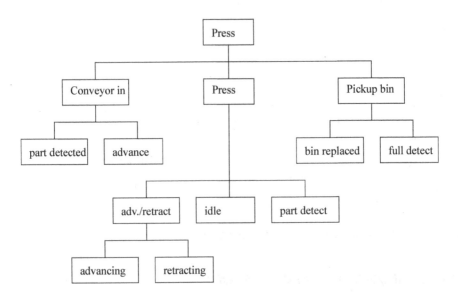

Figure 30.4   *Functional Diagram for Press Control*

Each block in the functional diagram can be written as a separate subroutine. A higher level *executive* program will call these subroutines as needed. The executive program can also be broken into smaller parts. This keeps the main program more compact, and reduces the overall execution time. And, because the subroutines only run when they should, the change of unexpected operation is reduced. This is the method promoted by methods such as SFCs and FBDs.

Each functional program should be given its' own block of memory so that there are no conflicts with shared memory. System wide data or status information can be kept in common areas. Typical examples include a flag to indicate a certain product type, or a recipe oriented system.

Testing should be considered during software planning and writing. The best scenario is that the software is written in small pieces, and then each piece is tested. This is important in a large system. When a system is written as a single large piece of code, it becomes much more difficult to identify the source of errors.

The most disregarded statement involves documentation. All documentation should be written when the software is written. If the documentation can be written first, the software is usually more reliable and easier to write. Comments should be entered when ladder logic is entered. This often helps to clarify thoughts and expose careless errors. Documentation is essential on large projects where others are likely to maintain the system. Even if you maintain it, you are likely to forget what your original design intention was.

Some of the common pitfalls encountered by designers on large projects are listed below.

- Amateur designers rush through design to start work early, but details they missed take much longer to fix when they are half way implemented.
- Details are not planned out and the project becomes one huge complex task instead of groups of small simple tasks.
- Designers write one huge program, instead of smaller programs. This makes proof reading much harder, and not very enjoyable.

- Programmers sit at the keyboard and debug by trial and error. If a programmer is testing a program and an error occurs, there are two possible scenarios. First, the programmer knows what the problem is, and can fix it immediately. Second, the programmer only has a vague idea, and often makes no progress doing trial-and-error debugging. If trial-and-error programming is going on the program is not understood, and it should be fixed through replanning.
- Small details are left to be completed later. These are sometimes left undone, and lead to failures in operation.
- The design is not frozen, and small refinements and add-ons take significant amounts of time, and often lead to major design changes.
- The designers use unprofessional approaches. They tend to follow poor designs, against the advice of their colleagues. This is often because the design is their *child*
- Designers get a good dose of the *not invented here* syndrome. Basically, if we didn't develop it, it must not be any good.
- Limited knowledge will cause problems. The saying goes "If the only tool you know how to use is a hammer every problem looks like a nail."
- Biting off more than you can chew. some projects are overly ambitious. Avoid adding wild extras, and just meet the needs of the project. Sometimes an unnecessary extra can take more time than the rest of the project.

30.4.2 Program Verification and Simulation

After a program has been written it is important to verify that it works as intended, before it is used in production. In a simple application this might involve running the program on the machine, and looking for improper operation. In a complex application this approach is not suitable. A good approach to software development involves the following steps in approximate order:

1. Structured design - design and write the software to meet a clear set of objectives.
2. Modular testing - small segments of the program can be written, and then tested individually. It is much easier to debug and verify the operation of a small program.
3. Code review - review the code modules for compliance to the design. This should be done by others, but at least you should review your own code.
4. Modular building - the software modules can then be added one at a time, and the system tested again. Any problems that arise can then be attributed to interactions with the new module.
5. Design confirmation - verify that the system works as the design requires.
6. Error proofing - the system can be tested by trying expected and unexpected failures. When doing this testing, irrational things should also be considered. This might include unplugging sensors, jamming actuators, operator errors, etc.
7. Burn-in - a test that last a long period of time. Some errors won't appear until a machine has run for a few thousand cycles, or over a period of days.

Program testing can be done on machines, but this is not always possible or desireable. In these cases simulators allow the programs to be tested without the actual machine. The use of a simulator typically follows the basic steps below.

1. The machine inputs and outputs are identified.
2. A basic model of the system is developed in terms of the inputs and outputs. This might include items such as when sensor changes are expected, what effects actuators should have, and expected operator inputs.
3. A system simulator is constructed with some combination of specialized software and hardware.
4. The system is verified for the expect operation of the system.
5. The system is then used for testing software and verifying the operation.

A detailed description of simulator usage is available [Kinner, 1992].

30.5 DOCUMENTATION

Poor documentation is a common complaint lodged against control system designers. Good documentation is developed as a project progresses. Many engineers will leave the documentation to the end of a project as an afterthought. But, by that point many of the details have been forgotten. So, it takes longer to recall the details of the work, and the report is always lacking.

A set of PLC design forms are given in Figure 30.1 to Figure 30.1. These can be used before, during and after a controls project. These forms can then be kept in design or maintenance offices so that others can get easy access and make updates at the controller is changed. Figure 30.1 shows a design cover page. This should be completed with information such as a unique project name, contact person, and controller type. The list of changes below help to track design, redesign and maintenance that has been done to the machine. This cover sheet acts as a quick overview on the history of the machine. Figure 30.1 to Figure 30.1 show sheets that allow free form planning of the design. Figure 30.1 shows a sheet for planning the input and output memory locations. Figure 30.1 shows a sheet for planning internal memory locations, and finally Figure 30.1 shows a sheet for planning the ladder logic. The sheets should be used in the order they are given, but they do not all.need to be used. When the system has been built and tested, a copy of the working ladder logic should be attached to the end of the bundle of pages.

PLC Project Sheet

Project ID:_____

Start Date: _____

Contact Person: _____

PLC Model: _____

Attached Materials/Revisions:

Date	Name	# Sheets	Reason

Figure 30.5   **Design Cover Page**

Project Notes

Project ID:_____   Date:_____

☐ System Description   Page_____of_____
☐ I/O Notes
☐ Power Notes   Name: _____
☐ Other Notes_____

Figure 30.6   **Project Note Page**

Design Notes

Project ID:_____ Date:_____

☐ State Diagram   ☐ Truth Table   Page_____of_____
☐ Flow Chart   ☐ Safety
☐ Sequential Function Chart ☐ Communications   Name: _____
☐ Boolean Equations   ☐ Other Notes_____

Figure 30.7   ***Project Diagramming Page***

Application Notes

Project ID:_____ Date:_____

☐ Test Plan
☐ Electrical I/O
☐ PLC Modules
☐ Other Notes _____

Page_____of_____

Name:_____

Figure 30.8   Project Diagramming and Notes Page

Input/Output Card   Page ____of____

Project I.D. _____ Name_____ Date_____

Card Type _____ Rack # _____ Slot # _____

Notes:

input/output	JIC symbol	Description

Vin

00

01

02

03

04

05

06

07

08/10

09/11

10/12

11/13

12/14

13/15

14/16

15/17

com

Figure 30.9   *IO Planning Page*

Internal Locations

Page_____ of_____

Project I.D. _____ Name _____ Date _____

Register or Word / Internal Word	Description

Figure 30.10   *Internal Memory Locations Page*

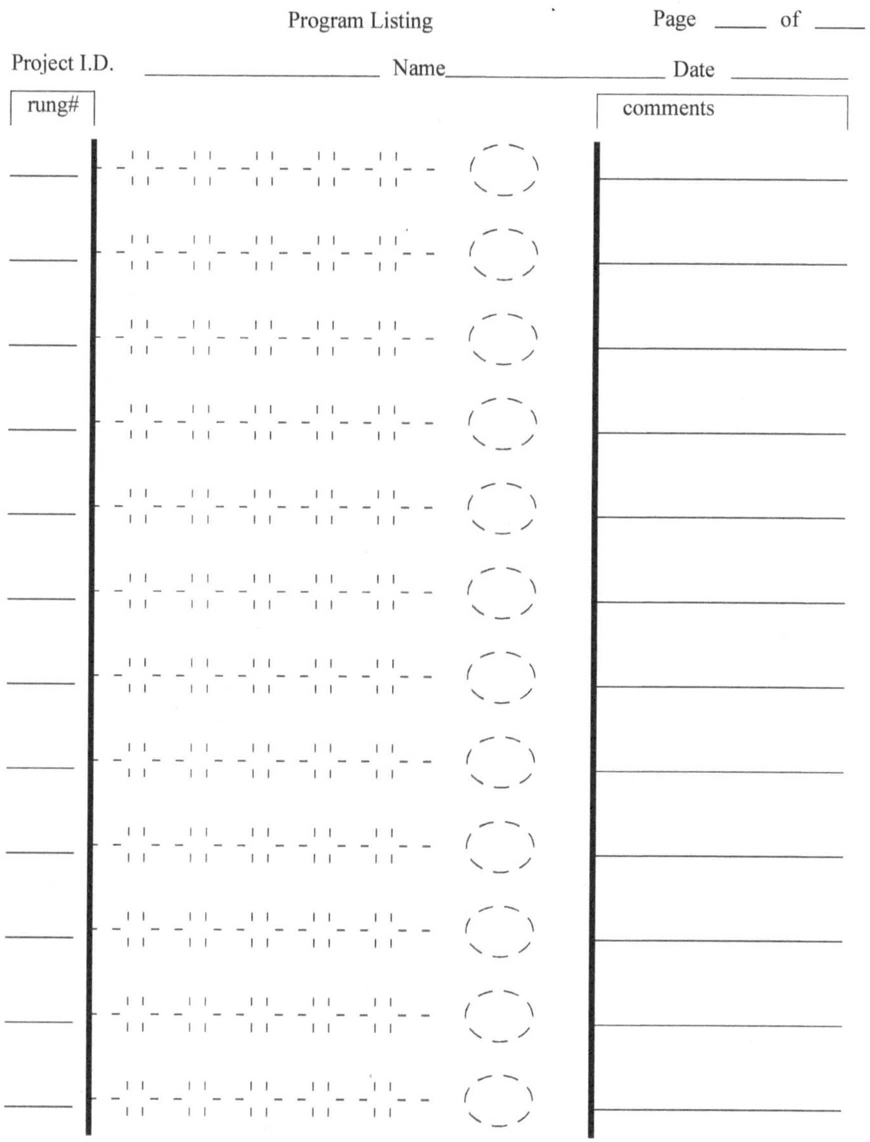
Program Listing
Page ____ of ____
Project I.D.
Name
Date
rung#
comments

Figure 30.11 ***Ladder Logic Page***

These design sheets are provided as examples. PLC vendors often supply similar sheets. Many companies also have their own internal design documentation procedures. If you are in a company without standardized design formats, you should consider implementing such a system.

30.6 COMMISIONING

When a new machine is being prepared for production, or has been delivered by a supplier, it is normal to go through a set of commissioning procedures. Some typical steps are listed below.

1. Visual inspection
    • verify that the machine meets internal and external safety codes
        - electrical codes
        - worker safety codes (e.g., OSHA)
    • determine if all components are present
2. Mechanical installation
    • physically located the machine
    • connect to adjacent machines
    • connect water, air and other required services
3. Electrical installation
    • connect grounds and power
    • high potential and ground fault tests
    • verify sensor inputs to the PLC
4. Functional tests
    • start the machine and test the emergency stops
    • test for basic functionality
5. Process verification
    • run the machine and make adjustments to produce product
    • collect process capability data
    • determine required maintenance procedures
6. Contract/specification verification
    • review the contact requirements and check off individually on each one
    • review the specification requirements and check off each one individually
    • request that any non-compliant requirements are corrected
7. Put into production
    • start the process in the production environment and begin normal use

30.7 SAFETY

Previous chapters have discussed safety as an inherent part of the design process - AS IT MUST BE. These include elements such as making stops normally closed, starts normally open, systems return to safe inherently, programs that are predictable, etc. There are a variety of systems for evaluating and managing safety issues from a range of groups including NASA, DOD, and many more.

30.7.1 IEC 61508/61511 safety standards

These standards cover electrical, electronic, and programmable electronic systems. Three categories of software languages covered by the standard

FPL (Fixed Programming Language) - a very limited approach to programming. For example the system is programmed by setting parameters.
LVL (Limited Variability Language) - a language with a strict programming model, such as ladder logic.
FVL (Full Variability Language) - a language that gives full access to a systems, such as C.

Dangers can be estimated using a Safety Integrity Level (SIL) - the safety requirements for a function in a system. These estimates of failure can be calculated using individual component reliability, historical data, and some careful consideration. Typical probabilities of failure by type of system are given below.

Low demand:
    level 4: $P = 10^{-4}$ to 10^{-5}
    level 3: $P = 10^{-3}$ to 10^{-4}
    level 2: $P = 10^{-2}$ to 10^{-3}
    level 1: $P = 10^{-1}$ to 10^{-2}
High Demand or continuous mode:
    level 4: $P = 10^{-8}$ to 10^{-9}
    level 3: $P = 10^{-7}$ to 10^{-8}
    level 2: $P = 10^{-6}$ to 10^{-7}

System safety levels are defined with one of the three categories below.

SIF (Safety Instrumented Function) - The SIL for each function is chosen to ensure an overall system functionality.

SIS (Safety Instrumented System) - A combined system with one or more logic processors.

SFF (Safe Failure Fraction) - the ratio of safe and dangerous detected failures to the total failures.

To calculate the SFF and/or overall system failure probability there is a need to refer to the relevant statistical theory. Some of the common approaches are listed.

- do an FMEA for each system component in the system
- classify failure modes as safe or dangerous
- calculate the probabilities of safe/dangerous failures (S/D [0, 1])
- estimate the fraction of the failures that can be detected (F [0, 1])

30.8 LEAN MANUFACTURING

Lean manufacturing focuses on the elimination of waste, formalizing a common source approach to machine design. Some general concepts to use when designing lean machines include,

- setups should be minimized or eliminated
- product changeovers should be minimized or eliminated
- make the tool fit the job, not the other way. If necessary, design a new tool
- design the machine be faster than the needed cycle time to allow flexibility and excess capacity - this does seem contradictory, but it allows better use of other resources. For example, if a worker takes a bathroom break, the production can continue with fewer workers.
- allow batches with a minimum capacity of one.
- people are part of the process and should integrate smoothly - the motions or workers are often described as dance like.
    - eliminate wasted steps, all should go into making the part
    - work should flow smoothly to avoid wasted motion
    - do not waste motion by spacing out machines
- make-one, check-one
- design "decouplers" to allow operations to happen independently.
- eliminate material waste that does not go into the product
- pull work through the cell
- design the product so that it is easy to manufacture
- use methods that are obvious, so that anybody can understand - this makes workers portable and able to easily cover for others.
- use poke-yoke
- design tools to reduce the needs for guards.

To achieve these principles the complexity of a machine may be increased to decrease the manual labor and setup issues. In lean manufacturing the ideal batch size is one with no switchover time.

30.9 SUMMARY

• Debugging and forcing are signs of a poorly written program.
• Process models can be used to completely describe a process.
• When programming large systems, it is important to subdivide the project into smaller parts.
• Documentation should be done at all phases of the project.

30.9.1 References

Eberhard, A, "Safety in Programmable Applications", Automation World, Nov., 2004.

Kenner, R. H., "The Use of Simulation Within a PLC to Improve Program Development and Testing", Proceedings of the First Automation Fair, Philadelphia, 1990.

McCrea-Steele, R., "Proven-in-Use: Making the right choices for process safety", Hotlinks, Invensys, Summer 2003.

Paques, Joseph-Jean, "Basic Safety Rules for Using Programmable Controllers", ISA Transactions, Vol. 29, No. 2, 1990.

30.10 PRACTICE PROBLEMS

(Note: Problem solutions are available at http://sites.google.com/site/automatedmanufacturingsystems/)

1. List 5 advantages of using structured design and documentation techniques.

30.11 ASSIGNMENT PROBLEMS

1. What documentation is requires for a ladder logic based controller? Are comments important? Why?

2. When should inputs and outputs be assigned when planning a control system?

3. Discuss when I/O placement and wiring documentation should be updated?

4. Should you use output forces?

5. Find web addresses for 10 PLC vendors. Investigate their web sites to determine how they would be as suppliers.

30.1 PRACTICE PROBLEM SOLUTIONS

1. more reliable programs - less debugging time - more routine - others can pick up where you left off - reduces confusion

31SELECTING A PLC

Topics:

- The PLC selection process
- Estimating program memory and time requirements
- Selecting hardware

Objectives:

- Be able to select a hardware and software vendor.
- Be able to size a PLC to an application
- Be able to select needed hardware and software.

After the planning phase of the design, the equipment can be ordered. This decision is usually based upon the required inputs, outputs and functions of the controller. The first decision is the type of controller; rack, mini, micro, or software based. This decision will depend upon the basic criteria listed below.

- Number of logical inputs and outputs.
- Memory - Often 1K and up. Need is dictated by size of ladder logic program. A ladder element will take only a few bytes, and will be specified in manufacturers documentation.
- Number of special I/O modules - When doing some exotic applications, a large number of special add-on cards may be required.
- Scan Time - Big programs or faster processes will require shorter scan times. And, the shorter the scan time, the higher the cost. Typical values for this are 1 microsecond per simple ladder instruction
- Communications - Serial and networked connections allow the PLC to be programmed and talk to other PLCs. The needs are determined by the application.
- Software - Availability of programming software and other tools determines the programming and debugging ease.

The process of selecting a PLC can be broken into the steps listed below.

1. Understand the process to be controlled (Note: This is done using the design sheets in the previous chapter).
    - List the number and types of inputs and outputs.
    - Determine how the process is to be controlled.
    - Determine special needs such as distance between parts of the process.
2. If not already specified, a single vendor should be selected. Factors that might be considered are, (Note: Vendor research may be needed here.)
    - Manuals and documentation
    - Support while developing programs
    - The range of products available
    - Support while troubleshooting
    - Shipping times for emergency replacements
    - Training
    - The track record for the company
    - Business practices (billing, upgrades/obsolete products, etc.)
3. Plan the ladder logic for the controls. (Note: Use the standard design sheets.)
4. Count the program instructions and enter the values into the sheets in Figure 31.1 and Figure 31.1. Use the instruction times and memory requirements for each instruction to determine if the PLC has sufficient memory, and if the response time will be adequate for the process. Samples of scan times and memory are given in Figure 31.1 and Figure 31.1.

PLC MEMORY TIME ESTIMATES - Part A

Project ID:_____

Name: _____

Date: _____

Instruc- tion Type	Time Max (us)	Time Min. (us)	Instruc- tion Memory (words)	Instruc- tion Data (words)	Instruc- tion Count (number)	Total Memory (words)	Min. Time (us)	Max. Time (us)
contacts								
outputs								
timers								
					Total			

Figure 31.1   **Memory and Time Tally Sheet**

PLC MEMORY TIME REQUIREMENTS - Part B

Project ID:_____

Name: _____

Date: _____

TIME

Input Scan Time	_____us	
Output Scan Time	_____us	
Overhead Time	_____us	
Program Scan Time	_____us	
Communication Time	_____us	
Other Times	_____us	
TOTAL		_____us

MEMORY

Total Memory	_____words	
Other Memory	_____words	
TOTAL	_____words	_____bytes

Figure 31.2   ***Memory and Timer Requirement Sheet***

5. Look for special program needs and check the PLC model. (e.g. PID)
6. Estimate the cost for suitable hardware, programming software, cables, manuals, training, etc., or ask for a quote from a vendor.

Typical values for an Allen-Bradley micrologix controller are,
input scan time 8us
output scan times 8us
housekeeping 180us
overhead memory for controller 280 words

Instruction Type	Time Max (us)	Time Min. (us)	Instruction Memory (words)	Instruction Data (words)
CTD - count down	27.22	32.19	1	3
CTU- count up	26.67	29.84	1	3
XIC - normally open contact	1.72	1.54	.75	0
XIO - normally closed contact	1.72	1.54	.75	0
OSR - one shot relay	11.48	13.02	1	0
OTE - output enable	4.43	4.43	.75	0
OTL - output latch	3.16	4.97	.75	0
OTU - output unlatch	3.16	4.97	.75	0
RES - reset	4.25	15.19	1	0
RTO - retentive on time	27.49	38.34	1	3
TOF - off timer	31.65	39.42	1	3
TON - on timer	30.38	38.34	1	3

Figure 31.3   *Typical Instruction Times and Memory Usage for a Micrologix Controller*

Typical values for an Allen-Bradley PLC-5 controller are,
input scan time ?us
output scan times ?us
housekeeping ?us
overhead memory for controller ? words

Instruction Type	Time Max (us)	Time Min. (us)	Instruction Memory (words)	Instruction Data (words)
CTD - count down	3.3	3.4	3	3
CTU- count up	3.4	3.4	3	3
XIC - normally open contact	0.32	0.16	1	0
XIO - normally closed contact	0.32	0.16	1	0
OSR - one shot relay	6.2	6.0	6	0
OTE - output enable	0.48	0.48	1	0
OTL - output latch	0.48	0.16	1	0
OTU - output unlatch	0.48	0.16	1	0
RES - reset	2.2	1.0	3	0
RTO - retentive on time	4.1	2.4	3	3
TOF - off timer	2.6	3.2	3	3
TON - on timer	4.1	2.6	3	3

Figure 31.4   *Typical Instruction Times and Memory Usage for a PLC-5 Controller*

31.1 SPECIAL I/O MODULES

Many different special I/O modules are available. Some module types are listed below for illustration, but the commercial selection is very large. Generally most vendors offer competitive modules. Some modules, such as fuzzy logic and vision, are only offered by a few supplier, such as Omron. This may occasionally drive a decision to purchase a particular type of controller.

PLC CPU's
- A wide variety of CPU's are available, and can often be used interchangeably in the rack systems. the basic formula is price/performance. The table below compares a few CPU units in various criteria.

PLC / FEATURE	Siemens S5-90U	Siemens S5-100U	Siemens S5-115U (CPU 944)	Siemens CPU03	AEG PC-A984-145
RAM (KB)	4	<= 20	96	20	8
Scan times (us) per basic instruc. overhead			0.8 2000		5
Package Power Supply Maximum Cards Maximum Racks Maximum Drops Distance	mini-module 24 VDC 6 with addon N/A	mini-module 24 VDC	card 24 VDC 2.5m or 3km	card 115/230VAC	
Counters Timers Flags			128 128 2048		
I/O - Digital on board maximum I/O - Analog on board maximum	16 208 0 16	0 448 0 32	0 1024 0 64	0 256 0 32	0 256
Communication network line human other	Sinec-L1	Sinec-L1	Sinec-L1, prop. printer, ASCII	Sinec-L1	Modbus/Modubs+
Functions PID			option	option	option

Legend:
prop. - proprietary technology used by a single vendor
option - the vendor will offer the feature at an additional cost

Figure 31.5   ***CPU Comparison Chart***

Programmers
- There are a few basic types of programmers in use. These tend to fall into 3 categories,
  1. PLC Software for Personal Computers - Similar to the specialized programming units, but the software runs on a multi-use, user supplied computer. This approach is typically preferred.

593

2. Hand held units (or integrated) - Allow programming of PLC using a calculator type interface. Often done using mnemonics.
3. Specialized programming units - Effectively a portable computer that allows graphical editing of the ladder logic, and fast uploading/downloading/monitoring of the PLC.

Ethernet/modem
- For communication with remote computers. This is now an option on many CPUs.

TTL input/outputs
- When dealing with lower TTL voltages (0-5Vdc) most input cards will not recognize these. These cards allow switching of these voltages.

Encoder counter module
- Takes inputs from an encoder and tracks position. This allows encoder changes that are much faster than the PLC can scan.

Human Machine Interface (HMI)
- A-B/Siemens/Omron/Modicon/etc offer human interface systems. The user can use touch screens, screen and buttons, LCD/LED and a keypad.

ASCII module
- Adds an serial port for communicating with standard serial ports RS-232/422.

IBM PC computer cards
- An IBM compatible computer card that plugs into a PLC bus, and allows use of common software.
- For example, Siemens CP580 the Simatic AT;
  - serial ports: RS-232C, RS-422, TTY
  - RGB monitor driver (VGA)
  - keyboard and mouse interfaces
  - 3.5" disk

Counters
- Each card will have 1 to 16 counters at speeds up to 200KHz.
- The counter can be set to zero, or up/down, or gating can occur with an external input.

Thermocouple
- Thermocouples can be used to measure temperature, but these low voltage devices require sensitive electronics to get accurate temperature readings.

Analog Input/Output
- These cards measure voltages in various ranges, and allow monitoring of continuous processes. These cards can also output analog voltages to help control external processes, etc.

PID modules
- There are 2 types of PID modules. In the first the CPU does the calculation, in the second, a second controller card does the calculation.
  - when the CPU does the calculation the PID loop is slower.
  - when a specialized card controls the PID loop, it is faster, but it costs less.
- Typical applications - positioning workpieces.

Stepper motor
- Allows control of a stepper motor from a PLC rack.

Servo control module
- Has an encoder and amplifier pair built in to the card.

Diagnostic Modules
- Plug in and they monitor the CPU status.

Specialty cards for IBM PC interface
- Siemens/Allen-Bradley/etc. have cards that fit into IBM buses, and will communicate with PLC's.

Communications
- This allows communications or networks protocols in addition to what is available on the PLC. This includes DH+, etc.

Thumb Wheel Module
- Numbers can be dialed in on wheels with digits from 0 to 9.

BCD input/output module
- Allows numbers to be output/input in BCD.

BASIC module
- Allows the user to write programs in the BASIC programming language.

Short distance RF transmitters
- e.g., Omron V600/V620 ID system
- ID Tags - Special "tags" can be attached to products, and as they pass within range of pickup sensors, they transmit an ID number, or a packet of data. This data can then be used, updated, and rewritten to the tags by the PLC. Messages are stored as ASCII text.

594

Voice Recognition/Speech
- In some cases verbal I/O can be useful. Speech recognition methods are still very limited, the user must control their speech, and background noise causes problems.

31.2 SUMMARY

- Both suppliers and products should be evaluated.
- A single supplier can be advantageous in simplifying maintenance.
- The time and memory requirements for a program can be estimated using design work.
- Special I/O modules can be selected to suit project needs.

31.3 ASSIGNMENT PROBLEMS

1. What is the most commonly used type of I/O interface?

2. What is a large memory size for a PLC?

3. What factors affect the selection of the size of a PLC.

APLC REFERENCES

SUPPLIERS

Asea Industrial Systems, 16250 West Glendale Dr., New Berlin, WI 53151, USA.
Adaptek Inc., 1223 Michigan, Sandpoint, ID 83864, USA.
Allen Bradley, 747 Alpha Drive, Highland Heights, OH 44143, USA.
Automation Systems, 208 No. 12th Ave., Eldridge, IA 52748, USA.
Bailey Controls Co., 29801 Euclid Ave., Wickliffe, OH 44092, USA.
Cincinatti Milacron, Mason Rd. & Rte. 48, Lebanon, OH 45036, USA.
Devilbiss Corp., 9776 Mt. Gilead Rd., Fredricktown, OH 43019, USA.
Eagle Signal Controls, 8004 Cameron Rd., Austin, TX 78753, USA.
Eaton Corp., 4201 North 27th St., Milwaukee, WI 53216, USA.
Eaton Leonard Corp., 6305 ElCamino Real, Carlsbad, CA 92008, USA.
Foxboro Co., Foxboro, MA 02035, USA.
Furnas Electric, 1000 McKee St., Batavia, IL 60510, USA.
GEC Automation Projects, 2870 Avondale Mill Rd., Macon, GA 31206, USA.
General Electric, Automation Controls Dept., Box 8106, Charlottesville, VA 22906, USA.
General Numeric, 390 Kent Ave., Elk Grove Village, IL 60007, USA.
Giddings & Lewis, Electrical Division, 666 South Military Rd., Fond du Lac, WI 54935-7258, USA.
Gould Inc., Programmable Control Division, PO Box 3083, Andover, MA 01810, USA.
Guardian/Hitachi, 1550 W. Carroll Ave., Chicago, IL 60607, USA.
Honeywell, IPC Division, 435 West Philadelphia St., York, PA 17404, USA.
International Cybernetics Corp., 105 Delta Dr., Pittsburgh, Pennsylvania, 15238, USA, (412) 963-1444.
Keyence Corp. of America, 3858 Carson St., Suite 203, Torrance, CA 90503, USA, (310) 540-2254.
McGill Mfg. Co., Electrical Division, 1002 N. Campbell St., Valparaiso, IN 46383, USA.
Mitsubishi Electric Automation Inc.(MEAU) 500 Corporate Woods Parkway, Vernon Hills, IL 60061
Modicon (AEG), 6630 Campobello Rd., Mississauga, Ont., Canada L5N 2L8, (905) 821-8200.
Modular Computer Systems Inc., 1650 W. McNabb Rd., Fort Lauderdale, FL 33310, USA.
Omron Electric, Control Division, One East Commerce Drive, Schaumburg, IL 60195, USA.
Reliance Electric, Centrl. Systems Division, 4900 Lewis Rd., Stone Mountain, GA 30083, USA.
Siemens, 10 Technology Drive, Peabody, MA 01960, USA.
Square D Co., 4041 N. Richards St., Milwaukee, WI 53201, USA.
Struthers-Dunn Systems Division, 4140 Utica Ridge Rd., Bettendorf, IA 52722, USA.
Telemechanique, 901 Baltimore Blvd., Westminster, MD 21157, USA.
Texas Instruments, Industrial Control Dept., PO Drawer 1255, Johnson City, IN 37605-1255, USA.
Toshiba, 13131 West Little York Rd., Houston, TX 77041, USA.
Transduction Ltd., Airport Corporate Centre, 5155 Spectrum Way Bldg., No. 23, Mississauga, Ont., Canada, L4W 5A1, (905) 625-1907.
Triconex, 16800 Aston St., Irvine, CA 92714, USA.
Westinghouse Electric, 1512 Avis Drive, Madison Heights, MI 48071.

PROFESSIONAL INTEREST GROUPS

American National Standards Committee (ANSI), 1420 Broadway, Ney York, NY 10018, USA.
Electronic Industries Association (EIA), 2001 I Street NW, Washington, DC 20006, USA.
Institute of Electrical and Electronic Engineers (IEEE), 345 East 47th St., New York, NY 10017, USA.
Instrument Society of America (ISA), 67 Alexander Drive, Research Triangle Park, NC 27709, USA.
International Standards Organization (ISO), 1430 Broadway, New York, NY 10018, USA.
National Electrical Manufacturers Association (NEMA), 2101 L. Street NW, Washington, DC 20037, USA.
Society of Manufacturing Engineers (SME), PO Box 930, One SME Drive, Dearborn, MI 48121, USA.

PLC/DISCRETE CONTROL REFERENCES - The table below gives a topic-by-topic comparison of some PLC books. (H=Good coverage, M=Medium coverage, L=Low coverage, Blank=little/no coverage).

Table 1: Published Books and Features

Author	Introduction/Overview	Wiring	Discrete Sensors/Actuators	Conditional Logic	Numbering	Timers/Counters/Latches	Sequential Logic Design	Advanced Functions	Structured Text Programming	Analog I/O	Continuous Sensors/Actuators	Continuous Control	Fuzzy Control	Data Interfacing/Networking	Implementation/Selection	Function Block Programming	pages on PLC topics
Filer...	H	M	L	H	M	H	M	H						M	M		303
Chang...	M	L		L	L		L		M					M	L		80
Petruzela	H	H	M	H	H	H	L	H		L	L	L		L			464
Swainston	H	L	L	L	L	M		H		M	M	M		M	M		294
Clements	H	M	L	L	L	L	L	L		L	L	M			H		197
Asfahl	L			H	L	L	L	L									86
Bollinger..	L		M	M	M	M	M			H	H	H					52
Boucher	M	L	M	L	M	M	H	L		L	M	M		H			59
Kirckof	L	L	L	M	L	M	H	L						M			202

Asfahl, C.R., "Robots and Manufacturing Automation", second edition, Wiley, 1992.

Batten, G.L., Programmable Controllers: Hardware, Software, and Applications, Second Edition, McGraw-Hill, 1994.

Batten, G.L., Batten, G.J., Programmable Controllers: Hardware, Software, and Applications,

*Bertrand, R.M., "Programmable Controller Circuits", Delmar, 1996.

Bollinger, J.G., Duffie, N.A., "Computer Control of Machines and Processes", Addison-Wesley, 1989.

Bolton, w., Programmable Logic Controllers: An Introduction, Butterworth-Heinemann, 1997.

Bryan, L.A., Bryan, E.A., Programmable Controllers, Industrial Text and VideoCompany, 1997.

Boucher, T.O., "Computer Automation in Manufacturing; An Introduction", Chapman and Hall, 1996.

*Bryan, L.A., Bryan, E.A., Programmable Controllers, Industrial Text Company, 19??.

*Carrow, R.A., "Soft Logic: A Guide to Using a PC As a Programmable Logic Controller", McGraw Hill, 1997.

Chang, T-C, Wysk, R.A., Wang, H-P, "Computer-Aided Manufacturing", second edition, Prentice Hall, 1998.

Clements-Jewery, K., Jeffcoat, W., "The PLC Workbook; Programmable Logic Controllers made easy", Prentice Hall, 1996.

*Cox, R., Technician's Guide to Programmable Controllers, Delmar Publishing, 19??.

?Crispin, A.J., "Programmable Logic Controllers and Their Engineering Applications", Books Britain, 1996.

*Dropka, E., Dropka, E., "Toshiba Medium PLC Primer", Butterworth-Heinemann, 1995.

*Dunning, G., "Introduction to Programmable Logic Controllers", Delmar, 1998.

Filer, R., Leinonen, G., "Programmable Controllers and Designing Sequential Logic", Saunders College Publishing, 1992.

**Hughes, T.A., "Programmable Controllers (Resources for Measuremwnt and Control Series)", Instrument Society of America, 1997.

?Johnson, D.G., "Programmable Controllers for Factory Automation", Marcel Dekker, 1987.

Kirckof, G., Cascading Logic; A Machine Control Methodology for Programmable Logic Controllers, The Instrumentation, Systems, and Automation Society, 2003.

*Lewis, R.W., "Programming Industrial Control Systems using IES1131-3",

*Lewis, R.W., Antsaklis, P.J., "Programming Industrial Control Systems Using IEC 1131-3 (lee Control Engineering, No. 59)", Inspec/IEE, 1995.

*Michel, G., Duncan, F., "Programmable Logic Controllers: Architecture and Application", John Wiley & Sons, 1990.

?Morriss, S.B., "Programmable Logic Controllers", pub??, 2000.

?Otter, J.D., "Programmable Logic Controllers: Operation, Interfacing and Programming", ???

Parr, E.A., Parr, A., Programmable Controllers: An Engineer's Guide, Butterworth-Heinemann, 1993.

*Parr, E.A., "Programmable Controllers", Butterworth-Heinemann, 1999.

Petruzella, F., Programmable Logic Controllers, Second Edition, McGraw-Hill Publishing Co., 1998.

*Ridley, J.E., "Introduction to Programmable Logic Controllers: The Mitsubishi Fx", John Wiley & Sons, 1997.

Rohner, P., PLC: Automation With Programmable Logic Controllers, International Specialized Book Service, 1996.

*Rosandich, R.G., "Fundamentals of Programmable Logic Controllers", EC&M Books, 1997.

*Simpson, C.D., "Programmable Logic Controllers", Regents/Prentice Hall, 1994.

Sobh, M., Owen, J.C., Valvanis, K.P., Gracanin, S., "A Subject-Indexed Bibliography of Discrete Event Dynamic Systems", IEEE Robotics and Applications Magazine, June 1994, pp. 14-20.

**Stenerson, J., "Fundamentals of Programmable Logic Controllers, Sensors and Communications", Prentice Hall, 1998.

Sugiyama, H., Umehara, Y., Smith, E., "A Sequential Function Chart (SFC) Language for Batch Control", ISA Transactions, Vol. 29, No. 2, 1990, pp. 63-69.

Swainston, F., "A Systems Approach to Programmable Controllers", Delmar, 1992.

Teng, S.H., Black, J. T., "Cellular Manufacturing Systems Modelling: The Petri Net Approach", Journal of Manufacturing Systems, Vol. 9, No. 1, 1988, pp. 45-54.

Warnock, I., Programmable Controllers: Operation and Application, Prentice Hall, 19??.

**Webb, J.W., Reis, R.A., "Programmable Logic Controllers, Principles and Applications", Prentice Hall, 1995.

Wright, C.P., Applied Measurement Engineering, Prentice-Hall, New Jersey, 1995.

BCOMBINED GLOSSARY OF TERMS

abort - the disruption of normal operation.

absolute pressure - a pressure measured relative to zero pressure.

absorption loss - when sound or vibration energy is lost in a transmitting or reflecting medium. This is the result of generation of other forms of energy such as heat.

absorptive law - a special case of Boolean algebra where A(A+B) becomes A.

AC (Alternating Current) - most commonly an electrical current and voltage that changes in a sinusoidal pattern as a function of time. It is also used for voltages and currents that are not steady (DC). Electrical power is normally distributed at 60Hz or 50Hz.

AC contactor - a contactor designed for AC power.

acceptance test - a test for evaluating a newly purchased system's performance, capabilities, and conformity to specifications, before accepting, and paying the supplier.

accumulator - a temporary data register in a computer CPU.

accuracy - the difference between an ideal value and a physically realizable value. The companion to accuracy is repeatability.

acidity - a solution that has an excessive number of hydrogen atoms. Acids are normally corrosive.

acoustic - another term for sound.

acknowledgement (ACK) - a response that indicates that data has been transmitted correctly.

actuator - a device that when activated will result in a mechanical motion. For example a motor, a solenoid valve, etc.

A/D - Analog to digital converter (see ADC).

ADC (Analog to Digital Converter) - a circuit that will convert an analog voltage to a digital value, also referred to as A/D.

ADCCP (Advanced Data Communications Procedure) - ANSI standard for synchronous communication links with primary and secondary functions.

address - a code (often a number) that specifies a location in a computers memory.

address register - a pointer to memory locations.

adsorption - the ability of a material or apparatus to adsorb energy.

agitator - causes fluids or gases to mix.

AI (Artificial Intelligence) - the use of computer software to mimic some of the cognitive human processes.

air dump valve - this valve will open to release system pressure when system power is removed.

algorithms - a software procedure to solve a particular problem.

aliasing - in digital systems there are natural limits to resolution and time that can be exceeded, thus aliasing the data. For example. an event may happen too fast to be noticed, or a point may be too small to be displayed on a monitor.

alkaline - a solution that has an excess of HO pairs will be a base. This is the compliment to an acid.

alpha rays - ions that are emitted as the result of atomic fission or fusion.

alphanumeric - a sequence of characters that contains both numbers and letters.

ALU (Arithmetic Logic Unit) - a part of a computer that is dedicated to mathematical operations.

AM (Amplitude Modulation) - a fixed frequency carrier signal that is changed in amplitude to encode a change in a signal.

ambient - normal or current environmental conditions.

ambient noise - a sort of background noise that is difficult to isolate, and tends to be present throughout the volume of interest.

ambient temperature - the normal temperature of the design environment.

amplifier - increased (or possibly decreases) the magnitude or power of a signal.

analog signal - a signal that has continuous values, typically voltage.

analysis - the process of review to measure some quality.

and - a Boolean operation that requires all arguments to be true before the result is true.

annealing - heating of metal to relieve internal stresses. In many cases this may soften the material.

annotation - a special note added to a design for explanatory purposes.

ANSI (American National Standards Institute) - a developer of standards, and a member of ISO.

APF (All Plastic Fibre cable) - fiber optic cable that is made of plastic, instead of glass.

API (Application Program Interface) - a set of functions, and procedures that describes how a program will use another service/library/program/etc.

APT (Automatically Programmed Tools) - a language used for directing computer controlled machine tools.

application - the task which a tool is put to, This normally suggests some level of user or real world interaction.

application layer - the top layer in the OSI model that includes programs the user would run, such as a mail reader.

arc - when the electric field strength exceeds the dielectric breakdown voltage, electrons will flow.

architecture - they general layout or design at a higher level.

armature - the central rotating portion of a DC motor or generator, or a moving part of a relay.

ARPA (Advanced Research Projects Agency) - now DARPA. Originally funded ARPANET.

ARPANET - originally sponsored by ARPA. A packet switching network that was in service from the early 1970s, until 1990.

ASCII (American Standard Code for Information Interchange) - a set of numerical codes that correspond to numbers, letters, special characters, and control codes. The most popular standard

ASIC (Application Specific Integrated Circuit) - a specially designed and programmed logic circuit. Used for medium to low level production of complex functions.

aspirator - a device that moves materials with suction.

assembler - converts assembly language into machine code.

assembly language - a mnemonic set of commands that can be directly converted into commands for a CPU.

associative dimensioning - a method for linking dimension elements to elements in a drawing.

associative laws - Boolean algebra laws A+(B+C) = (A+B)+C or A(BC) = (AB)C

asynchronous - events that happen on an irregular basis, and are not predictable.

asynchronous communications (serial) - strings of characters (often ASCII) are broken down into a series of on/off bits. These are framed with start/stop bits, and parity checks for error detection, and then send out one character at a time. The use of start bits allows the characters to be sent out at irregular times.

attenuation - to decrease the magnitude of a signal.

attenuation - as the sound/vibration energy propagates, it will undergo losses. The losses are known as attenuation, and are often measured in dB. For general specifications, the attenuation may be tied to units of dB/ft.

attribute - a nongraphical feature of a part, such as color.

audible range - the range of frequencies that the human ear can normally detect from 16 to 20,000 Hz.

automatic control - a feedback of a system state is compared to a desired value and the control value for the system is adjusted by electronics, mechanics and/or computer to compensate for differences.

automated - a process that operates without human intervention.

auxiliary power - secondary power supplies for remote or isolated systems.

AWG (American Wire Gauge) - specifies conductor size. As the number gets larger, the conductors get smaller.

B.1 B

B-spline - a fitted curve/surface that is commonly used in CAD and graphic systems.

backbone - a central network line that ties together distributed networks.

background - in multitasking systems, processes may be running in the background while the user is working in the foreground, giving the user the impression that they are the only user of the machine (except when the background job is computationally intensive).

background suppression - the ability of a sensing system to discriminate between the signal of interest, and background noise or signals.

backplane - a circuit board located at the back of a circuit board cabinet. The backplane has connectors that boards are plugged into as they are added.

backup - a redundant system to replace a system that has failed.

backward chaining - an expert system looks at the results and looks at the rules to see logically how to get there.

band pressure Level - when measuring the spectrum of a sound, it is generally done by looking at frequencies in a certain bandwidth. This bandwidth will have a certain pressure value that is an aggregate for whatever frequencies are in the bandwidth.

base - 1. a substance that will have an excess of HO ions in solution form. This will react with an acid. 2. the base numbering system used. For example base 10 is decimal, base 2 is binary

baseband - a network strategy in which there is a single carrier frequency, that all connected machines must watch continually, and participate in each transaction.

BASIC (Beginner's All-purpose Symbolic Instruction Code) - a computer language designed to allow easy use of the computer.

batch processing - an outdated method involving running only one program on a computer at once, sequentially. The only practical use is for very intensive jobs on a supercomputer.

battery backup - a battery based power supply that keeps a computer (or only memory) on when the master power is off.

BAUD - The maximum number of bits that may be transmitted through a serial line in one second. This also includes some overhead bits.

baudot code - an old code similar to ASCII for teleprinter machines.

BCC (Block Check Character) - a character that can check the validity of the data in a block.

BCD (Binary Coded Decimal) - numerical digits (0 to 9) are encoded using 4 bits. This allows two numerical digits to each byte.

beam - a wave of energy waves such as light or sound. A beam implies that it is not radiating in all directions, and covers an arc or cone of a few degrees.

bearing - a mechanical support between two moving surfaces. Common types are ball bearings (light weight) and roller bearings (heavy weight), journal bearings (rotating shafts).

beats - if two different sound frequencies are mixed, they will generate other frequencies. if a 1000Hz and 1001Hz sound are heard, a 1Hz (=1000-1001) sound will be perceived.

benchmark - a figure to compare with. If talking about computers, these are often some numbers that can be use to do relative rankings of speeds, etc. If talking about design, we can benchmark our products against our competitors to determine our weaknesses.

Bernoulli's principle - a higher fluid flow rate will result in a lower pressure.

beta ratio - a ratio of pipe diameter to orifice diameter.

beta rays - electrons are emitted from a fission or fusion reaction.

beta site - a software tester who is actually using the software for practical applications, while looking for bugs. After this stage, software will be released commercially.

big-endian - a strategy for storing or transmitting the most significant byte first.

BIOS (Basic Input Output System) - a set of basic system calls for accessing hardware, or software services in a computer. This is typically a level lower than the operating system.

binary - a base 2 numbering system with the digits 0 and 1.

bit - a single binary digit. Typically the symbols 0 and 1 are used to represent the bit value.

bit/nibble/byte/word - binary numbers use a 2 value number system (as opposed to the decimal 0-9, binary uses 0-1). A bit refers to a single binary digit, and as we add digits we get larger numbers. A bit is 1 digit, a nibble is 4 digits, a byte is 8 digits, and a word is 16 digits.

decimal(base 10)	binary(base 2)	octal(base 8)
0	0	0
1	1	1
2	10	2
3	11	3
4	100	4
5	101	5
6	110	6
7	111	7
8	1000	10
9	1001	11
10	1010	12
11	1011	13
.	.	.
.	.	.
.	.	.

e.g. differences

decimal
15 ... tens
3,052 ... thousands
1,000,365 ... millions

binary
1 ... bit
0110 .... nibble (up to 16 values)
10011101 ... byte (up to 256 values)
0101000110101011 ... work (up to 64,256 values)

Most significant bit

least significant bit

BITNET (Because It's Time NET) - An academic network that has been merged with CSNET.

blackboard - a computer architecture when different computers share a common memory area (each has its own private area) for sharing/passing information.

block - a group of bytes or words.

block diagrams - a special diagram for illustrating a control system design.

binary - specifies a number system that has 2 digits, or two states.

binary number - a collection of binary values that allows numbers to be constructed. A binary number is base 2, whereas normal numbering systems are base 10.

blast furnace - a furnace that generates high temperatures by blowing air into the combustion.

bleed nozzle - a valve or nozzle for releasing pressure from a system.

block diagram - a symbolic diagram that illustrates a system layout and connection. This can be used for analysis, planning and/or programming.

BOC (Bell Operating Company) - there are a total of 7 regional telephone companies in the U.S.A.

boiler - a device that will boil water into steam by burning fuel.

BOM (Bills Of Materials) - list of materials needed in the production of parts, assemblies, etc. These lists are used to ensure all required materials are available before starting an operation.

Boolean - a system of numbers based on logic, instead of real numbers. There are many similarities to normal mathematics and algebra, but a separate set of operators, axioms, etc. are used.

bottom-up design - the opposite of top-down design. In this methodology the most simple/basic functions are designed first. These simple elements are then combined into more complex elements. This continues until all of the hierarchical design elements are complete.

bounce - switch contacts may not make absolute contact when switching. They make and break contact a few times as they are coming into contact.

Bourdon tube - a pressure tube that converts pressure to displacement.

BPS (Bits Per Second) - the total number of bits that can be passed between a sender and listener in one second. This is also known as the BAUD rate.

branch - a command in a program that can cause it to start running elsewhere.

bread board - a term used to describe a temporary electronic mounting board. This is used to prototype a circuit before doing final construction. The main purpose is to verify the basic design.

breadth first search - an AI search technique that examines all possible decisions before making the next move.

breakaway torque - the start-up torque. The value is typically high, and is a function of friction, inertia, deflection, etc.

breakdown torque - the maximum torque that an AC motor can produce at the rated voltage and frequency.

bridge - 1. an arrangement of (typically 4) balanced resistors used for measurement. 2. A network device that connects two different networks, and sorts out packets to pass across.

broadband networks - multiple frequencies are used with multiplexing to increase the transmission rates in networks.

broad-band noise - the noise spectrum for a particular noise source is spread over a large range of frequencies.

broadcast - a network term that describes a general broadcast that should be delivered to all clients on a network. For example this is how Ethernet sends all of its packets.

brush - a sliding electrical conductor that conducts power to/from a rotor.

BSC (Binary Synchronous Communication) - a byte oriented synchronous communication protocol developed by IBM.

BSD (Berkeley Software Distribution) - one of the major versions of UNIX.

buffer - a temporary area in which data is stored on its way from one place to another. Used for communication bottlenecks and asynchronous connections.

bugs - hardware or software problems that prevent desired components operation.

bumpless transfer - a smooth transition between manual and automatic modes.

burn-in - a high temperature pre-operation to expose system problems.

burner - a term often used for a device that programs EPROMs, PALs, etc. or a bad cook.

bus - a computer has buses (collections of conductors) to move data, addresses, and control signals between components. For example to get a memory value, the address value provided the binary memory address, the control bus instructs all the devices to read/write, and to examine the address. If the address is valid for one part of the computer, it will put a value on the data bus that the CPU can then read.

byte - an 8 bit binary number. The most common unit for modern computers.

B.2 C

C - A programming language that followed B (which followed A). It has been widely used in software development in the 80s and 90s. It has grown up to become C++ and Java.

CAA (Computer Aided Analysis) - allows the user to input the definition of a part and calculate the performance variables.

cable - a communication wire with electrical and mechanical shielding for harsh environments.

CAD (Computer Aided Design) - is the creation and optimization of the design itself using the computer as a productivity tool. Components of CAD include computer graphics, a user interface, and geometric modelling.

CAD (Computer Aided Drafting) - is one component of CAD which allows the user to input engineering drawings on the computer screen and print them out to a plotter or other device.

CADD (Computer Aided Design Drafting) - the earliest forms of CAD systems were simple electronic versions of manual drafting, and thus are called CADD.

CAE (Computer Aided Engineering) - the use of computers to assist in engineering. One example is the use of Finite Element Analysis (FEA) to verify the strength of a design.

CAM (Computer Aided Manufacturing) - a family of methods that involves computer supported manufacturing on the factory floor.

capacitor - a device for storing energy or mass.

capacitance - referring to the ability of a device to store energy. This is used for electrical capacitors, thermal masses, gas cylinders, etc.

capacity - the ability to absorb something else.

carrier - a high/low frequency signal that is used to transmit another signal.

carry flag - an indication when a mathematical operator has gone past the limitations of the hardware/software.

cascade - a method for connecting devices to increase their range, or connecting things so that they operate in sequence. This is also called chaining.

CASE (Computer Aided Software Engineering) - software tools are used by the developer/programmer to generate code, track changes, perform testing, and a number of other possible functions.

cassette - a holder for audio and data tapes.

CCITT (Consultative Committee for International Telegraph and Telephone) - recommended X25. A member of the ITU of the United Nations.

CD-ROM (Compact Disc Read Only Memory) - originally developed for home entertainment, these have turned out to be high density storage media available for all platforms at very low prices (< $100 at the bottom end). The storage of these drives is well over 500 MB.

CE (Concurrent Engineering) - an engineering method that involves people from all stages of a product design, from marketing to shipping.

CE - a mark placed on products to indicate that they conform to the standards set by the European Common Union.

Celsius - a temperature scale the uses 0 as the freezing point of water and 100 as the boiling point.

centrifugal force - the force on an orbiting object the would cause it to accelerate outwards.

centripetal force - the force that must be applied to an orbiting object so that it will not fly outwards.

channel - an independent signal pathway.

character - a single byte, that when displayed is some recognizable form, such as a letter in the alphabet, or a punctuation mark.

checksum - when many bytes of data are transmitted, a checksum can be used to check the validity of the data. It is commonly the numerical sum of all of the bytes transmitted.

chip - a loose term for an integrated circuit.

chromatography - gases or liquids can be analyzed by how far their constituent parts can migrate through a porous material.

CIM (Computer Integrated Manufacturing) - computers can be used at a higher level to track and guide products as they move through the facility. CIM may or may not include CAD/CAM.

CL (Cutter Location) - an APT program is converted into a set of x-y-z locations stored in a CL file. In turn these are sent to the NC machine via tapes, etc.

clear - a signal or operation to reset data and status values.

client-server - a networking model that describes network services, and user programs.

clipping - the automatic cutting of lines that project outside the viewing area on a computer screen.

clock - a signal from a digital oscillator. This is used to make all of the devices in a digital system work synchronously.

clock speed - the rate at which a computers main time clock works at. The CPU instruction speed is usually some multiple or fraction of this number, but true program execution speeds are loosely related at best.

closed loop - a system that measures system performance and trims the operation. This is also known as feedback. If there is no feedback the system is called open loop.

CMOS (Complimentary Metal Oxide Semi-conductor) - a low power microchip technology that has high noise immunity.

CNC (Computer Numerical Control) - machine tools are equipped with a control computer, and will perform a task. The most popular is milling.

coalescing - a process for filtering liquids suspended in air. The liquid condenses on glass fibers.

coaxial cable - a central wire contains a signal conductor, and an outer shield provides noise immunity. This configuration is limited by its coaxial geometry, but it provides very high noise immunity.

coax - see coaxial cable.

cogging - a machine steps through motions in a jerking manner. The result may be low frequency vibration.

coil - wire wound into a coil (tightly packed helix) used to create electromagnetic attraction. Used in relays, motors, solenoids, etc. These are also used alone as inductors.

collisions - when more than one network client tries to send a packet at any one time, they will collide. Both of the packets will be corrupted, and as a result special algorithms and hardware are used to abort the write, wait for a random time, and retry the transmission. Collisions are a good measure of network overuse.

colorimetry - a method for identifying chemicals using their colors.

combustion - a burning process generating heat and light when certain chemicals are added.

command - a computer term for a function that has an immediate effect, such as listing the files in a directory.

commission - the typical name for getting equipment operational after delivery/installation.

communication - the transfer of data between computing systems.

commutative laws - Booleans algebra laws A+B = B+A and AB=BA.

compare - a computer program element that examines one or more variables, determines equality/inequality, and then performs some action, sometimes a branch.

compatibility - a measure of the similarity of a design to a standard. This is often expressed as a percentage for software. Anything less than 100% is not desirable.

compiler - a tool to change a high level language such as C into assembler.

compliment - to take the logical negative. TRUE becomes false and vice versa.

component - an interchangeable part of a larger system. Components can be used to cut down manufacturing and maintenance difficulties.

compressor - a device that will decrease the volume of a gas - and increase the pressure.

computer - a device constructed about a central instruction processor. In general the computer can be reconfigured (software/firmware/hardware) to perform alternate tasks.

Computer Graphics - is the use of the computer to draw pictures using an input device to specify geometry and other attributes and an output device to display a picture. It allows engineers to communicate with the computer through geometry.

concentric - a shared center between two or more objects.

concurrent - two or more activities occur at the same time, but are not necessarily the same.

concurrent engineering - all phases of the products life are considered during design, and not later during design review stages.

condenser - a system component that will convert steam to water. Typically used in power generators.

conduction - the transfer of energy through some medium.

configuration - a numbers of multifunction components can be connected in a variety of configurations.

connection - a network term for communication that involves first establishing a connection, second data transmission, and third closing the connection. Connectionless networking does not require connection.

constant - a number with a value that should not vary.

constraints - are performance variables with limits. Constraints are used to specify when a design is feasible. If constraints are not met, the design is not feasible.

contact - 1. metal pieces that when touched will allow current to pass, when separated will stop the flow of current. 2. in PLCs contacts are two vertical lines that represent an input, or internal memory location.

contactor - a high current relay.

continuous Noise - a noise that is ongoing, and present. This differentiates from instantaneous, or intermittent noise sources.

continuous Spectrum - a noise has a set of components that are evenly distributed on a spectral graph.

control relay - a relay that does not control any external devices directly. It is used like a variable in a high level programming language.

control variable - a system parameter that we can set to change the system operation.

controls - a system that is attached to a process. Its purpose is to direct the process to some set value.

convection - the transfer of heat energy to liquid or gas that is moving past the surface of an object.

cook's constant - another name for the fudge factor.

core memory - an outdated term describing memory made using small torii that could be polarized magnetically to store data bits. The term lives on when describing some concepts, for example a 'core dump' in UNIX. Believe it or not this has not been used for decades but still appears in many new textbooks.

coriolis force - a force that tends to cause spinning in moving frames of reference. Consider the direction of the water swirl down a drain pipe, it changes from the north to the south of the earth.

correction factor - a formal version of the 'fudge factor'. Typically a value used to multiply or add another value to account for hard to quantify values. This is the friend of the factor of safety.

counter - a system to count events. This can be either software or hardware.

cps (characters per second) - This can be a good measure of printing or data transmission speed, but it is not commonly used, instead the more confusing 'baud' is preferred.

CPU (Central Processing Unit) - the main computer element that examines machine code instructions and executes results.

CRC (Cyclic Redundancy Check) - used to check transmitted blocks of data for validity.

criteria - are performance variables used to measure the quality of a design. Criteria are usually defined in terms of degree - for example, lowest cost or smallest volume or lowest stress. Criteria are used to optimize a design.

crosstalk - signals in one conductor induce signals in other conductors, possibly creating false signals.

CRT (Cathode Ray Tubes) - are the display device of choice today. A CRT consists of a phosphor-coated screen and one or more electron guns to draw the screen image.

crucible - 1. a vessel for holding high temperature materials 2.

CSA (Canadian Standards Association) - an association that develops standards and does some product testing.

CSMA/CD (Carrier Sense Multiple Access with Collision Detection) - a protocol that causes computers to use the same communication line by waiting for turns. This is used in networks such as Ethernet.

CSNET (Computer+Science NETwork) - a large network that was merged with BITNET.

CTS (Clear To Send) - used to prevent collisions in asynchronous serial communications.

current loop - communications that use a full electronic loop to reduce the effects of induced noise. RS-422 uses this.

current rating - this is typically the maximum current that a designer should expect from a system, or the maximum current that an input will draw. Although some devices will continue to work outside rated values, not all will, and thus this limit should be observed in a robust system. Note: exceeding these limits is unsafe, and should be done only under proper engineering conditions.

current sink - a device that allow current to flow through to ground when activated.

current source - a device that provides current from another source when activated.

cursors - are movable trackers on a computer screen which indicate the currently addressed screen position, or the focus of user input. The cursor is usually represented by an arrow, a flashing character or cross-hair.

customer requirements - the qualitative and quantitative minimums and maximums specified by a customer. These drive the product design process.

cycle - one period of a periodic function.

cylinder - a piston will be driven in a cylinder for a variety of purposes. The cylinder guides the piston, and provides a seal between the front and rear of the piston.

B.3 D

daisy chain - allows serial communication of devices to transfer data through each (and every) device between two points.

darlington coupled - two transistors are ganged together by connecting collectors to bases to increase the gain. These increase the input impedance, and reduce the back propagation of noise from loads.

DARPA (Defense Advanced Research Projects Agency) - replaced ARPA. This is a branch of the US department of defence that has participated in a large number of research projects.

data acquisition - refers to the automated collection of information collected from a process or system.

data highway - a term for a communication bus between two separated computers, or peripherals. This term is mainly used for PLC's.

data link layer - an OSI model layer

data logger - a dedicated system for data acquisition.

data register - stores data values temporarily in a CPU.

database - a software program that stores and recalls data in an organized way.

DARPA (Defense Advanced Research Projects Agency) -

DC (Direct Current) - a current that flows only in one direction. The alternative is AC.

DCA (Defense Communications Agency) - developed DDN.

DCD (Data Carrier Detect) - used as a handshake in asynchronous communication.

DCE (Data Communications Equipment) - A term used when describing unintelligent serial communications clients. An example of this equipment is a modem. The complement to this is DTE.

DCE (Distributed Computing Environment) - applications can be distributed over a number of computers because of the use of standards interfaces, functions, and procedures.

DDN (Defense Data Network) - a group of DoD networks, including MILNET.

dead band - a region for a device when it no longer operates.

dead time - a delay between an event occurring and the resulting action.

debounce - a switch may not make sudden and complete contact as it is closes, circuitry can be added to remove a few on-off transitions as the switch mechanically bounces.

debug - after a program has been written it undergoes a testing stage called debugging that involves trying to locate and eliminate logic and other errors. This is also a time when most engineers deeply regret not spending more time on the initial design.

decibel (dB) - a logarithmic compression of values that makes them more suited to human perception (for both scaleability and reference)

decision support - the use of on-line data, and decision analysis tools are used when making decisions. One example is the selection of electronic components based on specifications, projected costs, etc.

DECnet (Digital Equipment Corporation net) - a proprietary network architecture developed by DEC.

decrement - to decrease a numeric value.

dedicated computer - a computer with only one task.

default - a standard condition.

demorgan's laws - Boolean laws great for simplifying equations $\sim(AB) = \sim A + \sim B$, or $\sim(A+B) = \sim A \sim B$.

density - a mass per unit volume.

depth first search - an artificial intelligence technique that follows a single line of reasoning first.

derivative control - a control technique that uses changes in the system of setpoint to drive the system. This control approach gives fast response to change.

design - creation of a new part/product based on perceived needs. Design implies a few steps that are ill defined, but generally include, rough conceptual design, detailed design, analysis, redesign, and testing.

design capture - the process of formally describing a design, either through drafted drawings, schematic drawings, etc.

design cycle - the steps of the design. The use of the word cycle implies that it never ends, although we must at some point decide to release a design.

design Variables - are the parameters in the design that describe the part. Design variables usually include geometric dimensions, material type, tolerances, and engineering notes.

detector - a device to determine when a certain condition has been met.

device driver - controls a hardware device with a piece of modular software.

DFA (Design For Assembly) - a method that guides product design/redesign to ease assembly times and difficulties.

DFT (Design for Testability) - a set of design axioms that generally calls for the reduction of test steps, with the greatest coverage for failure modes in each test step.

diagnostic - a system or set of procedures that may be followed to identify where systems may have failed. These are most often done for mission critical systems, or industrial machines where the user may not have the technical capability to evaluate the system.

diaphragm - used to separate two materials, while allowing pressure to be transmitted.

differential - refers to a relative difference between two values. Also used to describe a calculus derivative operator.

differential amplifier - an amplifier that will subtract two or more input voltages.

diffuse field - multiple reflections result in a uniform and high sound pressure level.

digital - a system based on binary on-off values.

DIN (the Deutsches Institut for Normung) - a German standards institute.

diode - a semiconductor device that will allow current to flow in one direction.

DIP switches - small banks of switches designed to have the same footprint as an integrated circuit.

distributed - suggests that computer programs are split into parts or functions and run on different computers

distributed system - a system can be split into parts. Typical components split are mechanical, computer, sensors, software, etc.

DLE (Data Link Escape) - An RS-232 communications interface line.

DMA (Direct Memory Access) - used as a method of transferring memory in and out of a computer without slowing down the CPU.

DNS (Domain Name System) - an internet method for name and address tracking.

documentation - (don't buy equipment without it) - one or more documents that instruct in the use, installation, setup, maintenance, troubleshooting, etc. for software or machinery. A poor design supported by good documentation can often be more useful than a good design unsupported by poor documentation.

domain - the basic name for a small or large network. For example (unc.edu) is the general extension for the University on North Carolina.

doppler shift - as objects move relative to each other, a frequency generated by one will be perceived at another frequency by the other.

DOS (Disk Operating System) - the portion of an operating system that handles basic I/O operations. The most common example is Microsoft MS-DOS for IBM PCs.

dotted decimal notation - the method for addressing computers on the internet with IP numbers such as '129.100.100.13'.

double pole - a double pole switch will allow connection between two contacts. These are useful when making motor reversers. see also single pole.

double precision - a real number is represented with 8 bytes (single precision is 4) to give more precision for calculations.

double throw - a switch or relay that has two sets of contacts.

download - to retrieve a program from a server or higher level computer.

downtime - a system is removed from production for a given amount of downtime.

drag - a force that is the result of a motion of an object in a viscous fluid.

drop - a term describing a short connection to peripheral I/O.

drum sequencer - a drum has raised/lowered sections and as it rotates it opens/closes contacts and will give sequential operation.

dry contact - an isolated output, often a relay switched output.

DSP (Digital Signal Processor) - a medium complexity microcontroller that has a build in floating point unit. These are very common in devices such as modems.

DSR (Data Set Ready) - used as a data handshake in asynchronous communications.

DTE (Data Terminal Equipment) - a serial communication line used in RS-232

DTR (Data Terminal Ready) - used as a data handshake in asynchronous communications to indicate a listener is ready to receive data.

dump - a large block of memory is moved at once (as a sort of system snapshot).

duplex - serial communication that is in both directions between computers at the same time.

dynamic braking - a motor is used as a brake by connecting the windings to resistors. In effect the motor becomes a generator, and the resistors dissipate the energy as heat.

dynamic variable - a variable with a value that is constantly changing.

dyne - a unit of force

B.4 E

EBCDIC (Extended Binary-Coded Decimal Information Code) - a code for representing keyboard and control characters.

eccentric - two or more objects do not have a common center.

echo - a reflected sound wave.

ECMA (European Computer Manufacturer's Associated) -

eddy currents - small currents that circulate in metals as currents flow in nearby conductors. Generally unwanted.

EDIF (Electronic Design Interchange Format) - a standard to allow the interchange of graphics and data between computers so that it may be changed, and modifications tracked.

EEPROM (Electrically Erasable Programmable Read Only Memory) -

effective sound pressure - the RMS pressure value gives the effective sound value for fluctuating pressure values. This value is some fraction of the peak pressure value.

EIA (Electronic Industries Association) - A common industry standards group focusing on electrical standards.

electro-optic isolator - uses optical emitter, and photo sensitive switches for electrical isolation.

electromagnetic - a broad range term referring to magnetic waves. This goes from low frequency signals such as AM radio, up to very high frequency waves such as light and X-rays.

electrostatic - devices that used trapped charge to apply forces and caused distribution. An example is droplets of paint that have been electrically charged can be caused to disperse evenly over a surface that is oppositely charged.

electrostatics discharge - a sudden release of static electric charge (in nongrounded systems). This can lead to uncomfortable electrical shocks, or destruction of circuitry.

email (electronic mail) - refers to messages passed between computers on networks, that are sent from one user to another. Almost any modern computer will support some for of email.

EMI (ElectroMagnetic Interference) - transient magnetic fields cause noise in other systems.

emulsify - to mix two materials that would not normally mix. for example an emulsifier can cause oil and water to mix.

enable - a digital signal that allows a device to work.

encoding - a conversion between different data forms.

energize - to apply power to a circuit or component.

energy - the result of work. This concept underlies all of engineering. Energy is shaped, directed and focused to perform tasks.

engineering work stations - are self contained computer graphics systems with a local CPU which can be networked to larger computers if necessary. The engineering work station is capable of performing engineering synthesis, analysis, and optimization operations locally. Work stations typically have more than 1 MByte of RAM, and a high resolution screen greater than 512 by 512 pixels.

EOH (End of Header) - A code in a message header that marks the end of the header block.

EOT (End Of Transmission) - an ASCII code to indicate the end of a communications.

EPROM (Erasable Programmable Read Only Memory) - a memory type that can be programmed with voltages, and erased with ultraviolet light.

EPS (Encapsulated PostScript) - a high quality graphics description language understood by high end printers. Originally developed by Adobe Systems Limited. This standard is becoming very popular.

error signal - a control signal that is the difference between a desired and actual position.

ESD - see electrostatic discharge.

esters - a chemical that was formed by a reaction between alcohol and an acid.

ETX (End Of Text) - a marker to indicate the end of a text block in data transmission.

even parity - a checksum bit used to verify data in other bits of a byte.

execution - when a computer is under the control of a program, the program is said to be executing.

expansion principle - when heat is applied a liquid will expand.

expert systems - is a branch of artificial intelligence designed to emulate human expertise with software. Expert systems are in use in many arenas and are beginning to be seen in CAD systems. These systems use rules derived from human experts.

B.5 F

fail safe - a design concept where system failure will bring the system to an idle or safe state.

false - a logical negative, or zero.

Faraday's electromagnetic induction law - if a conductor moves through a magnetic field a current will be induced. The angle between the motion and the magnetic field needs to be 90 deg for maximum current.

Farenheit - a temperature system that has 180 degrees between the freezing and boiling point of water.

fatal error - an error so significant that a software/hardware cannot continue to operate in a reliable manner.

fault - a small error that may be recoverable, or may result in a fatal error.

FAX (facsimile) - an image is scanned and transmitted over phone lines and reconstructed at the other end.

FCS (Frame Check Sequence) - data check flag for communications.

FDDI (Fibre Distributed Data Interface) - a fibre optic token ring network scheme in which the control tokens are counter rotating.

FDX (Full Duplex) - all characters that are transmitted are reflected back to the sender.

FEA (Finite Element Analysis) - is a numerical technique in which the analysis of a complex part is subdivided into the analysis of small simple subdivisions.

feedback - a common engineering term for a system that examines the output of a system and uses is to tune the system. Common forms are negative feedback to make systems stable, and positive feedback to make systems unstable (e.g. oscillators).

fetch - when the CPU gets a data value from memory.

fiberoptics - data can be transmitted by switching light on/off, and transmitting the signal through an optical fiber. This is becoming the method of choice for most long distance data lines because of the low losses and immunity to EMI.

FIFO (First In First Out) - items are pushed on a stack. The items can then be pulled back off last first.

file - a concept of a serial sequence of bytes that the computer can store information in, normally on the disk. This is a ubiquitous concept, but file is also used by Allen Bradley to describe an array of data.

filter - a device that will selectively pass matter or energy.

firmware - software stored on ROM (or equivalent).

flag - a single binary bit that indicates that an event has/has not happened.

flag - a single bit variable that is true or not. The concept is that if a flag is set, then some event has happened, or completed, and the flag should trigger some other event.

flame - an email, or netnews item that is overtly critical of another user, or an opinion. These are common because of the ad-hoc nature of the networks.

flange - a thick junction for joining two pipes.

floating point - uses integer math to represent real numbers.

flow chart - a schematic diagram for representing program flow. This can be used during design of software, or afterwards to explain its operation.

flow meter - a device for measuring the flow rate of fluid.

flow rate - the volume of fluid moving through an area in a fixed unit of time.

fluorescence - incoming UV light or X-ray strike a material and cause the emission of a different frequency light.

FM (Frequency Modulation) - transmits a signal using a carrier of constant magnitude but changing frequency. The frequency shift is proportional to the signal strength.

force - a PLC output or input value can be set on artificially to test programs or hardware. This method is not suggested.

format - 1. a physical and/or data structure that makes data rereadable, 2. the process of putting a structure on a disk or other media.

forward chaining - an expert system approach to examine a set of facts and reason about the probable outcome.

fragmentation - the splitting of an network data packet into smaller fragments to ease transmission.

frame buffers - store the raster image in memory locations for each pixel. The number of colors or shades of gray for each pixel is determined by the number of bits of information for each pixel in the frame buffer.

free field - a sound field where none of the sound energy is reflected. Generally there aren't any nearby walls, or they are covered with sound absorbing materials.

frequency - the number of cycles per second for a sinusoidally oscillating vibration/sound.

friction - the force resulting from the mechanical contact between two masses.

FSK (Frequency Shift Keying) - uses two different frequencies, shifting back and forth to transmit bits serially.

FTP (File Transfer Protocol) - a popular internet protocol for moving files between computers.

fudge factor - a number that is used to multiply or add to other values to make the experimental and theoretical values agree.

full duplex - a two way serial communication channel can carry information both ways, and each character that is sent is reflected back to the sender for verification.

fuse - a device that will destruct when excessive current flows. It is used to protect the electrical device, humans, and other devices when abnormally high currents are drawn. Note: fuses are essential devices and should never be bypassed, or replaced with fuses having higher current rating.

B.6 G

galvanometer - a simple device used to measure currents. This device is similar to a simple DC motor.

gamma rays - high energy electromagnetic waves resulting from atomic fission or fusion.

gate - 1. a circuit that performs on of the Boolean algebra function (i.e., and, or, not, etc.) 2. a connection between a runner and a part, this can be seen on most injection molded parts as a small bump where the material entered the main mold cavity.

gateway - translates and routes packets between dissimilar networks.

Geiger-Mueller tube - a device that can detect ionizing particles (eg, atomic radiation) using a gas filled tube.

global optimum - the absolute best solution to a problem. When found mathematically, the maximum or minimum cost/utility has been obtained.

gpm (gallons per minute) - a flow rate.

grafcet - a method for programming PLCs that is based on Petri nets. This is now known as SFCs and is part of the IEC 1131-3 standard.

gray code - a modified binary code used for noisy environments. It is devised to only have one bit change at any time. Errors then become extremely obvious when counting up or down.

ground - a buried conductor that acts to pull system neutral voltage values to a safe and common level. All electrical equipment should be connected to ground for safety purposes.

GUI (Graphical User Interface) - the user interacts with a program through a graphical display, often using a mouse. This technology replaces the older systems that use menus to allow the user to select actions.

B.7 H

half cell - a probe that will generate a voltage proportional to the hydrogen content in a solution.

half duplex - see HDX

handshake - electrical lines used to establish and control communications.

hard copy - a paper based printout.

hardware - a mechanical or electrical system. The 'functionality' is 'frozen' in hardware, and often difficult to change.

HDLC (High-level Data Link Control) - an ISO standard for communications.

HDX (Half Duplex) - a two way serial connection between two computer. Unlike FDX, characters that are sent are not reflected back to the sender.

head - pressure in a liquid that is the result of gravity.

hermetic seal - an airtight seal.

hertz - a measure of frequency in cycles per second. The unit is Hz.

hex - see hexadecimal.

hexadecimal - a base 16 number system where the digits are 0 to 9 then A to F, to give a total of 16 digits. This is commonly used when providing numbers to computers.

high - another term used to describe a Boolean true, logical positive, or one.

high level language - a language that uses very powerful commands to increase programming productivity. These days almost all applications use some form of high level language (i.e., basic, fortran, pascal, C, C++, etc.).

horsepower - a unit for measuring power

host - a networked (fully functional) computer.

hot backup - a system on-line that can quickly replace a failed system.

hydraulic - 1. a study of water 2. systems that use fluids to transmit power.

hydrocarbon - a class of molecules that contain carbon and hydrogen. Examples are propane, octane.

hysteresis - a sticking or lagging phenomenon that occurs in many systems. For example, in magnetic systems this is a small amount of magnetic repolarization in a reversing field, and in friction this is an effect based on coulomb friction that reverses sticking force.

Hz - see hertz

B.8 I

IAB (internet Activities Board) - the developer of internet standards.

IC (Integrated Circuit) - a microscopic circuit placed on a thin wafer of semiconductor.

IEC (International Electrical Commission) - A Swiss electrical standards group.

IEEE (Institute of Electrical and Electronics Engineers) -

IEEE802 - a set of standards for LANs and MANs.

IGES (Initial Graphics Exchange Specification) - a standard for moving data between various CAD systems. In particular the format can handle basic geometric entities, such as NURBS, but it is expected to be replaced by PDES/STEP in the near future.

impact instrument - measurements are made based by striking an object. This generally creates an impulse function.

impedance - In electrical systems this is both reactive and real resistance combined. This also applies to power transmission and flows in other types of systems.

impulse Noise - a short duration, high intensity noise. This type of noise is often associated with explosions.

increment - increase a numeric value.

inductance - current flowing through a coil will store energy in a magnetic field.

inductive heating - a metal part is placed inside a coil. A high frequency AC signal is passed through the coil and the resulting magnetic field melts the metal.

infrared - light that has a frequency below the visible spectrum.

inertia - a property where stored energy will keep something in motion unless there is energy added or released.

inference - to make a decision using indirect logic. For example if you are wearing shoes, we can infer that you had to put them on. Deduction is the complementary concept.

inference engine - the part of an expert system that processes rules and facts using forward or backward chaining.

Insertion Loss - barriers, hoods, enclosures, etc. can be placed between a sound source, and listener, their presence increases reverberant sound levels and decreases direct sound energy. The increase in the reverberant sound is the insertion loss.

instruction set - a list of all of the commands that available in a programmable system. This could be a list of PLC programming mnemonics, or a list of all of the commands in BASIC.

instrument - a device that will read values from external sensors or probes, and might make control decision.

intake stroke - in a piston cylinder arrangement this is the cycle where gas or liquid is drawn into the cylinder.

integral control - a control method that looks at the system error over a long period of time. These controllers are relatively immune to noise and reduce the steady state error, but the do not respond quickly.

integrate - to combine two components with clearly separable functions to obtain a new single component capable of more complex functions.

intelligence - systems will often be able to do simple reasoning or adapt. This can mimic some aspects of human intelligence. These techniques are known as artificial intelligence.

intelligent device - a device that contains some ability to control itself. This reduces the number of tasks that a main computer must perform. This is a form of distributed system.

interface - a connection between a computer and another electrical device, or the real world.

interlock - a device that will inhibit system operation until certain conditions are met. These are often required for safety on industrial equipment to protect workers.

intermittent noise - when sounds change level fluctuate significantly over a measurement time period.

internet - an ad-hoc collection of networks that has evolved over a number of years to now include millions of computers in every continent, and by now every country. This network will continue to be the defacto standard for personal users. (commentary: The information revolution has begun already, and the internet has played a role previously unheard of by overcoming censorship and misinformation, such as that of Intel about the Pentium bug, a military coup in Russia failed because they were not able to cut off the flow of information via the internet, the Tianneman square massacre and related events were widely reported via internet, etc. The last stage to a popular acceptance of the internet will be the World Wide Web accessed via Mosiac/Netscape.)

internet address - the unique identifier assigned to each machine on the internet. The address is a 32 bit binary identifier commonly described with the dotted decimal notation.

interlacing - is a technique for saving memory and time in displaying a raster image. Each pass alternately displays the odd and then the even raster lines. In order to save memory, the odd and even lines may also contain the same information.

interlock - a flag that ensures that concurrent streams of execution do not conflict, or that they cooperate.

interpreter - programs that are not converted to machine language, but slowly examined one instruction at a time as they are executed.

interrupt - a computer mechanism for temporarily stopping a program, and running another.

inverter - a logic gate that will reverse logic levels from TRUE to/from FALSE.

I/O (Input/Output) - a term describing anything that goes into or out of a computer.

IOR (Inclusive OR) - a normal OR that will be true when any of the inputs are true in any combinations. also see Exclusive OR (EOR).

ion - an atom, molecule or subatomic particle that has a positive or negative charge.

IP (internet Protocol) - the network layer (OSI model) definitions that allow internet use.

IP datagram - a standard unit of information on the internet.

ISDN (Integrated Services Digital Network) - a combined protocol to carry voice, data and video over 56KB lines.

ISO (International Standards Organization) - a group that develops international standards in a wide variety of areas.

isolation - electrically isolated systems have no direct connection between two halves of the isolating device. Sound isolation uses barriers to physically separate rooms.

isolation transformer - a transformer for isolating AC systems to reduce electrical noise.

B.9 J

JEC (Japanese Electrotechnical Committee) - A regional standards group.

JIC (Joint International Congress) - an international standards group that focuses on electrical standards. They drafted the relay logic standards.

JIT (Just in Time) - a philosophy when setting up and operating a manufacturing system such that materials required arrive at the worksite just in time to be used. This cuts work in process, storage space, and a number of other logistical problems, but requires very dependable supplies and methods.

jog - a mode where a motor will be advanced while a button is held, but not latched on. It is often used for clearing jams, and loading new material.

jump - a forced branch in a program

jumper - a short wire, or connector to make a permanent setting of hardware parameters.

B.10 K

k, K - specifies magnitudes. 1K = 1024, 1k = 1000 for computers, otherwise 1K = 1k = 1000. Note - this is not universal, so double check the meanings when presented.

Kelvin - temperature units that place 0 degrees at absolute zero. The magnitude of one degree is the same as the Celsius scale.

KiloBaud, KBaud, KB, Baud - a transmission rate for serial communications (e.g. RS-232C, TTY, RS-422). A baud = 1bit/second, 1 Kilobaud = 1KBaud = 1KB = 1000 bits/second. In serial communication each byte typically requires 11 bits, so the transmission rate is about 1Kbaud/11 = 91 Bytes per second when using a 1KB transmission.

Karnaugh maps - a method of graphically simplifying logic.

kermit - a popular tool for transmitting binary and text files over text oriented connections, such as modems or telnet sessions.

keying - small tabs, prongs, or fillers are used to stop connectors from mating when they are improperly oriented.

kinematics/kinetics - is the measure of motion and forces of an object. This analysis is used to measure the performance of objects under load and/or in motion.

B.11 L

label - a name associated with some point in a program to be used by branch instructions.

ladder diagram - a form of circuit diagram normally used for electrical control systems.

ladder logic - a programming language for PLCs that has been developed to look like relay diagrams from the preceding technology of relay based controls.

laminar flow - all of the particles of a fluid or gas are travelling in parallel. The complement to this is turbulent flow.

laptop - a small computer that can be used on your lap. It contains a monitor ad keyboard.

LAN (Local Area Network) - a network that is typically less than 1km in distance. Transmission rates tend to be high, and costs tend to be low.

latch - an element that can have a certain input or output lock in. In PLCs these can hold an output on after an initial pulse, such as a stop button.

LCD (Liquid Crystal Display) - a fluid between two sheets of light can be polarized to block light. These are commonly used in low power displays, but they require backlighting.

leakage current - a small amount of current that will be present when a device is off.

LED (Light Emitting Diode) - a semiconductor light that is based on a diode.

LIFO (Last In First Out) - similar to FIFO, but the last item pushed onto the stack is the first pulled off.

limit switch - a mechanical switch actuated by motion in a process.

line printer - an old printer style that prints single lines of text. Most people will be familiar with dot matrix style of line printers.

linear - describes a mathematical characteristic of a system where the differential equations are simple linear equations with coefficients.

little-endian - transmission or storage of data when the least significant byte/bit comes first.

load - In electrical system a load is an output that draws current and consumes power. In mechanical systems it is a mass, or a device that consumes power, such as a turbine.

load cell - a device for measuring large forces.

logic - 1. the ability to make decisions based on given values. 2. digital circuitry.

loop - part of a program that is executed repeatedly, or a cable that connects back to itself.

low - a logic negative, or zero.

LRC (Linear Redundancy Check) - a block check character

LSB (Least Significant Bit) - This is the bit with the smallest value in a binary number. for example if the number 10 is converted to binary the result is 1010. The most significant bit is on the left side, with a value of 8, and the least significant bit is on the right with a value of 1 - but it is not set in this example.

LSD (Least Significant Digit) - This is the least significant digit in a number, found on the right side of a number when written out. For example, in the number $1,234,567 the digit 7 is the least significant.

LSI (Large Scale Integration) - an integrated circuit that contains thousands of elements.

LVDT (Linear Variable Differential Transformer) - a device that can detect linear displacement of a central sliding core in the transformer.

B.12 M

machine language - CPU instructions in numerical form.

macro - a set of commands grouped for convenience.

magnetic field - a field near flowing electrons that will induce other electrons nearby to flow in the opposite direction.

MAN (Metropolitan Area Network) - a network designed for municipal scale connections.

manifold - 1. a connectors that splits the flow of fluid or gas. These are used commonly in hydraulic and pneumatic systems. 2. a description for a geometry that does not have any infinitely small points or lines of contact or separation. Most solid modelers deal only with manifold geometry.

MAP (Manufacturers Automation Protocol) - a network type designed for the factory floor that was widely promoted in the 1980s, but was never widely implemented due to high costs and complexity.

mask - one binary word (or byte, etc) is used to block out, or add in digits to another binary number.

mass flow rate - instead of measuring flow in terms of volume per unit of time we use mass per unit time.

mass spectrometer - an instrument that identifies materials and relative proportions at the atomic level. This is done by observing their deflection as passed through a magnetic field.

master/slave - a control scheme where one computer will control one or more slaves. This scheme is used in interfaces such as GPIB, but is increasingly being replaced with peer-to-peer and client/server networks.

mathematical models - of an object or system predict the performance variable values based upon certain input conditions. Mathematical models are used during analysis and optimization procedures.

matrix - an array of numbers

MB MByte, KB, KByte - a unit of memory commonly used for computers. 1 KiloByte = 1 KByte = 1 KB = 1024 bytes. 1 MegaByte = 1 MByte = 1MB = 1024*1024 bytes.

MCR (Master Control Relay) - a relay that will shut down all power to a system.

memory - binary numbers are often stored in memory for fast recall by computers. Inexpensive memory can be purchased in a wide variety of configurations, and is often directly connected to the CPU.

memory - memory stores binary (0,1) patterns that a computer can read or write as program or data. Various types of memories can only be read, some memories lose their contents when power is off.

> RAM (Random Access Memory) - can be written to and read from quickly. It requires power to preserve the contents, and is often coupled with a battery or capacitor when long term storage is required. Storage available is over 1MByte

> ROM (Read Only Memory) - Programs and data are permanently written on this low cost ship. Storage available is over 1 MByte.

> EPROM (ELECTRICALLY Programmable Read Only Memory) - A program can be written to this memory using a special programmer, and erased with ultraviolet light. Storage available over 1MByte. After a program is written, it does not require power for storage. These chips have small windows for ultraviolet light.

> EEPROM/E2PROM (Electronically Erasable Programmable Read Only Memory) - These chips can be erased and programmed while in use with a computer, and store memory that is not sensitive to power. These can be slower, more expensive and with lower capacity (measured in Kbytes) than other memories. But, their permanent storage allows system configurations/data to be stored indefinitely after a computer is turned off.

memory map - a listing of the addresses of different locations in a computer memory. Very useful when programming.

menu - a multiple choice method of selecting program options.

message - a short sequence of data passed between processes.

microbar - a pressure unit (1 dyne per sq. cm)

microphone - an audio transducer (sensor) used for sound measurements.

microprocessor - the central control chip in a computer. This chip will execute program instructions to direct the computer.

MILNET (MILitary NETwork) - began as part of ARPANET.

MMI (Man Machine Interface) - a user interface terminal.

mnemonic - a few characters that describe an operation. These allow a user to write programs in an intuitive manner, and have them easily converted to CPU instructions.

MODEM (MOdulator/DEModulator) - a device for bidirectional serial communications over phone lines, etc.

module - a part o a larger system that can be interchanged with others.

monitor - an operation mode where the computer can be watched in detail from step to step. This can also refer to a computer screen.

motion detect flow meter - a fluid flow induces measurement.

MRP (Material Requirements Planning) - a method for matching material required by jobs, to the equipment available in the factory.

MSD (Most Significant Digit) - the largest valued digit in a number (eg. 6 is the MSD in 63422). This is often used for binary numbers.

MTBF (Mean Time Between Failure) - the average time (hours usually) between the last repair of a product, and the next expected failure.

MTTR (Mean Time To Repair) - The average time that a device will out of use after failure before it is repaired. This is related to the MTBF.

multicast - a broadcast to some, but not necessarily all, hosts on a network.

multiplexing - a way to efficiently use transmission media by having many signals run through one conductor, or one signal split to run through multiple conductors and rejoined at the receiving end.

multiprocessor - a computer or system that uses more than one computer. Normally this term means a single computer with more than one CPU. This scheme can be used to increase processing speed, or increase reliability.

multivibrator - a digital oscillator producing square or rectangular waveforms.

B.13 N

NAK (Negative AKnowledgement) - an ASCII control code.

NAMUR - A european standards organization.

NAND (Not AND) - a Boolean AND operation with the result inverted.

narrowband - uses a small data transmission rate to reduce spectral requirements.

NC - see normally opened/closed

NC (Numerical Control) - a method for controlling machine tools, such as mills, using simple programs.

negative logic - a 0 is a high voltage, and 1 is a low voltage. In Boolean terms it is a duality.

NEMA (National Electrical Manufacturers Association) - this group publishes numerous standards for electrical equipment.

nephelometry - a technique for determining the amount of solids suspended in water using light.

nesting - a term that describes loops (such as FOR-NEXT loops) within loops in programs.

network - a connection of typically more than two computers so that data, email, messages, resources and files may be shared. The term network implies, software, hardware, wires, etc.

NFS (Network File System) - a protocol developed by Sun Microsystems to allow dissimilar computers to share files. The effect is that the various mounted remote disk drives act as a single local disk.

NIC (Network Interface Card) - a computer card that allows a computer to communicate on a network, such as ethernet.

NIH (Not Invented Here) - a short-lived and expensive corporate philosophy in which employees believe that if idea or technology was not developed in-house, it is somehow inferior.

NIST (National Institute of Standards and Technology) - formerly NBS.

NO - see normally opened

node - one computer connected to a network.

noise - 1. electrical noise is generated mainly by magnetic fields (also electric fields) that induce currents and voltages in other conductors, thereby decreasing the signals present. 2. a sound of high intensity that can be perceived by the human ear.

non-fatal error - a minor error that might indicate a problem, but it does not seriously interfere with the program execution.

nonpositive displacement pump - a pump that does not displace a fixed volume of fluid or gas.

nonretentive - when power is lost values will be set back to 0.

NOR (Not OR) - a Boolean function OR that has the results negated.

normally opened/closed - refers to switch types. when in their normal states (not actuated) the normally open (NO) switch will not conduct current. When not actuated the normally closed (NC) switch will conduct current.

NOT - a Boolean function that inverts values. A 1 will become a 0, and a 0 will become a 1.

NOVRAM (NOn Volatile Random Access Memory) - memory that does not lose its contents when turned off.

NPN - a bipolar junction transistor type. When referring to switching, these can be used to sink current to ground.

NPSM - American national standard straight pipe thread for mechanical parts.

NPT - American national standard taper pipe thread.

NSF (National Science Foundation) - a large funder of science projects in USA.

NSFNET (National Science Foundation NETwork) - funded a large network(s) in USA, including a high speed backbone, and connection to a number of super computers.

NTSC (National Television Standards Committee) - a Red-Green-Blue based transmission standard for video, and audio signals. Very popular in North America, Competes with other standards internationally, such as PAL.

null modem - a cable that connects two RS-232C devices.

B.14 O

OCR (Optical Character Recognition) - Images of text are scanned in, and the computer will try to interpret it, much as a human who is reading a page would. These systems are not perfect, and often rely on spell checkers, and other tricks to achieve reliabilities up to 99%

octal - a base 8 numbering system that uses the digits 0 to 7.

Octave - a doubling of frequency

odd parity - a bit is set during communication to indicate when the data should have an odd number of bits.

OEM (Original Equipment Manufacturer) - a term for a manufacturer that builds equipment for consumers, but uses major components from other manufacturers.

off-line - two devices are connected, but not communicating.

offset - a value is shifted away or towards some target value.

one-shot - a switch that will turn on for one cycle.

on-line - two devices are put into communications, and will stay in constant contact to pass information as required.

opcode (operation code) - a single computer instruction. Typically followed by one or more operands.

open collector - this refers to using transistors for current sourcing or sicking.

open loop - a system that does monitor the result. open loop control systems are common when the process is well behaved.

open-system - a computer architecture designed to encourage interconnection between various vendors hardware and software.

operand - an operation has an argument (operand) with the mnemonic command.

operating system - software that existing on a computer to allow a user to load/execute/develop their own programs, to interact with peripherals, etc. Good examples of this is UNIX, MS-DOS, OS/2.

optimization - occurs after synthesis and after a satisfactory design is created. The design is optimized by iteratively proposing a design and using calculated design criteria to propose a better design.

optoisolators - devices that use a light emitter to control a photoswitch. The effect is that inputs and outputs are electrically separate, but connected. These are of particular interest when an interface between very noisy environments are required.

OR - the Boolean OR function.

orifice - a small hole. Typically this is places in a fluid/gas flow to create a pressure difference and slow the flow. It will increase the flow resistance in the system.

oscillator - a device that produces a sinusoidal output.

oscilloscope - a device that can read and display voltages as a function for time.

OSF (Open Software Foundation) - a consortium of large corporations (IBM, DEC, HP) that are promoting DCE. They have put forth a number of popular standards, such as the Motif Widget set for X-Windows programming.

OSHA (Occupational safety and Health Act) - these direct what is safe in industrial and commercial operations.

OSI (Open System Interconnect) - an international standards program to promote computer connectivity, regardless of computer type, or manufacturer.

overshoot - the inertia of a controlled system will cause it to pass a target value and then return.

overflow - the result of a mathematical operation passes by the numerical limitations of the hardware logic, or algorithm.

B.15 P

parallel communication - bits are passed in parallel conductors, thus increasing the transmission rates dramatically.

parallel design process - evaluates all aspects of the design simultaneously in each iteration. The design itself is sent to all analysis modules including manufacturability, inspectibility, and engineering analysis modules; redesign decisions are based on all results at once.

parallel programs - theoretically, these computer programs do more than one thing simultaneously.

parity - a parity bit is often added to bytes for error detection purposes. The two typical parity methods are even and odd. Even parity bits are set when an even number of bits are present in the transmitted data (often 1 byte = 8 bits).

particle velocity - the instantaneous velocity of a single molecule.

Pascal - a basic unit of pressure

Pascal's law - any force applied to a fluid will be transmitted through the fluid and act on all enclosing surfaces.

PC (Programmable Controller) - also called PLC.

PCB (Printed Circuit Board) - alternate layers of insulating materials, with wire layout patterns are built up (sometimes with several layers). Holes thought the layers are used to connect the conductors to each other, and components inserted into the boards and soldered in place.

PDES (Product Data Exchange using Step) - a new product design method that has attempted to include all needed information for all stages of a products life, including full solids modeling, tolerances, etc.

peak level - the maximum pressure level for a cyclic variation

peak-to-peak - the distance between the top and bottom of a sinusoidal variation.

peer-to-peer - a communications form where connected devices to both read and write messages at any time. This is opposed to a master slave arrangement.

performance variables - are parameters which define the operation of the part. Performance variables are used by the designer to measure whether the part will perform satisfactorily.

period - the time for a repeating pattern to go from beginning to end.

peripheral - devices added to computers for additional I/O.

permanent magnet - a magnet that retains a magnetic field when the original magnetizing force is removed.

petri-net - an enhanced state space diagram that allows concurrent execution flows.

pH - a scale for determining is a solution is an acid or a base. 0-7 is acid, 7-4 is a base.

photocell - a device that will convert photons to electrical energy.

photoconductive cell - a device that has a resistance that will change as the number of incident photons changes.

photoelectric cell - a device that will convert photons to electrical energy.

photon - a single unit of light. Light is electromagnetic energy emitted as an electron orbit decays.

physical layer - an OSI network model layer.

PID (Proportional Integral Derivative) - a linear feedback control scheme that has gained popularity because of it's relative simplicity.

piezoelectric - a material (crystals/ceramics) that will generate a charge when a force is applied. A common transducer material.

ping - an internet utility that makes a simple connection to a remote machine to see if it is reachable, and if it is operating.

pink noise - noise that has the same amount of energy for each octave.

piston - it will move inside a cylinder to convert a pressure to a mechanical motion or vice versa.

pitch - a perceptual term for describing frequency. Low pitch means low frequency, high pitch means a higher frequency.

pitot tube - a tube that is placed in a flow stream to measure flow pressure.

pixels - are picture elements in a digitally generated and displayed picture. A pixel is the smallest addressable dot on the display device.

PLA (Programmable Logic Array) - an integrated circuit that can be programmed to perform different logic functions.

plane sound wave - the sound wave lies on a plane, not on a sphere.

PLC (Programmable Logic Controller) - A rugged computer designs for control on the factory floor.

pneumatics - a technique for control and actuation that uses air or gases.

PNP - a bipolar junction transistor type. When referring to switching, these can be used to source current from a voltage source.

poise - a unit of dynamic viscosity.

polling - various inputs are checked in sequence for waiting inputs.

port - 1. an undedicated connector that peripherals may be connected to. 2. a definable connection number for a machine, or a predefined value.

positive displacement pump - a pump that displaces a fixed volume of fluid.

positive logic - the normal method for logic implementation where 1 is a high voltage, and 0 is a low voltage.

potentiometer - displacement or rotation is measured by a change in resistance.

potting - a process where an area is filled with a material to seal it. An example is a sensor that is filled with epoxy to protect it from humidity.

power level - the power of a sound, relative to a reference level

power rating - this is generally the maximum power that a device can supply, or that it will require. Never exceed these values, as they may result in damaged equipment, fires, etc.

power supply - a device that converts power to a usable form. A typical type uses 115Vac and outputs a DC voltage to be used by circuitry.

PPP (Point-to-Point Protocol) - allows router to router or host to network connections over other synchronous and asynchronous connections. For example a modem connection can be used to connect to the internet using PPP.

presentation layer - an OSI network model layer.

pressure - a force that is distributed over some area. This can be applied to solids and gases.

pressure based flow meter - uses difference in fluid pressures to measure speeds.

pressure switch - activated above/below a preset pressure level.

prioritized control - control operations are chosen on the basic of priorities.

procedural language - a computer language where instructions happen one after the other in a clear sequence.

process - a purposeful set of steps for some purpose. In engineering a process is often a machine, but not necessarily.

processor - a loose term for the CPU.

program - a sequential set of computer instructions designed to perform some task.

programmable controller - another name for a PLC, it can also refer to a dedicated controller that uses a custom programming language.

PROM (Programmable Read Only Memory) -

protocol - conventions for communication to ensure compatibility between separated computers.

proximity sensor - a sensor that will detect the presence of a mass nearby without contact. These use a variety of physical techniques including capacitance and inductance.

pull-up resistor - this is used to normally pull a voltage on a line to a positive value. A switch/circuit can be used to pull it low. This is commonly needed in CMOS devices.

pulse - a brief change in a digital signal.

purge bubbling - a test to determine the pressure needed to force a gas into a liquid.

PVC - poly vinyl chloride - a tough plastic commonly used in electrical and other applications.

pyrometer - a device for measuring temperature

B.16 Q

QA (Quality Assurance) - a formal system that has been developed to improve the quality of a product.

QFD (Quality Functional Deployment) - a matrix based method that focuses the designers on the significant design problems.

quality - a measure of how well a product meets its specifications. Keep in mind that a product that exceeds its specifications may not be higher quality.

quality circles - a team from all levels of a company that meets to discuss quality improvement. Each members is expected to bring their own perspective to the meeting.

B.17 R

rack - a housing for holding electronics modules/cards.

rack fault - cards in racks often have error indicator lights that turn on when a fault has occurred. This allows fast replacement.

radar () - radio waves are transmitted and reflected. The time between emission and detection determines the distance to an object.

radiation - the transfer of energy or small particles (e.g., neutrons) directly through space.

radiation pyrometry - a technique for measuring temperature by detecting radiated heat.

radix - the base value of a numbering system. For example the radix of binary is 2.

RAID (Redundant Array of Inexpensive Disks) - a method for robust disk storage that would allow removal of any disk drive without the interruption of service, or loss of data.

RAM (Random Access Memory) - Computer memory that can be read from, and written to. This memory is the main memory type in computers. The most common types are volatile - they lose their contents when power is removed.

random noise - there are no periodic waveforms, frequency and magnitude vary randomly.

random-scan devices - draw an image by refreshing one line or vector at a time; hence they are also called vector-scan or calligraphic devices. The image is subjected to flicker if there are more lines in the scene that can be refreshed at the refresh rate.

Rankine - A temperature system that uses absolute 0 as the base, and the scale is the same as the Fahrenheit scale.

raster devices - process pictures in parallel line scans. The picture is created by determining parts of the scene on each scan line and painting the picture in scan-line order, usually from top to bottom. Raster devices are not subject to flicker because they always scan the complete display on each refresh, independent of the number of lines in the scene.

rated - this will be used with other terms to indicate suggested target/maximum/minimum values for successful and safe operation.

RBOC (Regional Bell Operating Company) - A regional telephone company. These were originally created after a US federal court split up the phone company into smaller units.

Read/Write (R/W) - a digital device that can store and retrieve data, such as RAM.

reagent - an chemical used in one or more chemical reactions. these are often used for identifying other chemicals.

real-time - suggests a system must be able to respond to events that are occurring outside the computer in a reasonable amount of time.

reciprocating - an oscillating linear motion.

redundancy - 1. added data for checking accuracy. 2. extra system components or mechanisms added to decrease the chance of total system failure.

refreshing - is required of a computer screen to maintain the screen image. Phosphors, which glow to show the image, decay at a fast rate, requiring the screen to be redrawn or refreshed several times a second to prevent the image from fading.

regenerative braking - the motor windings are reverse, and in effect return power to the power source. This is highly efficient when done properly.

register - a high speed storage area that can typically store a binary word for fast calculation. Registers are often part of the CPU.

regulator - a device to maintain power output conditions (such as voltage) regardless of the load.

relay - an electrical switch that comes in may different forms. The switch is activated by a magnetic coil that causes the switch to open or close.

relay - a magnetic coil driven switch. The input goes to a coil. When power is applied, the coil generates a magnetic field, and pulls a metal contact, overcoming a spring, and making contact with a terminal. The contact and terminal are separately wired to provide an output that is isolated from the input.

reliability - the probability of failure of a device.

relief valve - designed to open when a pressure is exceeded. In a hydraulic system this will dump fluid back in the reservoir and keep the system pressure constant.

repeatability - the ability of a system to return to the same value time after time. This can be measured with a standard deviation.

repeater - added into networks to boost signals, or reduce noise problems. In effect one can be added to the end of one wire, and by repeating the signals into another network, the second network wire has a full strength signal.

reset - a signal to computers that restarts the processor.

resistance - this is a measurable resistance to energy or mass transfer.

resistance heating - heat is generated by passing a current through a resistive material.

resolution - the smallest division or feature size in a system.

resonant frequency - the frequency at which the material will have the greatest response to an applied vibration or signal. This will often be the most likely frequency of self destruction.

response time - the time required for a system to respond to a directed change.

return - at the end of a subroutine, or interrupt, the program execution will return to where it branched.

reverberation - when a sound wave hits a surface, part is reflected, and part is absorbed. The reflected part will add to the general (reverberant) sound levels in the room.

Reynolds number - a dimensionless flow value based on fluid density and viscosity, flow rate and pipe diameter.

RF (Radio Frequency) - the frequency at which a magnetic field oscillates when it is used to transmit a signal. Normally this range is from about 1MHz up to the GHz.

RFI (Radio Frequency Interference) - radio and other changing magnetic fields can generate unwanted currents (and voltages) in wires. The resulting currents and voltages can interfere with the normal operation of an electrical device. Filters are often used to block these signals.

RFS (Remote File System) - allows shared file systems (similar to NFS), and has been developed for System V UNIX.

RGB (Red Green Blue) - three additive colors that can be used to simulate the other colors of the spectrum. This is the most popular scheme for specifying colors on computers. The alternate is to use Cyan-Magenta-Yellow for the subtractive color scheme.

ripple voltage - when an AC voltage is converted to DC it is passed through diodes that rectify it, and then through capacitors that smooth it out. A small ripple still remains.

RISC (Reduced Instruction Set Computer) - the more standard computer chips were CISC (Complete Instruction Set Computers) but these had architecture problems that limited speed. To overcome this the total number of instructions were reduced, allowing RISC computers to execute faster, but at the cost of larger programs.

rlogin - allows a text based connection to a remote computer system in UNIX.

robustness - the ability of a system to deal with and recover from unexpected input conditions.

ROM (Read Only Memory) - a permanent form of computer memory with contents that cannot be overwritten. All computers contain some ROM to store the basic operating system - often called the BIOS in personal computers.

rotameter - for measuring flow rate with a plug inside a tapered tube.

router - as network packets travel through a network, a router will direct them towards their destinations using algorithms.

RPC (Remote Procedure Call) - a connection to a specific port on a remote computer will request that a specific program be run. Typical examples are ping, mail, etc.

RS-232C - a serial communication standard for low speed voltage based signals, this is very common on most computers. But, it has a low noise immunity that suggests other standards in harsh environments.

RS-422 - a current loop based serial communication protocol that tends to perform well in noisy environments.

RS-485 - uses two current loops for serial communications.

RTC (Real-Time Clock) - A clock that can be used to generate interrupts to keep a computer process or operating system running at regular intervals.

RTD (Resistance Temperature Detector) - as temperature is changed the resistance of many materials will also change. We can measure the resistance to determine the temperature.

RTS (Request To Send) - A data handshaking line that is used to indicate when a signal is ready for transmission, and clearance is requested.

rung - one level of logic in a ladder logic program or ladder diagram.

R/W (Read/Write) - A digital line that is used to indicate if data on a bus is to be written to, or read from memory.

B.18 S

safety margin - a factor of safety between calculated maximums and rated maximums.
SCADA (Supervisory Control And Data Acquisition) - computer remote monitoring and control of processes.
scan-time - the time required for a PLC to perform one pass of the ladder logic.
schematic - an abstract drawing showing components in a design as simple figures. The figures drawn are often the essential functional elements that must be considered in engineering calculations.
scintillation - when some materials are high by high energy particles visible light or electromagnetic radiation is produced
SCR (Silicon Controlled Rectifier) - a semiconductor that can switch AC loads.
SDLC (Synchronous Data-Link Control) - IBM oriented data flow protocol with error checking.
self-diagnosis - a self check sequence performed by many operation critical devices.
sensitivity - the ability of a system to detect a change.
sensor - a device that is externally connected to survey electrical or mechanical phenomena, and convert them to electrical or digital values for control or monitoring of systems.
serial communication - elements are sent one after another. This method reduces cabling costs, but typically also reduces speed, etc.
serial design - is the traditional design method. The steps in the design are performed in serial sequence. For example, first the geometry is specified, then the analysis is performed, and finally the manufacturability is evaluated.
servo - a device that will take a desired operation input and amplify the power.
session layer - an OSI network model layer.
setpoint - a desired value for a controlled system.
shield - a grounded conducting barrier that steps the propagation of electromagnetic waves.
Siemens - a measure of electrical conductivity.
signal conditioning - to prepare an input signal for use in a device through filtering, amplification, integration, differentiation, etc.
simplex - single direction communication at any one time.
simulation - a model of the product/process/etc is used to estimate the performance. This step comes before the more costly implementation steps that must follow.
single-discipline team - a team assembled for a single purpose.
single pole - a switch or relay that can only be opened or closed. See also single pole.
single throw - a switch that will only switch one line. This is the simplest configuration.
sinking - using a device that when active will allow current to flow through it to ground. This is complimented by sourcing.
SLIP (Serial Line internet Protocol) - a method to run the internet Protocol (IP) over serial lines, such as modem connections.
slip-ring - a connector that allows indefinite rotations, but maintains electrical contacts for passing power and electrical signals.
slurry - a liquid with suspended particles.
SMTP (Simple Mail Transfer Protocol) - the basic connection protocol for passing mail on the internet.
snubber - a circuit that suppresses a sudden spike in voltage or current so that it will not damage other devices.
software - a program, often stored on non-permanent media.
solenoid - an actuator that uses a magnetic coil, and a lump of ferrous material. When the coil is energized a linear motion will occur.
solid state - circuitry constructed entirely of semiconductors, and passive devices. (i.e., no gas as in tubes)
sonar - sound waves are emitted and travel through gas/liquid. they are reflected by solid objects, and then detects back at the source. The travel time determines the distance to the object.
sound - vibrations in the air travel as waves. As these waves strike the human ear, or other surfaces, the compression, and rarefaction of the air induces vibrations. In humans these vibrations induce perceived sound, in mechanical devices they manifest as distributed forces.
sound absorption - as sound energy travels through, or reflects off a surface it must induce motion of the propagating medium. This induced motion will result in losses, largely heat, that will reduce the amplitude of the sound.
sound analyzer - measurements can be made by setting the instrument for a certain bandwidth, and centre frequency. The measurement then encompasses the values over that range.
sound level - a legally useful measure of sound, weighted for the human ear. Use dBA, dBB, dBC values.
sound level meter - an instrument for measuring sound exposure values.

source - an element in a system that supplies energy.

sourcing - an output that when active will allow current to flow from a voltage source out to a device. It is complimented by sinking.

specific gravity - the ratio between the density of a liquid/solid and water or a gas and air.

spectrometer - determines the index of refraction of materials.

spectrophotometer - measures the intensities of light at different points in the spectrum.

spectrum - any periodic (and random) signal can be described as a collection of frequencies using a spectrum. The spectrum uses signal power, or intensity, plotted against frequency.

spherical wave - a wave travels outward as if on the surface of an expanding sphere, starting from a point source.

SQL (Structured Query Language) - a standard language for interrogating relational databases.

standing wave - if a wave travels from a source, and is reflected back such that it arrives back at the source in phase, it can undergo superposition, and effectively amplify the sound from the source.

static head - the hydrostatic pressure at the bottom of a water tank.

steady state - describes a system response after a long period of time. In other words the transient effects have had time to dissipate.

STEP (Standard for the Exchange of Product model data) - a standard that will allow transfer of solid model data (as well as others) between dissimilar CAD systems.

step response - a typical test of system behavior that uses a sudden step input change with a measured response.

stoichiometry - the general field that deals with balancing chemical equations.

strain gauge - a wire mounted on a surface that will be stretched as the surface is strained. As the wire is stretched, the cross section is reduced, and the proportional change in resistance can be measured to estimate strain.

strut - a two force structural member.

subroutine - a reusable segment of a program that is called repeatedly.

substrate - the base piece of a semiconductor that the layers are added to.

switching - refers to devices that are purely on or off. Clearly this calls for discrete state devices.

synchronous - two or more events happen at predictable times.

synchronous motor - an AC motor. These motors tend to keep a near constant speed regardless of load.

syntax error - an error that is fundamentally wrong in a language.

synthesis - is the specification of values for the design variables. The engineer synthesizes a design and then evaluates its performance using analysis.

system - a complex collection of components that performs a set of functions.

B.19 T

T1 - a 1.54 Mbps network data link.

T3 - a 45 Mbps network data link. This can be done with parallel T1 lines and packet switching.

tap - a connection to a power line.

tare - the ratio between unloaded and loaded weights.

TCP (Transmission Control Protocol) - a transport layer protocol that ensures reliable data communication when using IP communications. The protocol is connection oriented, with full duplex streams.

tee - a tap into a larger line that does not add any special compensation, or conditioning. These connectors often have a T-shape.

telnet - a standard method for logging into remote computers and having access if connect by a dumb terminal.

temperature - the heat stored in an object. The relationship between temperature and energy content is specific to a material and is called the specific heat.

temperature dependence - as temperature varies, so do physical properties of materials. This makes many devices sensitive to temperatures.

thermal conductivity - the ability of a material to transfer heat energy.

thermal gradient - the change in temperature as we move through a material.

thermal lag - a delay between the time heat energy is applied and the time it arrives at the load.

thermistor - a resistance based temperature measurement device.

thermocouple - a device using joined metals that will generate a junction potential at different temperatures, used for temperature measurement.

thermopiles - a series of thermocouples in series.

thermoresistors - a category including RTDs and thermistors.

throughput - the speed that actual data is transmitted/processed, etc.

through beam - a beam is projected over an opening. If the beam is broken the sensor is activated.

thumbwheel - a mechanical switch with multiple positions that allow digits to be entered directly.

TIFF (Tagged Image File Format) - an image format best suited to scanned pictures, such as Fax transmissions.

time-division multiplex - a circuit is switched between different devices for communication.

time-proportional control - the amount of power delivered to an AC device is varied by changing the number of cycles delivered in a fixed period of time.

timer - a device that can be set to have events happen at predetermined times.

titration - a procedure for determining the strength of a solution using a reagent for detection. A chemical is added at a slow rate until the reagent detects a change.

toggle switch - a switch with a large lever used for easy reviews of switch settings, and easy grasping.

token - an indicator of control. Often when a process receives a token it can operate, when it is done it gives it up.

TOP (Technical Office Protocol) - a network protocol designed for offices. It was promoted in conjunction with MAP in the 1980s, but never became widely used.

top-down design - a design is done by first laying out the most abstract functions, and then filling in more of the details as they are required.

topology - 1. The layout of a network. 2. a mathematical topic describing the connection of geometric entities. This is used for B-Rep models.

torque - a moment or twisting action about an axis.

torus - a donut shape

toroidal core - a torus shaped magnetic core to increase magnetic conductivity.

TPDDI (Twisted Pair Distributed Data Interface) - counter rotating token ring network connected with twisted pair medium.

TQC (Total Quality Control) - a philosophical approach to developing quality methods that reach all levels and aspects of a company.

transceiver (transmitter receiver) - a device to electrically interface between the computer network card, and the physical network medium. Packet collision hardware is present in these devices.

transducer - a device that will convert energy from one form to another at proportional levels.

transformations - include translation, rotation, and scaling of objects mathematically using matrix algebra. Transformations are used to move objects around in a scene.

transformer - two separate coils wound about a common magnetic coil. Used for changing voltage, current and resistance levels.

transient - a system response that occurs because of a change. These effects dissipate quickly and we are left with a steady state response.

transmission path - a system component that is used for transmitting energy.

transport layer - an OSI network model layer.

TRIAC (TRIode Alternating Current) - a semiconductor switch suited to AC power.

true - a logic positive, high, or 1.

truth table - an exhaustive list of all possible logical input states, and the logical results.

TTL (Transistor Transistor Logic) - a high speed for of transistor logic.

TTY - a teletype terminal.

turbine - a device that generates a rotational motion using gas or fluid pressure on fan blades or vanes.

turbulent flow - fluids moving past an object, or changing direction will start to flow unevenly. This will occur when the Reynold's number exceeds 4000.

twisted pair - a scheme where wires are twisted to reduce the effects of EMI so that they may be used at higher frequencies. This is casually used to refer to 10b2 ethernet.

TXD (Transmitted Data) - an output line for serial data transmission. It will be connected to an RXD input on a receiving station.

B.20 U

UART (Universal Asynchronous Receiver/Transmitter) -

UDP (User Datagram Protocol) - a connectionless method for transmitting packets to other hosts on the network. It is seen as a counterpart to TCP.

ultrasonic - sound or vibration at a frequency above that of the ear (> 16KHz typ.)

ultraviolet - light with a frequency above the visible spectrum.

UNIX - a very powerful operating system used on most high end and mid-range computers. The predecessor was Multics. This operating system was developed at AT & T, and grew up in the academic environment. As a result a wealth of public domain software has been developed, and the operating system is very well debugged.

UPS (Uninterruptable Power Supply) -

user friendly - a design scheme that simplifies interaction so that no knowledge is needed to operate a device and errors are easy to recover from. It is also a marketing term that is badly misused.

user interfaces - are the means of communicating with the computer. For CAD applications, a graphical interface is usually preferred. User friendliness is a measure of the ease of use of a program and

implies a good user interface.
UUCP (Unix to Unix Copy Program) - a common communication method between UNIX systems.

B.21 V

Vac - a voltage that is AC.

vacuum - a pressure that is below another pressure.

vane - a blade that can be extended to provide a good mechanical contact and/or seal.

variable - a changeable location in memory.

varistor - voltage applied changes resistance.

valve - a system component for opening and closing mass/energy flow paths. An example is a water faucet or transistor.

vapor - a gas.

variable - it is typically a value that will change or can be changed. see also constant.

VDT (Video Display Terminal) - also known as a dumb terminal

velocity - a rate of change or speed.

Venturi - an effect that uses an orifice in a flow to generate a differential pressure. These devices can generate small vacuums.

viscosity - when moved a fluid will have some resistance proportional to internal friction. This determines how fast a liquid will flow.

viscosity index - when heated fluid viscosity will decrease, this number is the relative rate of change with respect to temperature.

VLSI (Very Large Scale Integration) - a measure of chip density. This indicates that there are over 100,000(?) transistors on a single integrated circuit. Modern microprocessors commonly have millions of transistors.

volt - a unit of electrical potential.

voltage rating - the range or a maximum/minimum limit that is required to prevent damage, and ensure normal operation. Some devices will work outside these ranges, but not all will, so the limits should be observed for good designs.

volume - the size of a region of space or quantity of fluid.

volatile memory - most memory will lose its contents when power is removed, making it volatile.

vortex - a swirling pattern in fluid flow.

vortex shedding - a solid object in a flow stream might cause vortices. These vortices will travel with the flow and appear to be shed.

B.22 W

watchdog timer - a timer that expects to receive a pulse every fraction of a second. If a pulse is not received, it assumes the system is not operating normally, and a shutdown procedure is activated.

watt - a unit of power that is commonly used for electrical systems, but applies to all.

wavelength - the physical distance occupied by one cycle of a wave in a propagating medium.

word - 1. a unit of 16 bits or two bytes. 2. a term used to describe a binary number in a computer (not limited to 16 bits).

work - the transfer of energy.

write - a digital value is stored in a memory location.

WYSIWYG (What You See Is What You Get) - newer software allows users to review things on the screen before printing. In WYSIWYG mode, the layout on the screen matches the paper version exactly.

B.23 X

X.25- a packet switching standard by the CCITT.

X.400 - a message handling system standard by the CCITT.

X.500 - a directory services standard by the CCITT.

X rays - very high frequency electromagnetic waves.

X Windows - a window driven interface system that works over networks. The system was developed at MIT, and is quickly becoming the standard windowed interface. Personal computer manufacturers are slowly evolving their windowed operating systems towards X-Windows like standards. This standard only specifies low level details, higher level standards have been developed: Motif, and Openlook.

XFER - transfer.

XMIT - transmit.

xmodem - a popular protocol for transmitting files over text based connections. compression and error checking are included.

B.24 Y

ymodem - a popular protocol for transmitting files over text based connections. compression and error checking are included.

B.25 Z

zmodem - a protocol for transmitting data over text based connections.